MICROFLORAL AND FAUNAL INTERACTIONS IN NATURAL AND AGRO-ECOSYSTEMS

DEVELOPMENTS IN BIOGEOCHEMISTRY

CJM Kramer and JC Duinker, eds: Complexation of Trace Metals in Natural Waters. 1984. ISBN 90-247-2973-4

EM Thurman: Organic Geochemistry of Natural Waters. 1985. ISBN 90-247-3143-7

MJ Mitchell and JP Nakas, eds: Microfloral and Faunal Interactions in Natural and Agro-Ecosystems. 1986. ISBN 90-247-3246-8

Microfloral and faunal interactions in natural and agro-ecosystems

edited by

MYRON J. MITCHELL and JAMES P. NAKAS

State University of New York
College of Environmental Science and Forestry
Syracuse, New York, USA

1986 **MARTINUS NIJHOFF/DR W. JUNK PUBLISHERS**
a member of the KLUWER ACADEMIC PUBLISHERS GROUP
DORDRECHT / BOSTON / LANCASTER

Distributors

for the United States and Canada: Kluwer Academic Publishers, 190 Old Derby Street, Hingham, MA 02043, USA
for the UK and Ireland: Kluwer Academic Publishers, MTP Press Limited, Falcon House, Queen Square, Lancaster LA1 1RN, UK
for all other countries: Kluwer Academic Publishers Group, Distribution Center, P.O. Box 322, 3300 AH Dordrecht, The Netherlands

Library of Congress Cataloging in Publication Data

```
Microfloral and faunal interactions in natural and
   agro-ecosystems.

   (Developments in biogeochemistry)
   Includes index.
   1. Soil ecology.  2. Biogeochemical cycles.
3. Microbial ecology.  4. Agricultural ecology.
I. Mitchell, Myron J.  II. Nakas, James P.
III. Series.
QH541.5.S6M53  1986     574.5'26404     85-21768
```

ISBN-13: 978-94-010-8789-6   e-ISBN-13: 978-94-009-5173-0
DOI: 10.1007/978-94-009-5173-0

CONTENTS

INTRODUCTION

MYRON J. MITCHELL*

The biota of soils constitute an integral part of both
natural and agronomic ecosystems. The soil microflora and
fauna in conjunction with the belowground portion of the
Metaphyta or higher plants constitute the living milieu,
components of which are in intimate association with each
other as well as the abiotic constituents of the soil. Since
these associations or interactions are important in regulating both the flux and availability of energy and nutrients,
the central theme of the present book focuses on these
interactions. The effects of microfloral and faunal interactions with regard to overall ecosystem dynamics and
specific critical processes will be examined.

HISTORICAL ASPECTS

The coverage of this volume is an extension of a vast
body of literature which dates back to the 18th century. A
brief compendium of major books and reviews published from
1960 to 1983 is given chronologically in Tables 1 and 2,
respectively. Russell (1961) has reviewed work in the 1800's
during which some of the basic tenets on the relationships
between plant nutrition and soil properties became established. In this period agricultural science was founded and
the study of soil bacteriology began. The evolution of soil
biology up to the early 1970's has been described by
Satchell in the volume edited by Dickinson and Pugh (1974).

---

*Department of Environmental and Forest Biology, SUNY,
College of Environmental Science and Forestry, Syracuse,
NY 13210

Table 1. Selected Books on Soil Biology from 1960 to 1983

Author(s) or Editor(s)	Date	Title
Russell, E.W. McE.	1961	Soil Conditions and Plant Growth.
Kevan, D.K. McE.	1962	Soil Animals.
Doeksen, J. and J. van der Drift (eds.)	1963	Soil Organisms.
Baker, K.F. and W.C. Snyder (eds.)	1965	Ecology of Soil-Borne Pathogens.
Kononova, M.M.	1966	Soil Organic Matter.
Burges, A. and F. Raw (eds.)	1967	Soil Biology.
Graff, O. and J.E. Satchell (eds.)	1967	Progress in Soil Biology.
Gray, T.R.G. and D. Parkinson (eds.)	1967	Ecology of Soil Bacteria.
McLaren, A.D. and G.H. Petersen (eds.)	1967	Soil Biochemistry.
Harley, J.L.	1969	The Biology of Mycorrhiza.
Garrett, S.D.	1970	Pathogenic Root-Infecting Fungi.
Reichle, D.E. (ed.)	1970	Analysis of Temperate Forest Ecosystems.
Toussoun, T.A.; R.V. Bega and P.E. Nelson (eds.)	1970	Root Diseases and Soil-Borne Pathogens.
Wallwork, J.A.	1970	Ecology of Soil Animals.
Gray, T.R.G. and S.T. Williams	1971	Soil Microorganisms.
Phillipson, J.	1971	Methods of Study in Quantitative Soil Ecology; Population, production and energy flow.
Hardy, R.W.F. (ed.)	1977	A Treatise on Dinitrogen Fixation.
Postgate, J.R.	1971	The Chemistry and Biochemistry of Nitrogen Fixation.
Schnitzer, M. and S.U. Khan	1972	Humic Substances in the Environment.
Domsch, K.H. and W. Gams	1973	Fungi in Agricultural Soils.
Carson, E.W. (ed.)	1974	The Plant Root and its Environment.
Dickinson, C.H. and G.J.F. Pugh (eds.)	1974	Biology of Plant Litter Decomposition.
Quispel, A. (ed.)	1974	The Biology of Nitrogen Fixation.
Richards, B.N.	1974	Introduction to the Soil Ecosystem.
Giesseking, J.E. (ed.)	1975	Soil Components.
Kilbertus, G., O. Reisinger, A. Mourey and J.A. Camela (eds.)	1975	Proc. First Symp. Bio-degradation et Humification.
Walker, N.	1975	Soil Microbiology: A Critical Review.
Anderson, J.M. and A. Macfadyen (eds.)	1976	The Role of Terrestrial and Aquatic Organisms in Decomposition Processes.
Marshall, K.D.	1976	Interfaces in Microbial Ecology.
Alexander, M.	1977	Introduction to Soil Microbiology.
Lohm, U. and T. Persson (eds.)	1977	Soil Organisms as Components of Ecosystems.
Burns, R.G. (ed.)	1978	Soil Enzymes.
Edwards, C.H. and J.R. Lofty	1978	Biology of Earthworms.
Harley, J.L. and R.S. Russell (eds.)	1979	The Soil-Root Interface.
Swift, M.J., O.W. Heal and J.M. Anderson	1979	Decomposition in Terrestrial Ecosystems.
Dindal, D.L. (ed.)	1980	Soil Biology as Related to Land Use Practices.
Ellwood, D.C., J.N. Hedger, M.J. Latham and J.M. Lynch and J.H. Slater (eds.)	1980	Contemporary Microbial Ecology.
Payne, W.J.	1980	Microorganisms and Nitrogen Sources.
Broughton, W.J. (ed.)	1981	Nitrogen Fixation, Vol. 1, Ecology.
Clark, F.E. and T. Rosswall (eds.)	1981	Terrestrial Nitrogen Cycles.
Payne, W.J.	1981	Denitrification.
Broughton, W.J. (ed.)	1982	Nitrogen Fixation, Vol. 2, Rhizobium.
Bull, A.T. and J.H. Slater (eds.)	1982	Microbial Interactions and Communities.
Burns, R.G. and J.H. Slater	1982	Experimental Microbial Ecology.
Freckman, E. (ed.)	1982	Nematodes in Soil Ecosystems.
Slater, J.H., R. Whittenburg and J.W.T. Wimpenny	1983	Microbes in Their Natural Environment.
Spicer, C.C. and R.E.O. Williams	1983	Microbial Ecology.

Table 2. Selected Reviews in Journals on Soil Biology from 1960 to 1983.

Author	Date	Title
Rovira, A.D.	1965	Interactions between plant roots and soil microorganisms.
Barber, D.A.	1968	Microorganisms and the inorganic nutrition of higher plants.
Harley, J.L.	1971	Fungi in ecosystems.
Witkamp, M.	1971	Soils as components of ecosystems.
Payne, W.J.	1973	Reduction of nitrogenous oxides by microorganisms.
Hewitt, E.J.	1975	Assimilatory nitrate-nitrite reduction.
Kirk, T.K., W.J. Connors and J.G. Zeikus	1977	Advances in understanding the microbiological degradation of lignin.
Stout, J.D.	1980	The role of protozoa in nutrient cycling and energy flow.
Knowles, R.	1982	Denitrification.
Luxton, M.	1982	Quantitative utilization of energy by the soil fauna.
Petersen, H.	1982	The total soil fauna biomass and its composition.
Coleman, D.C., C.P. Reid and C.V. Cole	1983	Biological strategies of nutrient cycling in soil systems.
Seastedt, T.R.	1983	The role of microarthropods in decomposition and mineralization processes.

In this latter period the disciplines of soil microbiology (e.g. Winogradsky 1925, Waksman 1927, Gray and Parkinson 1967, Alexander 1977), soil zoology (e.g. Bornebush 1930, Doekson and van der Drift 1963, Burges and Raw 1967, Wallwork 1970), soil organic matter relationships (Waksman 1936, Kononova 1966, Gieseking 1975, Schnitzer and Khan 1972) and soil-root interactions (Harley 1969, Garrett 1970, Tousson et al. 1970) became firmly established.

In much of the early work on ascertaining the biotic relationships of belowground systems, emphasis was placed on describing the components of the soil community (e.g. Jacot 1940). As some of these community relationships have become better delineated, interest has focused on ascertaining how the biotic constituents interact and affect critical soil processes. One of the major focal points has been to ascertain how energy fluxes in the soil are influenced by the microorganisms and the fauna. The pioneering work of Bornebush (1930) was a precursor to later studies which described trophic and energy relationships in the soil community (Birch and Clark 1953, Burges and Raw 1967, Dickinson and Pugh 1974). The relative contributions of specific taxa or functional groups of organisms, the importance of substrate composition and the role of the physical environment of the soil in affecting organic matter transformation and catabolism have been studied. Such considerations have led to the recognition and description of the soil system in an ecosystem context (Fenton 1947, Olson 1963, Reichle 1970, Weigert et al. 1970, Richards 1979, Swift et al. 1979).

There has been a rapid advancement during the past twenty years in the assessment of the relative contribution of various elements of the soil biota in the direct catabolism of soil organic matter, including litter inputs; much of this work was a direct result of research stimulated by the International Biological Program in which the determination of ecosystem energy relationships was a primary goal (Luxton 1982, Petersen 1982, Heal et al. 1981). There is

presently available an extensive literature on the relative
contribution of various faunal taxa in the overall net
dissipation of energy in the soil via respiratory metabolism.
The general consensus is that the fauna generally contribute
less than 10% of the net energy flux in the soil (Luxton 1982).
With regards to the microflora the relative contribution of
the fungi and bacteria have been studied (i.e. Jenkenson and
Powlson 1976), but still detailed and accurate information on
the specific role of microfloral taxa on decomposition in the
field is often lacking. Furthermore, although energy fluxes
have been estimated, the understanding of the contribution of
both the fauna and microflora as regulators of decomposition
has only recently been a focal point (Coleman et al. 1978,
Santos and Whitford 1981). Studying the role of interactions
among the soil microfloral and faunal components in affecting
energy flux is critical to understanding the soil system
(Mitchell and Parkinson 1974, Coleman et al. 1978).

Most recently, attention has shifted to studying the
effects of microflora and fauna on soil-plant nutrient relationships, especially with regard to the macronutrients N, P
and S (Stout 1980, Clark and Roswall 1981, Coleman et al.
1983, Mitchell et al. 1983, Seastedt 1983). Such studies
have demonstrated that there is a dynamic interchange among
the three biotic components of the soil: microflora, fauna
and higher plants. Furthermore, the importance of the biota
in affecting nutrient dynamics within soils is readily
apparent, especially when it is recognized that the microflora and fauna affect both the transformation and transfer
of nutrients. Although it has been known for many years that
microorganisms, especially the bacteria, are important in the
nitrogen cycle, only more recently has the importance of both
the soil fauna and soil microflora in phosphorus and sulfur
dynamics been recognized. Both phosphorus and sulfur have
important organic forms, the formation of which is attributed
to the biota (David et al. 1983, Newman and Tate 1980). The
mineralization and immobilization of these elements are major
factors in determining their availability for uptake via the

roots of higher plants.

The study of the interactions among the roots of higher plants and surrounding soil is a critical area of investigation since these processes constitute direct linkages to belowground and aboveground processes. For example, although it has been generally recognized that belowground processes are critical to grassland ecosystem dynamics (Innis 1978), the role of belowground processes in forest systems with regard to root dynamics has also recently become more established (McClaugherty et al. 1982). Furthermore, the consideration of root-soil interactions requires implicit integration of the disciplines of physics, chemistry and biology since an understanding of these processes requires concomitant consideration of abiotic and biotic soil components (Carson 1975, Harley and Russell 1979). The study of roots has also served as a useful interface between basic and applied soil investigations. Although early studies concentrated almost entirely on pathogenic relationships, more recently general aspects of root-soil interactions, including nutrient dynamics and general biotic relationships have also been emphasized (Harley and Russell 1979, Krupa and Dommergues 1979).

To unravel the complexities of these intricate interactions, regarding both nutrients and energy, there has been an increased interest in experimental approaches both in the field and laboratory which often attempt to examine specific mechanisms or pathways in simplified or manipulated soil systems (Burns and Slater 1982, Coleman et al. 1983, Ellwood et al. 1980, Santos and Whitford 1981). Furthermore, as the understanding of the soil system has expanded there has been a concomitant need for the examination of specific processes or components which are critical to the functioning of these systems. Specifically detailed information on degradation processes and humification biochemical pathways (Kirk et al. 1977, McClaren and Peterson 1967), soil enzymes (Burns 1978) and nutrient relationships, especially with respect to N (Broughton 1981, Knowles 1982) has become available.

BOOK COVERAGE

Within this book the description of the soil will combine ecosystem approaches with the examination of the role of specific components or processes in affecting interactions in soil systems. The book will not be an exhaustive review, but will instead focus on a series of topics which are of contemporary importance in the understanding of the role of microfloral and faunal interactions in natural and agro-ecosystems. The information within the volume should be of interest to professionals who are either working directly with soils such as agronomists, soil scientists, and soil biologists or those whose area of interest has a strong soil component such as terrestrial ecologists, waste managers, and environmental scientists. In addition, this book should be useful to advanced undergraduate students and graduate students who are seeking to further their knowledge of soil ecology.

Nutrient cycling and decomposition in natural ecosystems are discussed by Heal and Dighton (Chapter 2). The general role of soils in natural ecosystems is reviewed and the importance of temporal and spatial variation emphasized. How successional patterns in ecosystems are reflected by variation in soil processes and components including resource quality are considered with special emphasis on boreal and arctic environments. Major ecological themes such as "r" and "K" strategies are integrated into a consideration of the role of microfloral and faunal interactions in decomposition. Comparisons of specific patterns among ecosystems demonstrate which factors are most critical in affecting soil structure and function.

For agro-ecosystems Juma and McGill (Chapter 3) utilize analytical and simulation modelling procedures to examine how agricultural practices influence the transfer and transformation of C and N. These procedures are used to test a series of hypotheses for examining which factors are most critical in regulating decomposition. Since both the quality and temporal variation in organic matter inputs can be readily

quantified for these systems, they serve as useful experimental
tools for analyzing processes affecting decomposition and
nutrient cycling. This approach also demonstrates how certain
methodological procedures may limit the understanding of
agro-ecosystems.

The biochemistry of organic constituents in soils is
considered by Burns and Martin (Chapter 4). They outline
both the physical and chemical structure of organic matter
inputs to soil and discuss the major biochemical constituents
in belowground systems including cellulose and lignin. Understanding the functional groups of microorganisms responsible
for the degradation of specific substrates is emphasized.
Recent advances in characterizing the formation and decomposition of humus and the use of model compounds of humic
materials are reviewed. The authors stress that the determination of pesticide degradation necessitates a clear understanding of biodegradation pathways.

Smucker and Safir (Chapter 5) emphasize methodological
procedures in quantifying root and rhizosphere relationships.
The importance of the allocation of plant C to belowground
components is a central theme of this chapter. Although they
concentrate their discussion on cultivars, their conclusions
are relevant to a broad range of systems and include findings
for both applied and basic problems on root dynamics. The
effects of various processes including environmental stress
on root exudation and root respiration with respect to crop
yields are discussed. The role of mycorrhizal fungi on CO_2
fixation and water relationships is reviewed.

The importance of N-fixation, denitrification, and nitrification in soils is outlined by Smith and Rice (Chapter 6).
The roles of specific microbial taxa in denitrification and
nitrification pathways are detailed. They emphasize how
abiotic and biotic factors such as substrate availability,
soil moisture and oxygen availability affect these processes,
and how information on individual processes serves as an
important foundation in the assessment of nitrogen dynamics
in ecosystems.

The role of microflora in terrestrial S cycling is described by Nakas (Chapter 7) by considering the major pathways
of S transformation in soil. The study of S within terrestrial ecosystems has recently accelerated due to increased
documentation of S deficiencies in some regions while other
areas are being subjected to high levels of S loading
associated with acidic deposition. Regardless of whether a
system has excess S or is deficient in this element, the role
of S in an ecosystem is a direct function of microbial processes which alter the form of S. The chemical form of a S
chemical species has a major influence on biotic processes
such as immobilization and mineralization as well as abiotic
processes such as leaching.

Coleman (Chapter 8) outlines the general relationships
among soil fauna, soil microflora, nutrient cycling and
decomposition processes. The basic linkages between the soil
fauna and microflora and the soil's physical environment are
considered. The utilization of microcosms in testing hypotheses on the form of these interactions is described. The
need in addressing the role of higher plants in affecting the
interactions among soil fauna and microflora is emphasized.
The employment of field manipulations including fumigation
and agronomic alterations as experimental tools in assessing
microfloral and faunal interactions in grassland soils is
demonstrated.

The role of management practices in affecting decomposer
communities and decomposition is considered by Curry
(Chapter 9). Vertebrate grazing alters both the quantity
and quality of organic matter inputs to a variety of agronomic systems. Mowing and fertilization, organic manure
additions, burning, cultivation, and pesticides exhibit a
range of effects on soil systems. The nature of these
effects is highly dependent on the amount and timing of
application. Both biotic and abiotic factors show differential responses to these manipulations.

Recent methodological advances in quantifying soil taxa
and soil processes are discussed by Freckman, Wallwork and

Cromack (Chapter 10). Contemporary procedures for isolating, identifying and quantifying specific faunal and microfloral taxa and functional groups are outlined. The advancement of the understanding of microfloral and faunal interactions in soil is dependent on these techniques. The use of experimental manipulations for changing the biotic structure of the soil and assessing the role of functional groups in decomposition and nutrient cycling is shown.

The final chapter (11) by Hunt and Parton describes the use of models in studying decomposition and nutrient dynamics of soils. The implementation of modelling approaches has helped to synthesize information from a variety of sources which must be considered if the dynamics of soil systems are to be understood from either a systems or process orientation. Simulation models have been especially useful in both generating and testing hypotheses on the importance of specific components of the soil in energy and nutrient flux. Using examples of models which describe N cycling in grassland systems, the benefits of the modelling approach are demonstrated.

ACKNOWLEDGEMENTS

I wish to thank Cathy Bellinger for help in much of the correspondence needed for the production of this volume. Ruth Piatoff helped in correcting final drafts of some chapters.

REFERENCES

1. Alexander, M. 1977. Introduction to Soil Microbiology. 2nd Edition. John Wiley and Sons, NY.
2. Anderson, J.M., and A. MacFadyen (eds.). 1976. The Role of Terrestrial and Aquatic Organisms in Decomposition Processes. Blackwell, Oxford.
3. Baker, K.F., and W.C. Snyder (eds.). 1965. Ecology of Soil-Borne Pathogens; Prelude to Biological Control. University of California Press, Berkeley.
4. Barber, D.A. 1968. Microorganisms and the inorganic nutrition of higher plants. Annual Rev. Plant Physiol. 19:71-88.

10

5. Birch, L.C., and D.P. Clark. 1953. Forest soil as an ecological community with special reference to the fauna. Quart. Rev. Biol. 28:13-36.

6. Bornebush, C.H. 1930. The fauna of forest soil. Forst. Forsøgsvaesenm Danm. 11:1-224.

7. Broughton, W.J. (ed.). 1981. Nigrogen Fixation, Vol. 1. Ecology. Oxford University Press, NY.

8. _____. 1982. Nitrogen Fixation, Vol. 2. Rhizobium. Oxford University Press, NY.

9. Bull, A.T., and J.H. Slater (eds.). 1982. Microbial Interactions and Communities. Academic Press, London.

10. Burges, A., and F. Raw (eds.). 1967. Soil Biology. Academic Press, London.

11. Burns, R.G. (ed.). 1978. Soil Enzymes. Academic Press. London.

12. _____, and J.H. Slater (eds.). 1982. Experimental Microbial Ecology. Blackwell Sci. Publ., Oxford.

13. Carson, E.W. (ed.). 1974. The Plant Root and Its Environment. University Press of Virginia, Charlottesville.

14. Clark, F.E., and T. Roswall (eds.). 1981. Terrestrial nitrogen cycles. Ecol. Bull. Swedish National Science Research Council 33.

15. Coleman, D.C., R.V. Anderson, C.V. Cole, E.T. Elliot, L. Woods, and M.K. Campion. 1978. Trophic interactions in soils as they affect energy and nutrient dynamics. IV. Flows of metabolic and biomass carbon. Microb. Ecol. 4:373-380.

16. _____, C.P. Reid, and C.V. Cole. 1983. Biological strategies of nutrient cycling in soil systems. Adv. Ecol. Res. 13:1-55.

17. David, M.B., S.C. Schindler, M.J. Mitchell, and J.E. Strick. 1983. Importance of organic and inorganic sulfur to mineralization processes in a forest soil. Soil Biol. Biochem. 15:671-677.

18. Dickinson, C.H., and G.J.F. Pugh (eds.). 1974. Biology of Plant Litter Decomposition. Academic Press, London.

19. Dindal, D.L. (ed.). 1980. Soil Biology as Related to Land Use Practices. USEPA, Washington, D.C.

20. Doekson, J., and J. van der Drift (eds.). 1963. Soil Organisms. North Holland, Amsterdam.

21. Domsch, K.H., and W. Gams. 1973. Fungi in Agricultural Soils. Halsted Press, Wiley, NY.

22. Edwards, C.A., and J.R. Lofty. 1978. Biology of Earthworms. 2nd Edition. Chapman and Hall, London.

23. Ellwood, D.C., J.N. Hedger, M.J. Latham, J.M. Lynch, and J.H. Slater (eds.). 1980. Contemporary Microbiol Ecology. Academic Press, NY.

24. Fenton, G.R. 1947. The soil fauna: with special reference to the ecosystem of forest soil. J. Anim. Ecology 16:76-93.

25. Freckman, D. (ed.). 1982. Nematodes in Soil Ecosystems. University of Texas Press, Austin.

26. Garrett, S.D. 1970. Pathogenic Root-Infecting Fungi. Cambridge University, Cambridge.

27. Gieseking, J.E. (ed.). 1975. Soil Components. Springer-Verlag, NY.

28. Graff, O., and J.E. Satchell (eds.). 1967. Progress in Soil Biology. North Holland, Amsterdam.

29. Gray, T.R.G., and D. Parkinson (eds.). 1967. Ecology of Soil Bacteria. Liverpool University Press, Liverpool.

30. _____, and S.T. Williams. 1971. Soil Microorganisms. Oliver and Boyd, Edinburgh.

31. Hardy, R.W.F. (ed.). 1977. A Treatise on Dinitrogen Fixation. John Wiley and Sons, NY.

32. Harley, J.L. 1969. The Biology of Mycorrhiza. 2nd Edition. Leonard Hill, London.

33. _____. 1971. Fungi in Ecosystems. J. Appl. Ecology 8:627-642.

34. _____, and R.S. Russell (eds.). 1979. The Soil-Root Interface. Academic Press, London.

35. Heal, O.W., P.W. Flanagan, D.D. French, and S.F. McClean. 1981. Decomposition and accumulation of organic matter in Tundra, pages 587-633, in L.C. Bliss, O.W. Heal, and J.J. Moore, (eds.). Analysis of Ecosystems: Tundra Biome. Cambridge University Press, Cambridge.

36. Hewitt, E.J. 1975. Assimilatory nitrate-nitrite reduction. Ann. Rev. Plant. Physiol. 26:73-100.

37. Innis, G.S. (ed.). 1978. Grassland Simulation Model. Springer-Verlag, NY.

38. Jacot, A.P. 1940. The fauna of the soil. Quart. Rev. Biol. 15:28-58.

39. Jenkinson, D.J. and D.S. Powlson. 1976. The effects of biocidal treatments on metabolism in soil. V. A method for measuring soil biomass. Soil Biol. Biochem. 8:209-213.

40. Kevan, D.K. McE. 1962. Soil Animals. H.F. and G. Witherby, London.

41. Kilbertus, G., O. Reisinger, A. Mourey, and J.A. Camela (eds.). 1975. Proc. First Symp. Biodegradation et Humification. University of Nancy, France.

42. Kirk, T.K., W.J. Connors, and J.G. Zeikus. 1977. Advances in understanding the microbiological degradation of lignin. Rec. Adv. Phytopath. 11:369-394.

43. Knowles, R. 1982. Denitrification. Microbiol. Rev. 46:43-70.

44. Kononova, M.M. 1966. Soil Organic Matter. Pergamon Press, Oxford.

45. Krupa, S.V., and Y.R. Dommergues. 1979. Ecology of Root Pathogens. Elsevier Sci. Publ. Co., NY.

46. Lohm, U., and T. Persson (eds.). 1977. Soil Organisms as Components of Ecosystems. Swedish Natural Research Council, Stockholm.

47. Luxton, M. 1982. Quantitative utilization of energy by the soil fauna. Oikos 39:342-354.

48. Marshall, K.D. 1976. Interfaces in Microbial Ecology. Harvard University Press, Cambridge.

49. McClaugherty, C.A., J.A. Aber, and J.M. Melillo. 1982. The role of fine roots in organic matter and nitrogen budgets of two forested ecosystems. Ecology 63:1481-1490.

50. McLaren, A.D., and G.H. Peterson (eds.). 1967. Soil Biochemistry. Marcel Dekker Inc., NY.

51. Mitchell, M.J., M.B. David, and C.R. Morgan. 1983. Importance of organic sulfur constituents of forest soils and the role of the soil macro fauna in affecting sulfur flux and transformation, pages 575-583, in Ph. Lebrun, H.M. Andre, A. DeMedts, C. Gregoire-Wibo, and G. Wauthy (eds.). New Trends in Soil Biology. Dieu-Brichart, Ottignies, Belgium.

52. _____, and D. Parkinson. 1976. Fungal feeding of oribatid mites (Acari: Cryptostigmata) in an aspen woodland soil. Ecology 57:302-312.

53. Newman, R.H., and K.R. Tate. 1980. Soil phosphorus characterization by ^{31}P nuclear magnetic resonance. Comm. Soil Sci. Plant Anal. 11:835-842.

54. Olson, J.S. 1963. Energy storage and balance of producers and decomposers in ecological systems. Ecology 44:322-331.

55. Payne, W.J. 1973. Reduction of nitrogenous oxides by microorganisms. Bacteriol. Rev. 37:409-452.

56. _____. 1980. Microorganisms and Nitrogen Sources. John Wiley and Sons, NY.

57. _____. 1981. Denitrification. Wiley-Interscience, NY.

58. Petersen, M. 1982. The total soil fauna biomass and its composition. Oikos 39:330-339.

59. Phillipson, J. 1971. Methods of Study in Quantitative Soil Ecology: Population, production and energy flow. IBP Handbook 18, Blackwell Sci. Publ., Oxford.

60. Postgate, J.R. 1971. Chemistry and Biochemistry of Nitrogen Fixation. Plenum Press, NY.

61. Quispel, A. (ed.). 1974. Biology of Nitrogen Fixation. North-Holland Publ. Co., Amsterdam.

62. Reichle, D.E. (ed.). 1970. Analysis of Temperate Forest Ecosystems. Springer-Verlag, NY.

63. Richards, B.N. 1974. Introduction to the Soil Ecosystem. Longman, London.

64. Rovira, A.D. 1965. Interactions between plant roots and soil microorganisms. Ann. Rev. Microbiol. 19:241-266.

65. Russell, E.W. 1961. Soil Conditions and Plant Growth. 9th Edition. Longman, Londan.

66. Santos, P.F., and W.G. Whitford. 1981. The effects of microarthropods on litter decomposition in a Chihnauhuan desert ecosystem. Ecology 62:654-663.

67. Schnitzer, M., and S.U. Khan. 1972. Humic Substances in the Environment. Marcel Dekker, NY.

68. Seastedt, T.R. 1983. The role of microarthropods in decomposition and mineralization processes. Ann. Rev. Entomology 29:25-46.

69. Slater, J.H., R. Whittenbury, and J.W.T. Wimpenny (eds.). 1983. Microbes in Their Natural Environments. 34th Symposium Soc. Gen. Microbiol. Cambridge University Press, Cambridge.

70. Spicer, C.C., and R.E.O. Williams (eds.). 1983. Microbial Ecology (Soc. Gen. Microbial 7th Symp.). Cambridge University Press, Cambridge.

71. Stout, J.D. 1980. The role of protozoa in nutrient cycling and energy flow. Adv. Microbial Ecol. 4:1-50.

72. Swift, M.J., O.W. Heal, and J.M. Anderson. 1979. Decomposition in terrestrial ecosystems. Blackwell Scientific Publ. Oxford.

73. Tousson, T.A., R.V. Bega, and P.E. Nelson (eds.). 1970. Root Diseases and Soilborne Pathogens. University of California Press, Berkeley.

74. Waksman, S.A. 1927. Principles of Soil Microbiology. Williams and Wilkins Co., Baltimore.

75. _____. 1936. Humus. Williams and Wilkins Co., Baltimore.

76. Walker, N. 1975. Soil Microbiology: A critical review. Halsted Press, Wiley, NY.

77. Wallwork, J.A. 1970. Ecology of Soil Animals. McGraw-Hill, London.

78. Wiegert, R.G., D.C. Coleman, and E.P. Odum. 1970. Energetics of the litter-soil subsystem, pages 93-98, in J. Phillipson (ed.) Methods of Study in Soil Ecology. Proc. Paris Sympos. UNESCO, Paris.

79. Winogradsky, S. 1925. Etudes sur la microbiologie du Sol 1. Sur la Methode. Annls. Inst. Pasteur (Paris) 39:299-354.

80. Witkamp, M. 1971. Soils as components of ecosystems. Annual Review of Ecology and Systematics 2:85-110.

NUTRIENT CYCLING AND DECOMPOSITION IN NATURAL TERRESTRIAL ECOSYSTEMS

O.W. HEAL AND J. DIGHTON*

INTRODUCTION

Ecosystem research is in its infancy, its roots are varied, it has a reasonable theoretical basis but it lacks the extensive data so essential for the development of the subject. The very complexity and scale of ecosystems makes analysis and comprehension difficult. However, the stimulus for research lies in the recognition that the ecosystem is a level of organization at which many organisms and processes interact. It is a unit which man manages for agriculture and forestry, and the flow of energy, carbon and nutrients provides a common currency for comparison of components and ecosystems.

It is in this context that microfloral and faunal interactions occur. The microflora are responsible for most of the heterotrophic activity through which organic matter is decomposed and nutrients mobilized. Fauna contribute less to direct heterotrophy but modify the physico-chemical environment and influence microbial activity through grazing. The details of microfloral and faunal interactions are considered in subsequent chapters; here we identify ecosystem structure and functions affecting and being influenced by these interactions. We will examine variations in nutrient dynamics among ecosystems and the changes occurring during primary succession and response to subsequent disturbance. We have not attempted a comprehensive review of the vast literature, but have

*Institute of Terrestrial Ecology, Merlewood Research Station, Grange-over-Sands, Cumbria, UK.

examined the main generalizations and illustrated these with a few examples.

The concept of 'ecosystems' has a long history (Major 1969) and is defined by Odum (1972) as any unit, including all of the organisms, in a given area interacting with the physical environment so that a flow of energy leads to a clearly defined trophic structure, biotic diversity, and material cycles within the system. An internal energy source, i.e. plant production, is envisaged and the system cannot develop homeostatic mechanisms if the main source is external.

The extent to which ecosystems exist as a functional unit of organization is arguable, the argument polarizing around the view of the 'super-organism' versus a collection of organisms responding to a common environment. It is difficult or even impossible to validate either view, and the argument may be irrelevant to the usefulness of the concept.

However, the key features of ecosystems are: (1) Reproducibility, i.e. the same ecosystem structure and function will be replicated in similar environments, (2) Responsiveness, i.e. the system is dynamic and will respond to changes in nutrient or energy flow or alteration in the environment, (3) Predictability, i.e. future states can be predicted from knowledge of the past and present state of the ecosystem.

Ecosystems can be considered as open, dissipative thermodynamic systems (Blackburn 1973) whose energy flow is controlled by interactions and feedback mechanisms, selected to maintain maximum biomass and to minimize the effects of fluctuations in environmental factors (Whittaker and Woodwell 1972, O'Neill et al. 1975). Four central properties characterize such dissipative structures (ecosystems):

1. Dependence on energy flow;
2. Homeostasis: development of self preservative negative feedback loops buffering the effect of moderate changes in external conditions;

3. Succession: change in properties until a stable state is reached in which parameters attain constant or cyclic values;

4. Limitation by rate of mass transfer: climax states are reached when as much matter as possible has been entrained in material cycles. Physiological and behavioural mechanisms mediate mass and energy flow through storage and homeostasis.

Many fields of research have contributed to the understanding of ecosystems. The concept of vegetation change in the succession to climax communities (Clements 1916) is a main base, especially when feedback mechanisms are identified (Slatyer 1977, Horn, 1981). These mechanisms include the link between primary production and decomposition subsystems and recognize the concept of a changing soil system which evolved from the holistic view developed by soil scientists (Jenny 1980). Analyses of energy flow through various trophic levels have resulted in classifications of the structure and flow of materials particularly through the fauna and the identification of energetic constraints to production (Lindeman 1942, McNeill and Lawton 1970, Humphreys 1977). Limitations to trophic level concepts have been identified (Rigler 1975) and alternative analyses of energy flow in ecosystems proposed based, for example, on organism size (Cousins 1980). When combined with primary production estimates, trophic relationships allow prediction of secondary production (Phillipson 1973, Heal and MacLean 1975) without adequately expressing the dynamics of population regulation (May 1981), one of the key feedback mechanisms by which ecosystem integrity can be maintained.

The dynamic nature of ecosystems, often obscured in short-term and single site studies, is reflected in the strongly theoretical research on stability and diversity (May 1981). Much of the interest has focused on species composition but, as recognized in research on growth strategies and r and K selection (MacArthur and Wilson 1967, Pianka 1981),the biological characteristics of the species represent the most important determinants of nutrient turnover.

The combination of organisms which constitute an ecosystem and any pattern of change in their growth characteristics will be reflected in the nutrient circulation within the ecosystem. The links between species and ecosystem processes are seen in the summary of trends over succession identified by Odum (1969).

Quantitative expression in mathematical models has focused on processes rather than organisms at the ecosystem level. This is largely due to the difficulty in obtaining comprehensive data at a realistic level of resolution, particularly for soil organisms. Many models have been produced which explore and synthesize information on stability characteristics, trophic structure, productivity, carbon and energy flow (O'Neill et al. 1975, Webster et al. 1975, Kirkwood and Lawton 1981, Clymo 1978, DeAngelis 1980, Bunnell and Scouller 1981). Ecosystem models increasingly highlight the interaction between plant growth and decomposition through nutrient availability and attempt to incorporate microbial population dynamics (McGill et al. 1981, Van Veen et al. 1981). However, microfloral and faunal interactions are still poorly represented even in the mechanistic models, although such detail can be realistically subsumed in the more general deterministic management models which are increasingly concerned with nutrient transfers (Aber et al. 1979, Swank and Waide 1980) or pollutants such as radionuclides (Frissel and Pennders 1983).

The study of primary and secondary successions has provided a fertile field for understanding vegetation dynamics and soil development. There is implied predictability of a sequence of events determined by external and internal responses leading to an approximate equilibrium state. There is, however, a chance (stochastic) element in early stages of succession influencing species composition and hence activity. The holistic or Clementsian view of succession is that the interaction between components of an ecosystem leads to an ordered structure with characteristics of production, energy and nutrient flow which direct the development of the system to select maximization of energy flow or nutrient conservation. The opposing, reductionist, view is of an association of organisms determined by chance (the stochastic element) and by their ability to survive and grow under the prevailing conditions (the deterministic element). Any resulting pattern reflects species characteristics and the ecosystem is the sum of the attributes of its component species.

Whether one takes the view of holistic or whole ecosystems there are marked changes in species characteristics, productivity and nutrient dynamics as ecosystems develop on previously uncolonized parent material (primary succession) or following disturbance (secondary succession). These changes provide the framework for examination of microflora, fauna and their interactions. The context is also important because soil biology has generally had little influence on,and been relatively uninfluenced by, considerations of ecosystem development, structure and function.

PRIMARY SUCCESSION

The succession of plant communities and soil development on bare ground has long been a fruitful area of research, particularly on sand dunes, glacial and lake deposits (Jenny 1980, Walker et al. 1981, Van Cleve et al. 1980) and on industrial wastes (Marrs et al. 1983, De Selm and Shanks 1963). The literature is vast and the conditions varied but a general pattern of development of nutrient pools emerges, as summarized by Reiners (1981) for N (Fig. 1).

Biomass and production

During succession the vegetation changes from herbaceous and relatively short lived species to longer lived shrubs and trees with greater biomass. Litter and soil organic matter accumulate, reflecting the increased input of woody and other fractions more resistant to decomposition. The pattern below ground is obscure but it probably follows that above ground, except that later stages dominated by woody plants will tend to have a lower proportion of biomass below ground than herb- or grass-dominated stages (Harris et al. 1980).

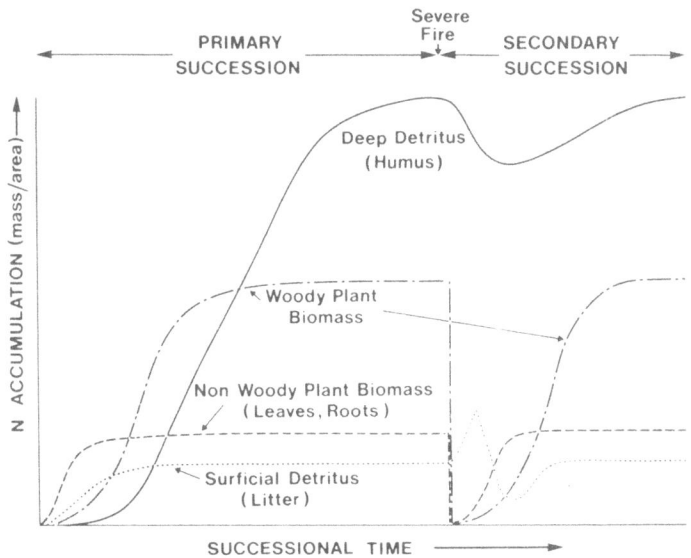

Figure 1. A generalized pattern for changes in organic nitrogen pools through primary succession, a consumptive fire disturbance, and secondary succession. The relative magnitudes and presence of pools, such as woody biomass, will differ between ecosystems but the relative rates of change should be consistent. These patterns are generalized from a wide range of examples (from Reiners 1981)

The development of organic matter pools, production and litter fall are broadly paralleled by nutrient increases as seen in the Tanana River floodplain succession in Alaska (Van Cleve and Viereck 1981) (Fig. 2). The extent to which the amount and state of the nutrient capital controls ecosystem development is uncertain, although Marrs et al. (1983) suggest that invasion by trees may be initiated only when the soil has developed a N capital of 500-1000 kg/ha. In some successions the pattern of nutrient cycling reflects that of the dominant species, as in Pinus nigra plantations on nutrient poor sands in Scotland (Miller 1981). There uptake rate and immobilization of nutrients in biomass increase through canopy closure, followed by a longer phase of reduced nutrient uptake and return as net production declines and woody litter fall increases, combined with greater internal recycling.

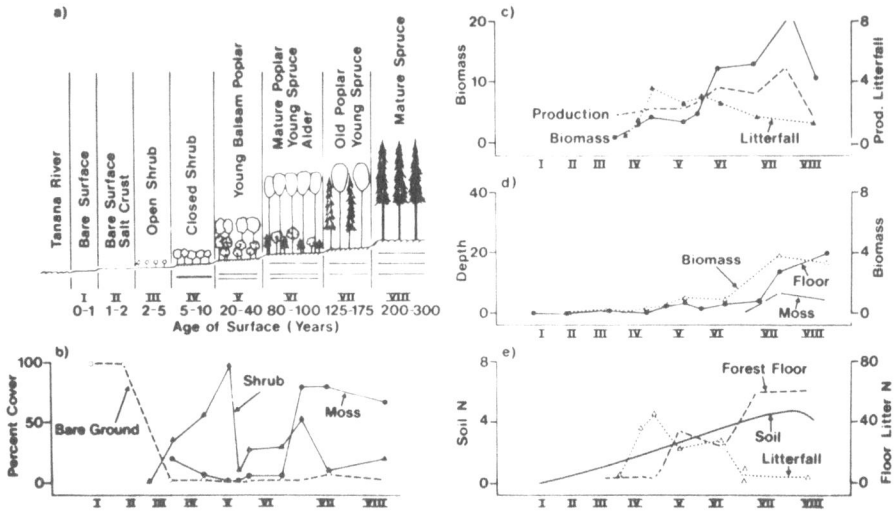

Figure 2. Changes in ecosystem processes during primary succession on the Tanana River floodplain, Alaska. (a) general vegetation sequence with time (b) percent cover of bare ground, shrubs and moss, (c) plant biomass ($g.m^{-2}.y^{-1}$ x 10^2), above ground production and litterfall ($g.m^{-2}.y^{-1}$ x 10^2), (d) forest floor biomass (g/m^2 x 10^3) and depth, and depth of moss cover (cm), (e) nitrogen in litter fall ($g.m^{-2}.y^{-1}$ x 10^{-1}), forest floor (g/m^2) and soil (g/m^2 x 10^2) (from Van Cleve et al. 1980)

Microclimate and soil

The Alaskan study (Van Cleve and Viereck 1981) also identifies the change in microclimate during succession with a decrease in the soil heat content as the increasing plant biomass and organic accumulation dampens the diurnal and seasonal temperature oscillations. Although less well documented, the moisture content of soil and litter layers also varies during succession. In the floodplain, there is some initial drying through evapotranspiration and although the process is enhanced later by increased biomass, there is a compensating moisture retention by an increasing moss layer. In drier, sandy soils moisture retention is likely to be increased by the development of vegetation and surface organic matter.

The rate of succession varies with climatic and soil conditions, taking as little as a decade to reach approximately stable nutritional status under warm, wet, fertile conditions, but thousands of years in some dune successions. The apparent biomass changes above ground correspond with alteration in soil properties and less obvious changes in soil processes. The pedogenic sequence includes movements of organic matter down the soil profile (Dickson and Crocker 1953), transformation of nutrients from inorganic to organic form (Walker and Syers 1974), decreasing bulk density and pH, and increasing pore size, aggregate size and cation exchange capacity (Cromack 1981, Jenny 1980). Mineral weathering is probably rapid in early successional stages and enhanced by disturbance but, like soil profile development, it occurs over much longer time scales than changes in vegetation structure and nutrient cycling (Gorham et al. 1979).

Decomposition

Despite its central position in the processes of nutrient cycling and soil development, surprisingly little is known of the variation in rates of organic matter decomposition during ecosystem development, but a pattern can be inferred from basic principles. Rate of decomposition is determined by climate (mainly temperature and moisture), resource quality (energy source, nutrient concentration, allelochemicals, physical structure) and soil physico-chemical properties. These factors, acting through the soil biota, control the three processes of decomposition: catabolism (mineralization of carbon and nutrients), comminution (reduction in particle size) and leaching (translocation of water soluble fractions) (Swift et al. 1979).

The evidence suggests that, especially under dry conditions, the environment for decomposition would improve during early successional stages and then decline with damping of temperature and gradual decrease in pH (Fig. 3a). Early litter input tends to be high in nutrients, simple sugars and cellulosic material but low in lignin and inhibiting compounds (Cromack 1981) resulting in a resource of high decomposition potential represented by a low C:N ratio. Litter quality tends to decline in later succession (Table 1) through increased wood production and, especially in nutrient limited situations, internal nutrient recycling.

The interaction of resource quality with environment is likely to produce an initial increase followed by a decline in decay rate (Table 1, Fig. 3b). Variability in decay rates will be highest in early and late succession through environmental heterogeneity, and in late stages, through resource variation. When resource decay is combined with the expected pattern of resource input, a double sigmoid curve of loss per unit area is predicted lagging behind the curve of litter fall (Fig. 3b). Additional information on biochemical

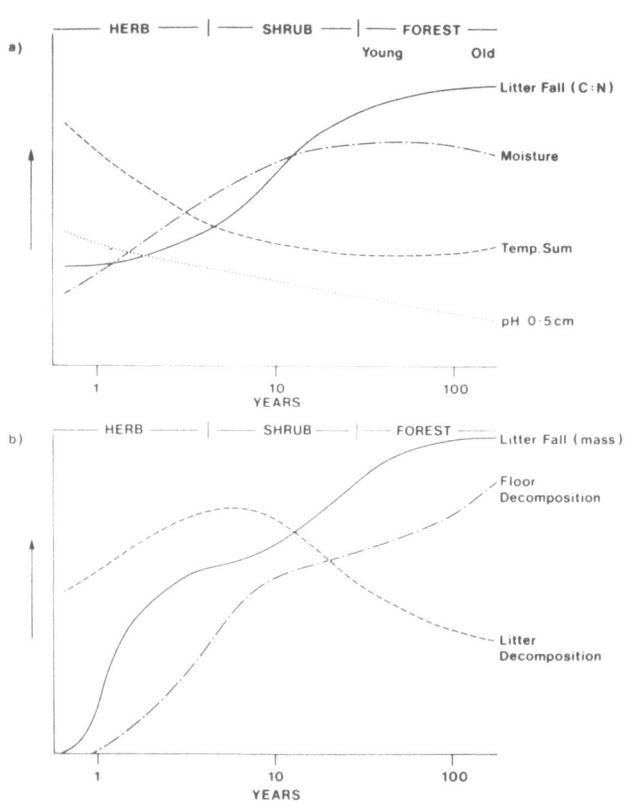

Figure 3. Hypothetical changes in climate, soil and resource quality (a) and in litter fall and decomposition rates (b) during a primary succession

aspects of biodegradation and transformation is presented in Chapter 4.

In general, nutrient release from decomposing litter is directly, but not linearly, related to decomposition (Swift et al. 1979). There tends to be a recurrent sequence in which K is rapidly leached, N and P are retained until late stages of decomposition, and other elements exhibit intermediate retention (Lousier and Parkinson 1979). Although the sequence of nutrient release (net mineralization) may be largely independent of litter type, the concentration at which release occurs varies directly with initial concentration in the litter (Berg and Staaf 1981). Thus it is expected that the pattern of nutrient release will follow the curve of decomposition during succession (Fig. 3). However, with the lower available nutrient concentration in later stage litters the amounts released per unit area will increasingly diverge from the decomposition curve.

Table 1. Nutrient composition and C substrate quality of foliage and needle litter of both early and late successional species in the H.J. Andrews Forest. ADF = acid detergent fraction (from Cromack 1981)

Species	N (%)	C (%)	C:N ratio	Lignin (%)	Cellu-lose (%)	ADF (%)	Annual weight loss (%)
Early succession							
Epilobium	0.65	46	71	10.1	14.9	74.6	52
Alnus (N fixer)	2.1	15	21	9.4	9.9	80.6	51
Ceanothus (N fixer)	0.81	45	55	10.0	11.0	77.3	48
Early & late succession							
Acer	0.70	44	63	8.5	14.7	73.9	59
Rhododendron	0.20	51	255	10.3	16.0	73.7	25
Castanopsis	0.32	50	156	22.8	21.1	55.9	37
Late succession							
Pseudotsuga							
Green needles	0.80	50	63	21.8	13.2	64.2	25
Senescent needles	0.50	50	100	24.1	14.6	59.5	20
Seed cones	0.24	49	204	44.2	34.1	21.2	8
Twigs	0.15	49	327	43.4	30.0	26.7	9
Bark	0.19	56	294	58.6	12.6	23.0	3

The importance of organic matter decomposition as a source of plant nutrients increases during succession. Increased efficiency of nutrient utilization by plants provides an important positive feedback mechanism on decomposition through production of resources poor in nutrients which further retards nutrient release (Gosz 1981, Van Cleve and Viereck 1981). Adaptation by plants to nutrient limitation is likely to include increased dependence on mycorrhizal associations enhancing P and N uptake (Alexander 1983). Shifts in selective uptake from NO_3^- to NH_4^+ and possibly soluble organic N may also occur with increasing competition for nutrients and decreasing N concentration in litters (Heal et al. 1982).

Decomposer microflora and fauna

As a result of the changes in physico-chemical environment and resource input during ecosystem succession, the composition and activity of the decomposer community will change. It may parallel the r-K life history strategies of the plant communities, with variations in response to stress or adverse conditions (Grime 1979), but few data are available to test this hypothesis (Usher and Parr 1977, Frankland 1982, Peterson and Luxton 1982).

Any observed pattern of change in decomposers during ecosystem succession will reflect the combination of successions on individual resources. For example, with increasing proportions of litters low in nutrients, during ecosystem succession the frequency of 'sugar fungi' is expected to decrease (Table 2). Basidiomycete frequency will increase with selection for their capacity to degrade lignin, conserve nutrients and to compete with other organisms (Rayner 1978, Pugh 1980, Frankland 1982). The characteristics of organisms in the resource succession indicates that over ecosystem succession there is a trend from species with an exploitation strategy (r) to those with an interaction (K) strategy (Table 2). Such a pattern is likely to occur under fertile conditions, but in nutrient poor conditions, species diversity will be low with decreased growth rates and maximum

Table 2. A summary of trends in resource succession with seral change in ecosystem succession (from Dickinson and Pugh 1974, Swift et al. 1979)

		Ecosystem succession			
		Increasing contribution of component to litter →			
	Lower plant	Herb- aceous plant	Angio- sperm leaves	Coniferous leaves	Wood
Chemical composition (%)					
Cellulose	16–35	20–37	6–22	20–31	36–63
Hemi.cell.	25–31	18–24	11–26	12–23	13–24
Sol.carb.	1–20	1–18	6–27	6–17	2–27
Lignin	7–36	3–30	9–42	20–58	17–35
Nitrogen	0.4–2	0.5–4	0.6–2	0.2–0.8	0.1
C:N	13–150	29–160	21–71	63–327	294–327
Decay rate (%/y)	20	30–70	40–60	3–50	1–90

Microbial succession

(Limited data) Sugar fungi Ascomycetes Fungi imperfecti	Yeasts Sugar fungi Ascomycetes Fungi imperfecti Basidio- mycetes		Ascomyc. Fungi imperf. Basidio- mycetes	Ascomyc. Fungi imperf. Basidio- mycetes

Importance of fauna

Relatively important   Increasing importance   Synergistic relationships →

Dominant fauna

(Enchy- traeids)	Enchy- traeids Oligo- chaetes Diptera	Oligo- chaetes Coll- embolla Acari	Acari Coll- embola Oligo- chaetes	Insects, other arthro- pods

adaptation for nutrient conservation including intimate symbiotic associations. Thus, species adopting an adversity (A) strategy due to poor growth conditions will predominate throughout the succession (Southwood 1977, Grime 1979, Pugh 1980, Greenslade 1982, Heal and Ineson 1984).

Although information is limited, Rusek (1978) has suggested a pattern of faunal succession related to primary succession of temperate ecosystems. Diversity and size of fauna increase through initial, pioneer and development stages to climax. Associated with the changes in flora and fauna, soil development passes from protorendsina to protoranker with microarthropod moder humus, through ranker and rendsina with ranging humus forms, to mull forest soil at climax. Increasing organism size, generation time and total biomass are expected during succession, a shift from r to K strategists exemplified by the trend from surface to subterranean earthworms (Dunger 1969, Satchell 1980). A shift in microfloral and faunal interactions is also expected, with rapid population fluctuations through opportunistic grazing in early seral stages moving to more damped population change and increasingly selective grazing as environmental variation and resource quality are reduced and diversity increases. As the size of soil fauna increases, the scale of organic matter comminution, organic matter redistribution and microbial dispersal will expand, as indicated in the role of Allolobophora caliginosa in polder soil profile development (van Rhee 1969) and mycorrhizal spore dispersal (McIlveen and Cole 1976, Maser et al. 1978). In the more nutrient poor resources in later seral stages e.g. wood, the concentration of nutrients by microflora may be an important precuser to faunal invasion and nutrient release (Swift 1977).

MATURE ECOSYSTEMS

Primary succession results in the accumulation of plant biomass and organic matter on and in the soil, with associated nutrients. A climax or stable state is rarely, if ever, reached. Rather the rate of change is reduced, there is little change in the accumulated live and dead organic matter, and nutrient uptake by the vegetation is balanced by nutrient release through decomposition. Thus the system represents a quasi-steady state with net ecosystem productivity near to zero, while nutrient input in wet and dry deposition and fixation tends to be balanced by outputs in gaseous and aqueous form as the biotic control of retention is relaxed. This concept and the corroborative evidence has been explored by Vitousek and Reiners (1975), Gorham et al. (1979) and Reiners (1981) and focuses on the biotic control of nutrient dynamics through the coupled processes of primary production and decomposition. Specific aspects of processes affecting N transformations are outlined in Chapter 6.

A comparison of mature ecosystems is given later but a brief summary of results and interpretation from one comprehensive study, a wet meadow tundra ecosystem in Alaska, indicates some central features as well as areas of uncertainty (Brown et al. 1980).

The meadow tundra at Point Barrow has an annual net primary production of about 2000 kg/ha from a standing crop of about 8000 kg/ha, production being severely limited by the length of the growing season. About 75% of the production, and even more of the live standing crop of vegetation, occurs below ground, indicative of the importance of maintaining nutrient absorptive and storage tissues. The distribution of nutrients closely follows the biomass, but an unusually small proportion of the nutrient capital of the ecosystem is contained in live biomass (1%). With a large nutrient store in dead organic matter and a limited nutrient supply from mineral soil because of permafrost, nutrient availability to plants is largely determined by release through decomposition. Evidence from fertilizer trials and nutrient pulses through lemmings tends to corroborate the budget interpretation of plant growth being restricted by N and P supply. The N uptake by plants of 23 $kg.ha^{-1}.y^{-1}$ is about 20-25% of the amount in the vegetation standing crop, and, as usual, is estimated from information on dry matter production. Physiological studies indicate that internal translocation of P within plants accounts for lack of coincidence between nutrient release and plant requirements within and to some extent between years.

The atmospheric input of nutrients is a much smaller fraction of the annual growth requirement and of the total nutrient reservoir than in many ecosystems. Precipitation input represents only about 1% of the inorganic N and 6% of the labile P in the rooting zone, but a greater input of N comes through fixation by algae, particularly Nostoc commune, the amount fixed being directly related to habitat moisture content. The fate of this fixed N is not known but it appears to be a key process which more than offsets losses in run-off and denitrification.

The cutting of vegetation and defecation by lemmings increase both quality and quantity of input to the decomposer subsystem and modify the microclimate resulting in increased decomposition rate. In particular 80-90% of the ingested P (about 1 $kgP.ha^{-1}.y^{-1}$) is returned to the soluble inorganic pool of about 0.01 kg/ha. Nutrient redistribution by lemmings between microhabitats (0.07 $kgP.ha^{-1}.y^{-1}$) could account for the observed differences in plant production and decomposition in the long-term development of the polygonal terrain.

Microbial biomass is about 180 kg/ha and production is estimated at 4-5 times the biomass. Nutrient turnover through microbial production approaches the annual plant uptake but the P requirement for microbial production necessitates considerable recirculation within the microbial population and hence greater competition with plants for this nutrient compared to C or N. However, much depends on microbial population dynamics and the sudden death of a large standing crop of microflora could release a pulse of P equivalent to the annual uptake by plants.

Such a pulse could be induced by faunal grazing but estimates indicate that only about 30% of the microbial production is consumed annually by soil fauna, consumption of fungi being three times that of bacteria. The fauna, with a biomass of 30 kg/ha and annual production of about 70 kg.ha^{-1}.y^{-1}, dominated by enchytraeids, ingests an equivalent of only 20% of net primary production.

The broad pattern of nutrient transfer within the meadow tundra shows the dominance of internal circulation within the vegetation, within the decomposer microflora and between vegetation and microflora. This evenly balanced tight cycling results from the primary control of climate on production and decomposition processes and the long-term positive feed back between production and decomposition. The nutrient budget or model allows limited interpretation but it is founded on the detailed understanding of the functioning of the processes. It is not the model itself, rather its construction, which is important because it focuses research on quantitative assessment of the main components and their interactions. For further discussions on the use of models consult Chapter 11.

As in virtually all such ecosystem studies there are two important weaknesses, (1) the errors associated with estimates are unspecified. With many transformation rates being small in relation to the nutrient concentrations of standing crops, quantifying the degree of imbalance of nutrients, e.g. between input and output is almost impossible; (2) despite the concerted research effort in such sites as Point Barrow, the functional relationships between nutrient release through decomposition and plant uptake, and between fauna and microflora, are still largely inferred and unquantified. For example, the extent to which a change in faunal population alters microbial populations and hence nutrient release and plant uptake remains to be quantified, at an ecosystem scale.

Although the mechanisms are not fully quantified, the main process relationships at Barrow are expressed in Fig. 4. The causal relationships and the budget are specific to wet meadow tundra at Barrow. In the theoretical terms expressed by Blackburn (1973) within the constraints of the severe physico-chemical environment, the biomass and turnover of organisms and materials have expanded to the maximum extent

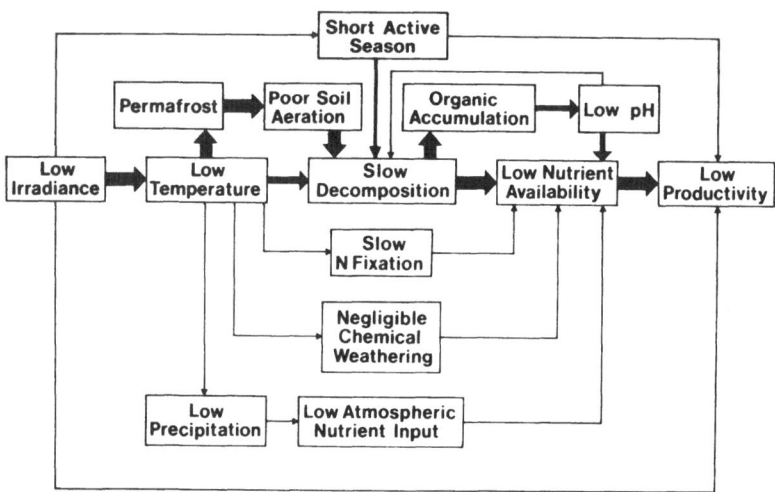

Figure 4. Causal relationships between low solar irradiance and low primary productivity of arctic coastal tundra in Alaska. Thickness of arrows indicates magnitude of effect (from Chapin et al. 1980)

allowed by the kinetics of the ecosystem. In other words the high environmental stress and low disturbance has selected for species with stress or adversity characteristics (Grime 1979, Pugh 1980, Greenslade 1982, Heal and Ineson 1984) which have combined to produce an ecosystem with an optimum balance of biomass or production of the different functional groups. As constraints of the physico-chemical environment are eased in more temperate or nutrient rich environments the organism structure and turnover expands but the degree of expansion and the balance of functions are limited by energy and nutrient kinetics.

SECONDARY SUCCESSION

The apparently stable or mature ecosystem state described in the previous section is rather one of minimal change. Gradual change may continue, e.g. podzolization in which elements are redistributed to lower horizons (possibly outside the rooting zone) or converted to unavailable forms. Such slow processes can, over centuries, lead to decreases in nutrient turnover and production, with associated changes in species composition. But how long does slow change continue and to what end? Evidence from most ecosystems indicates that sooner or later the period of relative stability is interrupted by an event leading to a rapid change in system characteristics. The event may be initiated by external factors (allogenic) but ecosystem characteristics often increase its susceptibility to change (autogenic). Some examples illustrate the main types of change.

1. Plagioclimax. Many ecosystem states are stabilized by man's management (plagioclimax) and alteration in management can produce relatively rapid but asymptotic change – a stepped sequence. Abandonment of arable farming leads to soil organic matter accumulation up to new stable grassland conditions under grazing (Whitehead 1970) or to woodland in the absence of grazing (Jenkinson 1971) with N accumulation of the order of 65 $kg.ha^{-1}.y^{-1}$. Reduction or removal of grazing in grasslands allows accumulation of above ground biomass, reduced return of nutrient rich organic matter and reduced decay rates as the succession moves through shrub to forest. These changes may be considered as secondary succession following disturbance of a relatively stable state or as continuation of a delayed primary succession. The difference is largely semantic.

2. Complete recycle or regeneration. A catastrophic event may result in the virtual destruction of the ecosystem and reversion to the equivalent of a primary succession.
    Although rare within the time frame of human experience, events occurring over millenia result in ecosystem regeneration. The apparently stable meadow tundra at Point Barrow is probably a stage in a long-term cycle with periodic flooding followed by a lake succession leading to a reformation of polygonal land (Chapin et al. 1980). In blanket bog ecosystems, accumulation of about 10% of the net primary production as peat under acid anaerobic conditions,

maintains a consistent ecosystem structure for periods of 10,000 years or so in northern Britain. Eventually erosion results in the destruction of the system, exposure of the parent material and development of a new ecosystem with initial characteristics of primary succession (Fig. 1). Such catastrophies are much less frequent than the third type of disturbance which produces secondary succession.

3. Partial regeneration. In many situations disturbance results in the loss of plant biomass followed by rapid reaccumulation of dry matter and nutrients through growth of populations with r characteristics. Death of dominant trees can provide a small scale mosaic of regeneration - the shifting mosaic steady state of Bormann and Likens (1979). Defoliation of boreal forest induces a redevelopment of dwarf shrub tundra (Kallio and Lehtonen 1973). Burning of old heather moor encourages regeneration of the same vegetation and pattern of nutrient cycling (Chapman and Webb 1978) but the burning may allow colonization by birch followed by a process of depodzolization and increased rates of nutrient circulation (Miles and Young 1980). Regeneration through fire causes loss of nitrogen and sulphur, but retention of other nutrients in surface soil horizons allows revegetation (Raison 1979). However, the frequency of fire determines the vegetation type (Fig. 5) and the distribution and circulation of nutrients (Noble and Slatyer 1981) especially where a high proportion of the nutrients are contained in plant biomass.

The effect of fire on soil processes is influenced by fire intensity and frequency, microclimate and chemical change (including deposition and leaching of nutrients in ash (Raison 1979)) and by the subsequent revegetation. In grasslands an initial volatilization of NH_3 and NH_4^+ after fire is followed by an explosion of microbial populations and NO_3^- production. The enhanced NO_3 concentration stimulates revegetation and decomposition followed by an increase in N_2 fixation as available N supplies are reduced (Woodmansee and Wallach 1981). However, microbial stimulation by fire does not always occur. Controlled burning of Eucalyptus forests

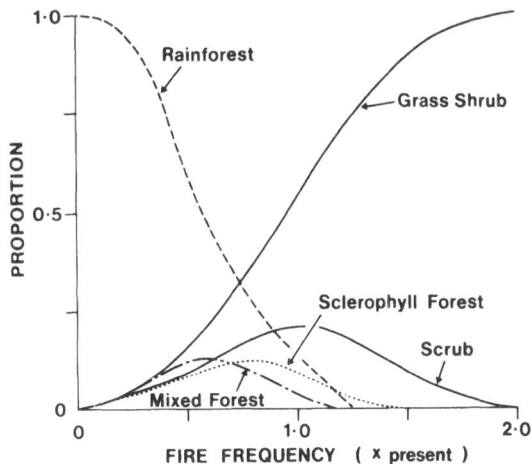

Figure 5. Changes in the proportion of vegetation types predicted to occur at steady state under different fire regimes in the tropics. The present fire regime is represented by a frequency of 1.0 (from Noble and Slatyer 1981)

in Australia on a short rotation (about 5-7 years) can markedly reduce litter decomposition, faunal species number and diversity (Springett 1976) during the inter-burn period. Thus a flush of microbial activity may be of short duration immediately after burning.

The general response of decomposer populations and activity to fire (Woodmansee and Wallach 1981) has similarities with the response to clear cutting (Fig. 6) the increase in mineralization varies with quality and quantity of fresh organic matter input, changes in microclimate and rate of re-establishment of plant uptake. It is the combination of these factors, plus the soil conditions, which determine N transformations, which are reflected in variation in amount and timing of NO_3^- output in leaching (Vitousek 1981).

Although the response to disturbance is variable, primary production is generally more severely disturbed than is the decomposer subsystem. The residual decomposer populations,

34

often stimulated by a pulse of organic matter input and
reduced competition from plant uptake of nutrients, respond
more rapidly than the primary producers due to shorter
generation times. The uncoupling of the balance between
production and decomposition processes, with negative net
ecosystem productivity, results in a net loss of nutrients
(Gorham et al. 1979, Fig. 6). The retention of the decomposer
subsystem also provides a buffer to disturbance and the base
for redevelopment of the ecosystem. However the degree and
type of change in the decomposer subsystem influence the
extent to which secondary succession recapitulates previous
successions. Change in pH and gain, loss or redistribution
of nutrients determine the reversibility of the successional
sequence in species composition but less obviously in terms
of the balance and rates of processes.

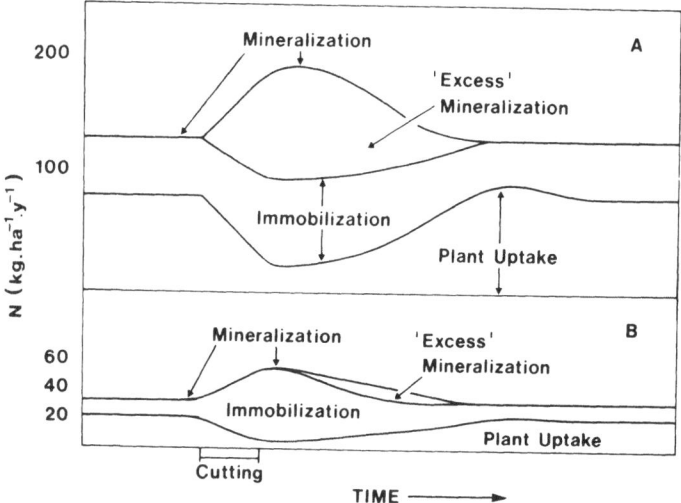

Figure 6. The effects of clear-cutting on the internal
nitrogen cycle in a deciduous (A) and a coniferous (B)
temperate forest. The 'excesses' of nitrogen mineralized
over nitrogen utilized can accumulate in the soil or be lost
to the system by several pathways. Succession eventually
re-establishes plant nitrogen uptake within the forest, and
excess nitrogen mineralization is eliminated (from Vitousek
1981)

PATTERNS WITHIN ECOSYSTEMS

Concealed within the seemingly stable state of an ecosystem, viewed at the macro level, a small scale mosaic of changing populations, energy and nutrient flow exists. Such mosaics are readily recognized in old forests as a result of death of individual trees, initiating new, localized successions, but the mosaic in space and time is a general phenomenon.

Spatial heterogeneity

Spatial heterogeneity may arise from local changes in microclimate, resource quality and vertical distribution of resources and nutrients in the soil profile. Although a curse to sampling programmes, such heterogeneity can be exploited to investigate factors influencing decomposition and nutrient cycling. For example on a scale of 0.25 m^2 in a wet, cool temperate north England blanket bog, moisture accounted for 86% of the variation in decomposition of standard Rubus leaves under different vegetation (Heal et al. 1978). Although mean annual temperatures in Eriophorum tussocks were 1°C higher (daily maxima up to 10°C higher) than Calluna or Sphagnum, weight loss of Rubus in each vegetation was 34%, 38% and 52%, respectively (Heal et al. 1978). Similar effects of topography on decomposition, acting through microclimate, are clearly shown in the polygonal tundra (Flanagan and Bunnell 1980).

Spatial variation in quality of resources derived from the same or different plant species (Table 1) may be compounded with microclimate variation. The mosaic of vegetation, tree stumps, dung pats, etc. form foci for different rates of decomposition each acting as nutrient sources and sinks which are separated in both space and time (Heal et al. 1982). Most litters show a short-term increase in N and sometimes P in the early stages of decomposition (Berg and Staaf 1981), possibly through faunal and/or fungal immigration. In decaying wood the increase may extend for a number of years, acting as a nutrient sink (Swift 1977, Foster and Lang 1982). The extensive network of rhizomorphic basidiomycete individuals such as Armillaria and Phanerochaete volutina, imply the capacity to transport nutrients for distances up to 50 m (Thompson and Boddy 1983). A reverse transport of energy sources also occurs with photosynthates from plants to mycorrhizal fungi (Reid et al. 1983).

Vertical separation or redistribution of organic matter and element pools is clearly marked in many soil profiles, but analysis of nutrient flows through canopy, surface organic matter and mineral horizons provides a picture of the dynamics with differential flow of elements (Table 3). The cations Na, K, Ca, Mg are leached from the canopy and surface horizons but retained within the mineral horizons. While N input, as both NH_4^+ and NO_3^-, is retained in the forest floor. Surface N retention probably indicates the zone of high N demand by roots and microflora, even though roots were absent in the lysimeters of Bringmark (1980). However, movement of an element, whilst reflecting its availability and solubility also depends on other elements. Cation transport is matched by anion movement (not fully shown in Table 3) in maintaining electroneutrality and the balance of element solubility and transport shifts during seasonal and longer term acidification whether natural or man-made (Bache

Table 3. Ion flows during April 25–December 1 1975 at different levels in a Pinus sylvestris forest. Depth figures refer to a zero level at the top of the mineral soil. Increase of fluxes after passage of a layer are marked (+) and (++) for the 5% and 1% levels of significance. Decreases of fluxes are marked correspondingly with (−) and (−−). n.s. = not significant (from Bringmark 1980)

Ion flow (mmole m^{-2})

	H	Na	K	Ca	Mg	NH_4	NO_3
Incoming rain	12.2	3.4	1.3	3.2	1.2	6.4	6.6
	n.s	+	+	+	+	−	n.s
Throughfall	11.1	5.7	5.5	5.9	2.6	4.2	6.3
	n.s	n.s	n.s	++	+	n.s	−−
S-layer outflow	11.7	6.0	9.2	13.3	4.4	5.6	1.4
	n.s.	++	n.s.	−	n.s.	−−	−−
6 cm depth	7.6	31.6	7.4	7.2	7.5	1.1	0.55
	−	−−	−	−−	−−	−−	−−
27 cm depth		13.9	0.96	2.2	0.88	0.15	0.16
		n.s.	−	−	n.s.	n.s.	n.s.
80 cm depth	0.38	9.8	0.30	0.65	0.55	0.07	0.08

1980, Ulrich 1980). Selective uptake of NH_4^+ or NO_3^- by plants and release during decomposition can shift the H^+ and OH^- balance both spatially and over time (Nilsson et al. 1982).

Thus, the spatial heterogeneity in microclimate, resource quality and physico-chemical profile provides microbial and faunal habitats with the capacity for a wide range of community structure and function. The microhabitat variation within many ecosystems may be as great as the variation between major ecosystem types so that a wide spectrum of microfloral-faunal interactions is expected within an ecosystem.

Temporal heterogeneity

Decomposition, nutrient release and plant uptake are variable over short and long periods of time, partly through interdependence but also in response to independent factors. The consequences of, and adaptations to, temporal variation in processes of nutrient cycling have received little attention. Within the seasonal pattern of climate, a synchrony of decomposer release and plant uptake of nutrients is indicated by the winter increase in nitrate into stream waters (Caporali et al. 1981). Release of nutrients from plants as litter and root exudates is usually markedly seasonal, as is nutrient uptake; however, uptake at different times of year may allow co-existence of species in a sward (Rorison et al. 1983). Within decomposer communities microbial biomass tends to peak in spring and autumn (Gray et al. 1974, Soderstrom 1979, Clarholm and Rosswall 1980). Much seasonal variation in processes is causally related to climate but it is probable that there is selection for species with differing time patterns of activity as well as for opportunism.

Pulses of net immobilization and mineralization may be related to faunal grazing on microflora where the addition of a simple carbon source causes initial immobilization by increased microbial biomass followed by increased mineralization as a result of faunal grazing (Ausmus et al.

1976, Coleman et al. 1977, Clarholm 1981, Woods et al. 1982). Varying degrees of competition within microflora, and between microflora and roots, are expected over time. The mathematical analysis of Bosatta (1981) indicates that variations in pulses of immobilization and mineralization can result in a state of dynamic equilibrium or of root death. The short-term functional relationships between decomposition and nutrient uptake, linked with microfloral-faunal interactions, are thus areas of particular interest which are amenable to experimental approaches including microcosms in both laboratory and field (Anderson and Ineson 1982).

Long term timing of pulses of nutrient uptake and release may be most clearly seen in single species even-aged forest plantations where asynchrony of production and decomposition processes may be greatest (Heal 1979). Development of a forest stand shows three stages of nutrient availability: (I) Supply from the soil during establishment, (II) Circulation within the crop and through the forest floor following canopy closure, and (III) Potential late deficiency resulting from immobilization of nutrients in biomass and humus (Miller 1981, Cole 1981). To what extent is nutrient uptake related to a sequence in the decomposer system? In Stage I nutrient release from decomposing ground vegetation acts as a source, while accumulation of needle litter acts as a sink. Following canopy and root closure the forest floor acts increasingly as a source as litter input and decomposition come into balance (II). The increased input of wood in Stage III acts as a nutrient sink. The slow turnover of woody fractions can extend beyond the crop rotation. At harvesting the boles remaining on site provide a further sink for N and P which is largely outweighed by enhanced decomposition of fresh residues, thus returning to Stage I. Slow turnover of the large mass of resistant soil organic matter provides a more constant supply of nutrients than shorter term pulses, but the soil organic fraction itself changes with time.

Within the intimate feedback between tree growth and decomposition, it is the rate of tree growth and successional stages which reflect changes in nutrient availability from resident organic matter and recent input. Quantitative assessment of growth limitation due to short term temporal changes in sources and sinks of nutrients and to long term effects of nutrient removal in harvests remains a major challenge which needs to be explored through a variety of approaches, including mathematical models (Aber et al. 1979, Swank and Waide 1980, Bosatta et al. 1980, Bosatta and Staaf 1982). The forest manager or silviculturist takes a more pragmatic approach by controlling nutrient supply through site preparation, fertilization and probably through thinning.

Succession on resource input

Within an ecosystem, spatial and temporal variation in quality and quantity of plant litter input, provide selection mechanisms which maintain microbial and faunal diversity. Resource quality is one of the key variables and analysis of microbial and faunal successions focuses attention on interactions which are developed in communities under different resource availabilities.

The succession of microflora has been studied on many resources (Dickinson and Pugh 1974; Rayner and Todd 1979). Unlike ecosystem succession which tends towards stability in the climax community, resource succession tends to extinction through decomposition of the resource. Fresh resources and invasion by microorganisms reinitiates this cyclical event (Frankland 1981).

Species composition during succession is often related to Webster's 'nutritional hypothesis' (Webster 1970), with composition determined by the type and availability of carbohydrates. These substrates exert selective pressure for the expression of different enzyme complements by the soil microflora. Fungal successions (Table 4) indicate that cellulolytic and lignolytic activity of basidiomycetes dominate later in succession following the earlier removal of readily available carbohydrates. The sequence on wood differs in that lignin-exploiting basidiomycetes are more important in early stages.

Table 4. Difference in species composition of soil microflora between forest litters and their change in vertical distribution. 0 = oak, As = aspen, Al = alder, Sp = spruce and species which are resource-specific in a horizon are shown in heavy type (from Kjøller and Struwe 1982)

	Litter				F				Humus			
	O	As	Al	Sp	O	As	Al	Sp	O	As	Al	Sp
Cryptococcus	O											
Epicoccum	O											
Mycena galopus	O											
Sporobolomyces	O											
Aureobasidium	O		Al	Sp	O							
Cladosporium	O	As	Al	Sp	O		Al		O		Al	
Sterile dark		**As**										
Beauveria		**As**										
Phialophora		**As**										
Phoma			**Al**		O	As	Al			As	Al	
Cylindrocladium			**Al**									
Yeast (yellow)			**Al**									
Alternaria			**Al**				**Al**					
Cylindrocarpon			**Al**				**Al**			**As**		
Hormonema				**Sp**								
Centrospora				**Sp**								
Trichoderma				**Sp**	O	As		Sp	O	As		
Mucor					O				O			
Penicillium					O	As			O	As	Al	Sp
Discula		**As**				**As**						
Mortierella		**As**				As	Al	Sp		As	Al	Sp
Absidia						**As**				**As**		
Acremonium						**As**				**As**		
Verticillium								**Sp**				
Volutella										**As**		
Trichosporiella											**Al**	
Sterile white											**Al**	
Mycelium sterilum												**Sp**
Oidiodendron												**Sp**
Mycelium rad. atro.												**Sp**

Consideration of growth rate, colonization ability and possibly antibiotic production as agencies in competition (Rayner 1978, Frankland 1981, Rai and Srivastava 1983), when linked to the 'nutritional hypothesis', give more validity to concepts of r and K strategists (Gerson and Chet 1981). These categories broadly correspond to Winogradsky's (1924) 'allochthonous' and 'autochthonous' groups, respectively. Further recognition of selection for morphological, physiological and phenological characteristics by 'stress' or 'adversity' factors helps to clarify successional patterns (Grime 1979, Pugh 1980). Adverse conditions are found when resources are very resistant to decomposition due to low nutrients, resistant carbohydrates and inhibitory secondary compounds. Such resources, or stages in resource decomposition, select for relatively slow growth, nutrient conservation mechanisms, intra-specific competition and intimate symbiotic relationships (Heal and Ineson 1984). Most of the information on microbial succession on different substrates is derived from fungal research. The more comprehensive analyses (Remacle 1971), including fungal-bacterial interactions, which can improve these hypotheses remain to be developed.

Although faunal succession on decomposing litters has received less emphasis than fungi, the results indicate some of the main features of microfloral-faunal interactions which are important in decomposition and nutrient release. Faunal succession during wood decomposition shows recurrent features: (1) The importance of temperature and moisture in determining patterns of colonization; (2) Symbiosis between insects and microflora in early succession; (3) The change in food resources causes the replacement of xylophagous insects by fungivores; (4) The increasing diversity associated with physical breakdown of substrates and invasion of forest floor species (Kaarik 1974, Ausmus 1977).

Analysis of fauna during decomposition of leaf litter has concentrated on particular taxa and feeding habits. High polyphenol content of some litters can inhibit colonization and palatability until it is degraded by fungi (Heath and Arnold 1966, Satchell and Lowe 1967, Harding and Stuttard 1974, Anderson 1975). Following polyphenol degradation there is ingestion of Pinus needles by phthiracarid mites and succession of Cryptostigmata on Fagus and Castanaea litter. Early mite colonizers are small, unspecialized feeders, while in later stages larger fungal feeders predominate with large macrophytophagus mites concluding the succession. Similar microarthropod sequences were shown by Luxton (1972) and Hagvar and Kjondal (1981).

Twinn (1974) showed increasing nematode biomass, correlated with hyphal length, on decomposing Quercus litter. The succession of feeding types in nematodes varies with litter type. On *Quercus*, early dominance of fungivores is followed by equal numbers of fungivores and bacterivores, while on *Fagus* bacterivores are replaced by fungivores. Within a wider range of fauna, the importance of mobility, dispersal and rapid growth rate in initial colonizers compared with later stages is recognized in various resources ranging from *Agrostis* and *Festuca* leaves (Curry 1969) to sewage sludge (Huhta et al. 1979).

Consumption by macrofauna is related to resource quality and stage of decomposition as exemplified by earthworms (Satchell and Lowe 1967, Edwards and Lofty 1977). These may be initial colonizers which consume freshly fallen leaf litters, such as *Fraxinus* and *Ulmus*, and later successional organisms in wood decomposition.

The brief comments here mask the morphological and physiological diversity in soil fauna related to feeding and the flexibility in feeding habits (Swift et al. 1979) When combined with taxonomic variety and response to the physical environment, the complexity within soil fauna inhibits all but superficial generalizations. However, the analysis of habitats in relation to ecological strategies by Southwood (1977) provides a series of principles applicable to soil zoological analyses (Fig. 7). In terms of decomposition, the selection for rapid exploitation (r-strategies) followed by more stable interactions (K-strategies) is found on resources of high resource quality. When resource quality is poor, certain characteristics of the soil biota will be selected to overcome adversity (e.g. resistance to inhibitory compounds, conservation of nutrients, synthesis of depolymerization enzymes) rather than for r or K strategies. Low quality resources tend to have high durational stability and therefore select for some of the species characteristics represented by K strategies. Resource quality, however, varies during decomposition, and the quality of adjacent resources can modify the intensity of selection caused by a particular resource. The situation is certainly complex, but

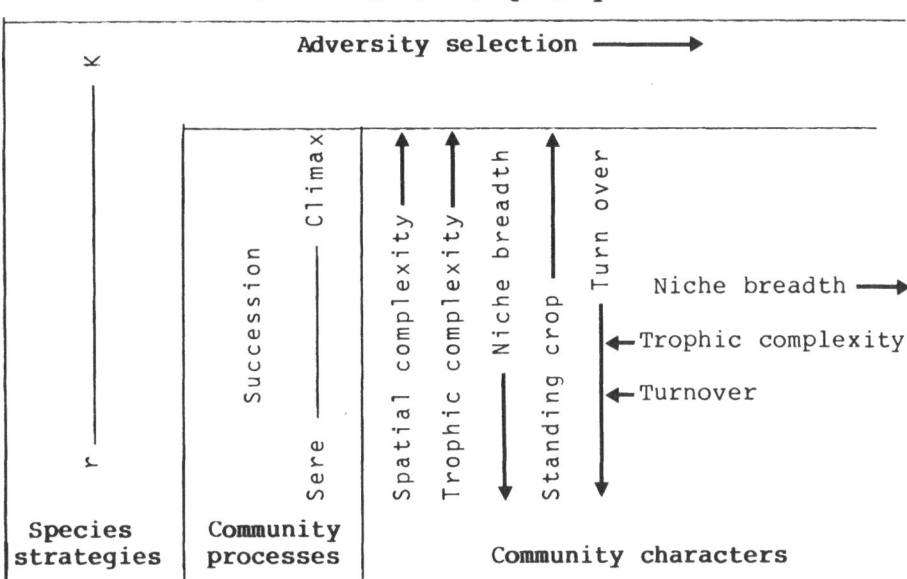

Figure 7. Ecological strategies of fauna related particularly to decomposition of plant litters (Modified from Southwood 1977)

further definition of the combinations of species characteristics selected in response to the physical environment, resource quality and other organisms, offers an opportunity to explain the activities of fauna in decomposition and their relationships with microflora.

Relationships between fauna, microflora and decomposition processes

Microfloral-faunal interactions can enhance C utilization above that of a system containing either only microbial or faunal components. The direct contribution of fauna to heterotrophic respiration is small, usually less than 15% in any ecosystem, but experiments using faunal elimination, exclusion or addition show fauna increase respiration by stimulation of microbial activity (Reichle et al. 1973, Addison and Parkinson 1978, Anderson et al. 1981). Depending on the turnover rate of organisms, nutrient mineralization may also be enhanced, but the strategies adopted by the fauna and microflora in response to resource quality will determine both the degree and the mechanism of enhancement. Although there is stimulation of microbial activity with increased fauna populations there tends to be an optimum density above which fauna decrease activity. Reduction may occur through 'over-grazing' of the microflora by specifically microbial feeders, while in larger fauna over-grazing may be combined with physical disruption of the microbial population (Anderson and Ineson 1983).

The hypothetical model of C utilization and nutrient release in grazed and ungrazed microbial systems (Anderson et al. 1981) (Fig. 8) is exemplified for C in Quercus harvardii litter decay (Elkins and Whitford 1982). In this study the use of insecticide decreased nematophagous mite populations which caused overgrazing of microflora with a concomitant suppression of organic matter loss, indicating hierarchical population control of decomposition. McBrayer (1977)

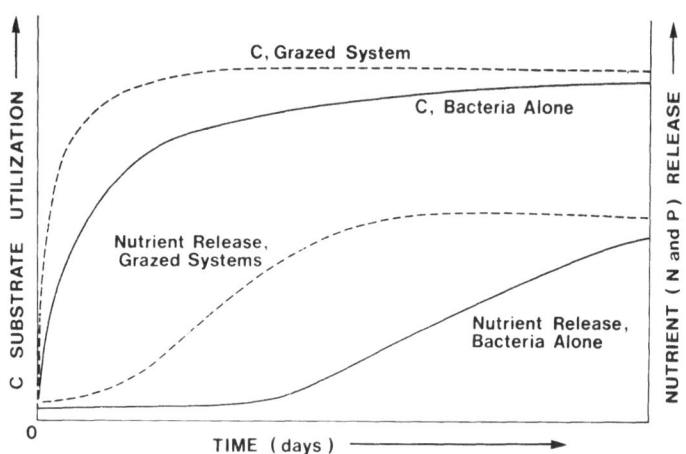

Figure 8. Hypothetical curves of resource utilization and nutrient mineralization with and without grazers (from Anderson et al. 1981)

emphasized the temporal control of nutrient release through immobilization in microbial, or faunal biomass and subsequent release after death. The sequential release of different elements can be highly variable (Anderson and Ineson 1983). Seastedt and Crossley (1980) demonstrated that fauna can enhance P release from litter while retention of Ca, K and Mg was increased. Grazing intensity can influence not only the timing of response but also whether there is stimulation (light grazing) or suppression (heavy grazing) of the microflora as shown for the Collembola, Folsomia candida, grazing on the basidiomycete, Coriolus versicolor(Hanlon and Anderson 1979).

The direct interaction of grazing is amenable to experiment, but, although the preference of individual faunal species for particular fungi (Parkinson et al. 1979), bacteria, or other microflora (Heal and Felton 1970) has been shown repeatedly, there is yet no consensus of general principles or patterns identifying the faunal and microbial characteristics which determine food selection. Nevertheless, the result of grazing can modify the distribution and abundance of microbial species as shown by Newell (1984 a,b). Through a combination of laboratory and field studies, she demonstrated that a change in the outcome of competition between the fast growing but palatable Marasmius androsaceus and the slower growing, less palatable Mycena galopus, could be induced by selective grazing by the collembolan Onychiurus latus.

The mechanisms responsible for increased decomposition in the presence of both fauna and microflora may be chemical, as in the increased palatability of litter to enchytraeids following bleaching by Marasmius (Latter 1977), or physical, as in increased substrate surface area through faunal comminution, classically demonstrated by van der Drift and Witkamp (1951). However the interaction may be a more intimate, symbiotic relationship as exemplified by cellulose decomposition by gut microflora and fauna in various insects (Breznak 1982), or the cultivation of the fungus Termitomyces by termites (Macrotermitinae) (Collins 1982). Such intimate associations may be selected by particularly unfavorable conditions of resource quality or climate, or in highly competitive situations.

The consequence of many microfloral-faunal interactions appears to be an increase in the mineralization and decay rates of organic matter. The latter has been shown chemically, (Coleman et al. 1977) but its significance in plant growth has rarely been quantified. Clarholm (1983) has demonstrated that in experimental systems grazing of bacterial populations by amoebae can enhance N uptake and dry matter production of wheat by 50% compared with ungrazed systems. With larger fauna, microbial stimulation and physical and chemical changes in litter and soil, can similarly enhance plant growth (Hoogerkamp et al. 1983, Syers and Springett 1983). More direct effects on plant growth have been demonstrated due to grazing by nematodes on ericoid mycorrhizae (Shafer et al. 1981) and with nematode and collembolan grazing in agricultural systems (Riffle 1971, Warnock et al. 1982).

Thus, the range of interactions between fauna and microflora is wide but the general results can be summarized as increased decomposition, nutrient release and plant growth. Although only a few examples have been given here the interactions can be categorized as originally presented by Mignolet (1972): (1) predation or grazing, (2) mutualism including comminution and dissemination of propagules, (3) commensalism such as the intimate association of gut microflora, (4) competition for substrates, (5) parasitism and (6) other interactions such as production of fungal toxins. These qualitative interactions are reflected in the conceptual model of Seastedt and Crossley (1980, Fig. 9) and other chapters in this book, especially Chapter 8.

Leaf grazing: a potential influence on decomposition and nutrient cycling

Decomposition and nutrient cycling in soil are influenced in canopy lichens. Microbial succession begins on live leaf surfaces and up to 40% decay of branches can occur in the

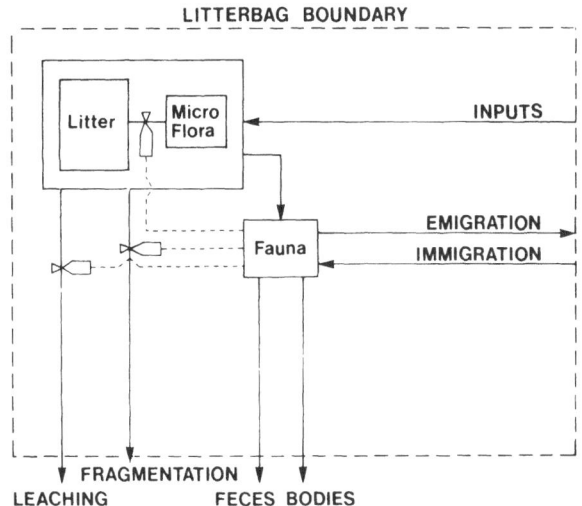

Figure 9. Conceptual model of elemental contents and flux in forest litter. Inputs include particulates, rainfall, throughfall and possible transfers by fungal hyphae. ———— = direct flows, ---- = indirect control of flows (from Seastedt and Crossley 1980)

canopy (Swift et al. 1976). Population outbreaks of insects (Kallio and Lehtonen 1973) result in major pulses of organic matter and nutrients to the decomposer subsystem as litter or frass, but these exceptional events tend to mask the more continuous influence of leaf grazing on decomposition (Larsson and Tenow 1980).

The carrying capacity of some woodland ecosystems for canopy dwelling insect herbivores may be very large. Rafes (1971) found up to 2.5×10^6 larvae/ha of Gypsy moth (Lymantria dispar) on Quercus, depending on stand age. Although litter quantity in infested and non-infested stands was similar, the proportion of entomogenous material and nutrient concentration increased with tree age. Similarly, Tortrix frass from Quercus, showed increases in N, P and K and a decrease of C:N ratio from 50 in leaves to 20 in the frass (Carlisle et al. 1966). Small non-leafy material and frass accounted for only 14% of the litter weight, but about 30% of the N, 40% of the P and 25% of the K, and it fell mainly in spring and early summer.

The proportion of sap sucking insects may be greater than leaf chewers in temperate regions (Moran and Southwood 1982) and this may influence decomposition processes. The production of sugar excretions by aphids of up to 1000 g/m^2 (Llewellyn 1972) can increase the rate of N fixation in soil with possible evolutionary advantage to the tree (Owen 1978; Petelle 1980). Melezitose, a major constituent of aphid secretions (Dighton 1978), may be implicated in this process, although fructose supports higher N fixation (Petelle 1980). A different function is cited by Owen and Weigert (1976) and Owen (1980) for cercopids (Ptyellus) feeding in African forest and savanna. There, excretion of copious quantities of water allows decomposition and nutrient uptake to occur, even during the dry season.

Contrary to the hypotheses of Owen (1980), microcosm experiments indicate that readily available carbohydrates can increase microbial immobilization of nutrients and root-microbe competition, thus reducing growth and survival of seedlings of Pinus sylvestris (Baath et al. 1978, Bosatta 1981). There are many possible explanations for the apparent conflict in results, since carbohydrate concentration, availability of alternative nutrient sources and time scale of response can all influence immobilization. However, most data indicate that faunal activity within the plant canopy influences decomposition processes and has significant effects on plant growth.

PATTERNS BETWEEN ECOSYSTEMS

Community structure and nutrient cycling

Despite considerable variation within mature ecosystem types, and the rather arbitrary nature of the divisions between types, the main pattern of primary productivity is clearly directly related to the combined controls of temperature and rainfall, a theme developed by Leith (1975) and others.

One major difference between ecosystems which still is not adequately quantified is the proportion of net primary production which is allocated to below ground growth. The proportion of biomass which is below ground is often much greater in deserts, grasslands and tundra than in forests, probably as a response to grazing pressure, frequency of disturbance and climate (Lange et al. 1983). Most biomass measurements ignore fine roots which, in forests, may turnover annually (McClaugherty et al. 1982) or 5-10 times per season, with a production approximately equal to leaf litter (Jarvis and Leverenz 1983). With the addition of root sloughing and exudation, the differences between ecosystems may not be so great as appears from biomass data.

The quality of the input to decomposition from roots may also show less variation between ecosystems than is superficially apparent. Withdrawal of nutrients and soluble carbohydrates before death tend to leave a resource dominated by structural carbohydrate, with decay rates more similar to wood than to leaves (McClaugherty et al. 1982, Berg 1984). Unfortunately, most root decomposition studies have used live rather than naturally dead roots, probably resulting in over-estimates of their decay rates. In contrast to the input of particulate material, soluble organic fractions exuded from roots are rapidly decomposed.

The general pattern of decomposition, assumed from measured input and surface accumulation of organic matter, shows trends in control by temperature and moisture which are similar to that for primary production (Swift et al. 1979, Esser et al. 1982, Anderson and Swift 1983), with the implied parallel of turnover of nutrients. The pattern of decomposition of organic matter within the soil is less clear. Both quality and quantity of input and its rate of decomposition are poorly understood and estimates of the standing crop of soil organic matter are very variable, without marked distinctions between temperate and tropical or between forest and grassland ecosystems (Sanchez et al. 1982). Even within forests, soil organic matter standing crops show no clear evidence of a gradient from equatorial regions to near the Arctic Circle (Anderson and Swift 1983). The search for general patterns obscures, however, potentially important distinctions between ecosytems. For example, of greater interest than the total amount of soil organic matter, are the formation and decomposition of the functionally important fractions and in their regulation of nutrient supply to plants.

The climatically controlled pattern is more clear in the decomposition of individual litters or standard substrates (Meentemeyer 1978, Heal et al. 1981). However, superimposed on this pattern is the large variation resulting from differences in resource quality with major ecosystem types. Thus there is a considerable overlap in rate of litter decomposition between ecosystems, for example between temperate and tropical forests (Anderson and Swift 1983). Nutrient circulation broadly follows production and decomposition of organic matter but there are some marked differences related to plant growth characteristics and to nutrient conservation. Using N as an example, Clark and Rosswall (1981) indicated that N concentration in plant biomass and rates of nitrogen uptake by forests are more similar to non-forest ecosystems than are dry matter levels. Except for the low concentration in wood, internal resorbtion before litter fall and internal recycling can minimize annual uptake. The N efficiency of production, defined as above-ground dry matter production per unit uptake of N, in forests (168 ± 49 kg.kg^{-1}N.ha^{-1}.y^{-1}) is 2-3 times higher than in other ecosystems (Cole 1981).

Interpretation must be cautious because belowground production has not been fully accounted, and translocation to belowground parts in tundra, grasslands and deserts may be greater than estimated. The low uptake and high N efficiency in conifers compared with deciduous forests is related to needle retention and translocation and probably also applies to many evergreen dwarf shrubs and desert plants. The distinction between deciduous and evergreen trees occurs in soils of similar nutrient status indicating a phylogenetic characteristic as well as possible adaptation to nutrient shortage (Cole 1981). Certainly the efficiency of nutrient use within a species varies with nutrient supply (Gosz 1981, Miller 1981).

A less commonly recognized phenomenon is the potentially positive feedback effect of plant nutrient conservation. Increased translocation before litter fall reduces nutrient concentration in litter which, combined with greater proportions of structural and secondary compounds in perennial tissues, retards decomposition. This can exacerbate plant nutrient deficiency and may be a mechanism in succession and competition, and, as indicated by Gosz (1981) in the balance between mull and mor humus formation.

Microflora

Comparison of population densities, biomass and activity of microflora between ecosystems is severely limited by restriction of data to a few ecosystems, the lack of standardized techniques and the selectivity of both research workers and culture methods. From data compiled by Kjøller and Struwe (1982) it appears that maximum hyphal length (per g) is attained in temperate coniferous forests and tundra, and declines in deciduous forests and grasslands (Fig. 10a). With increasing bulk density of soil in tundra, through coniferous and deciduous forests to grasslands, the fungal weight per unit area increases in temperate deciduous forests and grassland. The limited data show that as little as 10% of hyphae is living.

The ordination analyses of Christensen (1981) indicate some specificity of habitat resulting in patterns of fungal communities associated with primary producers. The association of trees with basidiomycete mycorrhizal fungi, and the association of grassland vegetation with vesicular-arbuscular mycorrhizae, may reflect more general distribution patterns of basidiomycetes. There are indications that bacterial populations become increasingly important in deciduous woodlands and grasslands, possibly associated with trends in earthworm populations (Kozlovskaja 1969). Sundman (1970) and others have demonstrated taxonomic and physiological distinctions between bacterial populations of different soils with some evidence of decreasing

52

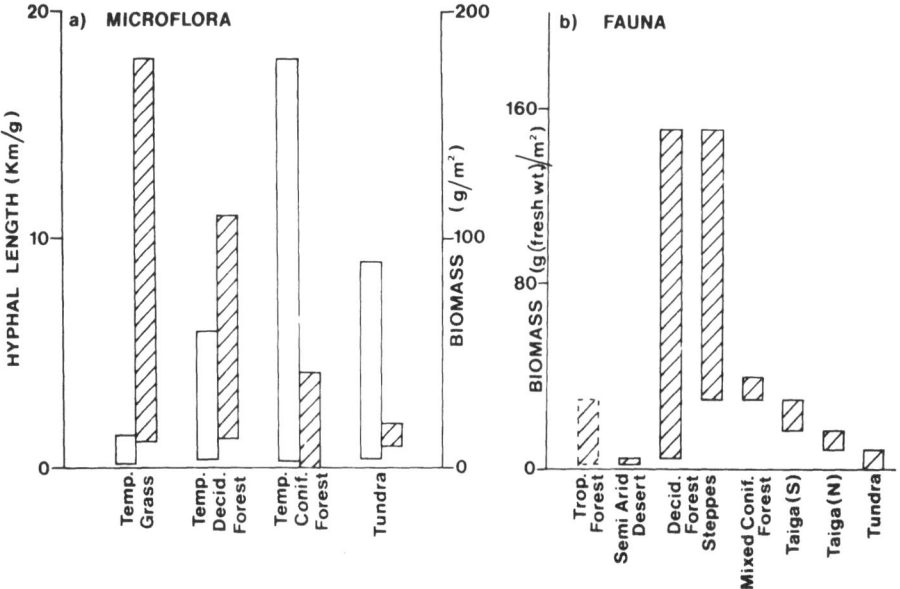

Figure 10. (a) Maximum and minimum estimates of fungal biomass in different ecosystems, expressed as hyphal length (Km/g) and mass (g/m²) (derived from Kjøller and Struwe 1982); (b) Maximum and minimum estimates of faunal biomass (g fresh weight/m²) in different ecosystems. The tropical forest has been expanded to include termites (from Petersen and Luxton 1982)

physiological versatility of the populations with decreasing soil fertility (Rosswall and Kvillner 1978). However there seems to be a greater similarity in the microflora between ecosystems than among other plants and animals.

Swift (1976) argued that there is little correlation among the gross features of difference between macrohabitats (ecosystems) and soil microflora. Data on woodland microflora (Kjøller and Struwe 1982) tend to support Swift's view that the quality of the primary resources within habitat is more important than differences in species composition of the primary producers.

Fauna

Collation of data on soil fauna (Petersen and Luxton 1982) supported by specific data from the Russian literature, suggests increasing faunal biomass from the northern tundra to deciduous temperate forests and grasslands and then a decline through semi-desert and desert ecosystems to tropical grasslands and forests where biomass increases again (Fig. 10b).

Within this change in biomass, correlated with change in quantity and quality of litter, moisture and temperature, change in the community structure also occurs (Table 5). These data, together with summaries from Wallwork (1976), suggest that in tundra and taiga ecosystems the microarthropods, enchytraeids and Diptera (tipulid) larvae predominate. Collembola show little change in population between non-wooded ecosystems, whereas Acari are more prominent in woodlands. The tendency for fungal domination in forests is probably causally related to the mite dominated fauna with a high Cryptostigmata component, many of which are mycophagous. Mesostigmatid mites may account for up to 20% of the mite fauna in temperate forests and do not exhibit the same population decline as other mite groups in tropical forests.

Table 5. Variation in dominant fauna, as estimated by biomass, of different ecosystems. Values are in mg/m^2. With ranges for mor[a] and mull[b] humus types in some cases data for tropical systems are given separately[c]. Dominant fauna in each ecosystem shown in bold type (modified from Petersen and Luxton 1982)

	Tundra	Forest Coniferous	Forest Deciduous	Grassland
Formicoidea	0	10	10	**100**
Diplopoda	0	50	**420**	**1000**
Oligochaeta	330	450	**200[a]–5300[b]**	**3100**
Gastropoda	0	20	**270**	**100**
Chilopoda	20	70	**130**	**140**
Nematoda	160	120	**330**	**440**
Acari	90	**500**	**300[b]–900[a]**	120
Diptera larvae	**470**	**260**	**330**	60
Enchytraeidae	**1800**	**480**	430	330
Collembola	**150**	80	**100[b]–130[a]**	90
Oligochaeta			**340[c]**	170[c]
Isoptera			**1000[c]**	**1000[c]**
Formicoidea			30[c]	**300[c]**

A decrease in the C:N ratio and soil acidity of deciduous forest and grassland litter is associated with increased dominance of larger saprophages, particularly earthworms. Earthworm burrowing activity in soils enhances the development of mull humus type with an increase in organo-mineral complexes, tending to promote an increase in bacterial populations (Kozlovskaja 1969) with an associated expansion of populations of Protozoa and Nematoda. In temperate grasslands Diplopoda, Chilopoda, Gastropoda and Formicoidea join large Oligochaeta as the dominant macro fauna but in tropical grasslands and forests the Isoptera (termites) become the dominant group. The latter tend to exploit niches occupied by a greater diversity of species in the temperate ecosystems. Hot and cold deserts have a low faunal species diversity but populations of groups of organisms may be as high as in less extreme environments, with associated physiological and behavioural survival strategies (Block 1980, Wallwork 1982, Greenslade 1982).

This superficial analysis of faunal variation between biomes is based on the most readily obtainable information of species, numbers and biomass, but is not easily related to ecosystem processes. One tentative hypothesis (Fig. 11) suggests that the faunal contribution to decomposition decreases with latitude, although severely constrained by climate in deserts. A more mechanistic view is taken by MacMahon (1981) who approached the comparison of ecosystems through the factors influencing successional processes (Table 6). Although his faunal analysis includes a strong element of the role of vertebrates, the general hypothesis has some validity for soil fauna and provides a framework for the development of a more functional approach.

GENERAL DISCUSSION

There is strong evidence for a general pattern of processes during ecosystem succession, with accumulation of nutrients in biomass and in dead organic matter. Development is towards a balance of input and output of nutrients and

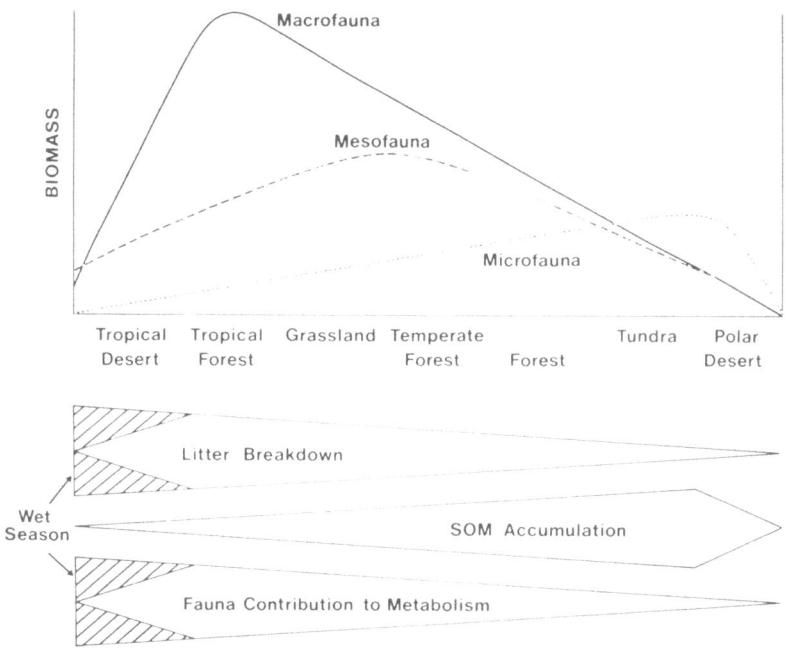

Figure 11. Hypothetical patterns of latitudinal variation in
the contribution of the macro-meso- and micro-fauna to total
soil fauna biomass. The effects on litter breakdown rates of
change in the relative importance of the three fauna size
groups are represented as a gradient together with the faunal
contribution to soil community metabolism. The favorability
of the soil environment for microbial decomposition is
represented by the cline of soil organic matter (SOM)
accumulation from the poles to the Equator (from Swift et
al. 1979)

increasing inter-dependence of production and decomposition,
resulting in increasing internal regulation of processes.
Ecosystem processes are however limited in their expansion by
the general climatic and soil regimes. Although the species
composition of succession is strongly influenced,
particularly in the initial stages, by random, opportunistic
events, there tends to be a recurrent pattern in different
ecosystems of growth characteristics which are common to the
vegetation, fauna and microflora. These characteristics are
common among organisms because:

1. Similarity of the influence of C and nutrient availability
2. Responses to the same environmental conditions, although there may be some differences in the micro-climatic regimes within vegetation, litter and soil, and in particular environmental factors relevant to specific organisms.
3. Responses to the same disturbances. Although disturbances tend to affect the decomposer system less directly and severely than the vegetation, the decomposer system also acts

Table 6. Successional processes involving animals among various biomes. The specfic comparisons summarized are nudation, the role of animals as agents of site disturbance; migration, the role of animals in dispersing plant propagules, only positive effects are summarized, ecesis, the role of animals in enhancing plant germination and establishment; biotic interaction, the importance of species-species interactions involving animals in determining the vector of succession; reaction, the degree to which animals are agents in altering the physical and chemical environment of a site; stabilization, the role of animals in determining and maintaining a given, quasi-equilibrium, plant species mix and in determining the cycling of nutrients (from MacMahon 1981)

	DESERT	TUNDRA	GRASSLAND	CONIFEROUS FOREST	DECIDUOUS FOREST	RAIN FOREST
NUDATION	Low	Moderate	Mod. high	High	Mod. low	Very high
MIGRATION	Low	Low	Moderate	Mod. high	High	Very high
ECESIS	Low	Low	Moderate	Moderate	Mod. high	High
BIOTIC INTERACTION	Low	Low	Low	Mod. low	Moderate	High
REACTION	Low	Mod. low	High	High	Moderate	Low
STABILIZATION (Sp. Comp.)	Low	Low	Moderate	Moderate	High	Very high
STABILIZATION (Nutrients)	Mod. high	Low	High	Mod. low	Mod. high	Moderate

as a buffer providing some continuity of conditions for vegetation regeneration.

The general increase in the energy flow and homeostasis during successional development is represented in Figure 12. A simplified linear sequence is represented for fertile climatic and soil conditions selecting for organisms adopting a sequence of r-K strategies. The link between the plant and decomposer sub-systems is a function of litter quality, which reflects the plant growth strategies and, in turn, selects for decomposer microflora and fauna with r-K characteristics. Thus, as indicated by MacMahon (1981), "animals, (and microorganisms), to a surprising degree, are decoupled from specific plant species but are coupled to the architecture of the plants".

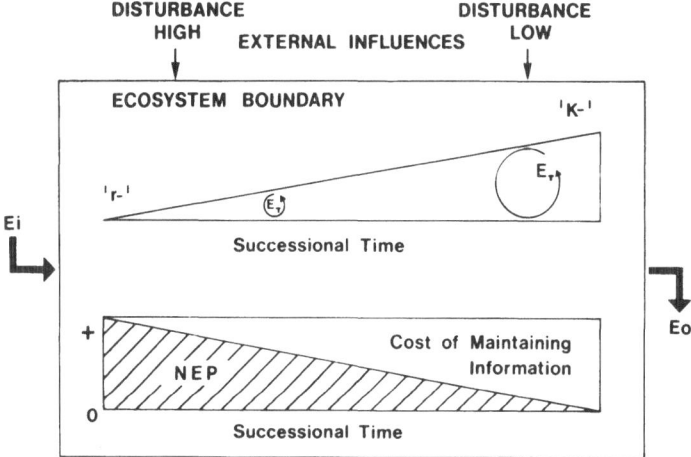

Figure 12. General concept of ecosystem development, with expansion of information reflecting accumulation of biomass, nutrients and biomass, nutrients and biotic structure during succession. Energy transformation (E_T) within the system function as homeostatic mechanisms. High and low frequencies of disturbance select for r and k strategies within the biota. The cost of maintaining biomass and activity in relation to energy fixation increases with time, hence net ecosystem productivity (NEP) moves from being positive in early stages to zero in late stages of succession. The balance of energy input (E_i) and output (E_o) changes during succession, with energy accumulation in early stages, but with equilibrium in later stages representing the higher net cost of homeostatic mechanisms

The variety of resource inputs tends to increase with succession, resulting in increased soil organism diversity. The pattern of diversity with macro-succession is paralleled on a microscale by the microbial and faunal succession on individual resources in response to changes in micro-physical and substrate conditions (Swift 1976). The resource succession, in contrast to ecosystem succession, results in the eventual destruction of the resource, the decomposer community being regenerated by fresh resource input, analogous to disturbance. Upon disturbance, the degree of reversion of the ecosystem in successional time is dependent on intensity of disturbance, but the decomposer sub-system in particular remains relatively intact, i.e. it retains information in terms of organisms, resources and nutrients (Fig. 12). The 'cost of maintaining information' is heterotrophic respiration which results in continued nutrient mineralization. Where disturbance drastically reduces primary production, maintenance cost briefly exceeds input and net ecosystem productivity becomes negative, with net mineralization exceeding uptake until the intimate coupling of production and decomposition is restored.

A third dimension not shown in Fig. 12 is that of stress or adversity in that some of the unfavourable environmental conditions such as extreme climatic variation or nutrient poverty. The concepts of stress (Grime 1979) or adversity (Southwood 1977, Greenslade 1982) strategies appear to apply to the soil microflora and fauna with implications to decomposition and nutrient cycling. In the highly fluctuating climatic conditions of tundra and desert, selection will favour organisms which can endure conditions of low productivity through slow growth rate and turnover coupled with mechanisms which allow exploitation of temporarily favourable conditions. Mechanisms may be dormancy or encystment, with reactivation in response to climatic events, movement to locally more favourable microclimates, or physiological tolerances to various stresses. This 'discontinuous adversity' selection contrasts with the 'continuous adversity' of resources such as wood which, while having climatic stability, is low in nutrients and readily available energy, and sometimes high in inhibitory secondary compounds. Conditions of poor resource quality will tend to favour specific enzymic capabilities and nutrient conservation, which are often linked to slow growth rates and large size (Heal and Ineson 1984).

The combined influence of disturbance and adversity (largely extrinsic factors) and competition (intrinsic), resulting in the selection of organisms with different characteristics, will have a marked effect on the type of interaction between fauna and microflora:

1. Under conditions of high disturbance and low adversity, the selection for small size, rapid growth and dispersal (r), will tend to produce rapid and large population fluctuations, dominance of direct grazing interactions, with little food specificity, and high rates of nutrient release.

2. Under conditions of low disturbance and low adversity, the K strategists will exhibit more stable populations, with density-dependent relationships which reflect more continuous grazing and a relatively high degree of food selection. Food chains may be long but intimate symbiotic associations involving nutrient conservation are likely to occur. Indirect interactions will tend to be common (e.g. large body size will tend to increase the importance of comminution), while long generation times will tend to be associated with development of biochemical 'defense' mechanisms.

3. Under conditions of discontinuous adversity the extreme climatic conditions tend to restrict diversity and hence increase potential niche breadth (Southwood 1977). This implies wider range of substrate use by the microflora and less selective grazing by fauna, with short food chains and irregular population dynamics influenced mainly by climate. Where moisture is severely limiting, intimate symbiosis as with gut microflora may serve as a mechanism maintaining microbial activity. In contrast, under conditions of continuous adversity of resource quality, associated with a relatively stable environment, a limited range of large, slow growing, organisms and specific grazing or symbiotic

associations, possibly with damped, density-dependent population fluctuations, will tend to occur.

The application of the hypotheses of selection of ecological strategies in response to disturbance and adversity, has yet to be fully explored for soil fauna and microflora (Grime 1979, Pugh 1980, Heal and Ineson 1984). However, it is argued that variations in resource quality, either between resources or within the same resource as it decomposes, have a major influence on the composition of the decomposer microfloral and faunal community. As a result, major variations in the type and intensity of interaction between microflora and fauna can be expected, related to resource quality but with some modification by climate. Although a wide variation in resource quality, and thus of microfloral-faunal interactions, occurs within most ecosystems, some differences in quality can be expected during primary and secondary succession as a result of selection for different plant growth strategies. Thus, while the general control of primary production, decomposition and nutrient cycling processes may be related to the physico-chemical environment, an understanding of the organism interactions is required to explain the subtleties of these processes.

REFERENCES

1. Aber, J.D., Botkin, D.B., and Melillo, J.M. 1979. Predicting the effects of different harvesting regimes on productivity and yield in northern hardwoods. Can. J. For. Res. 9: 10-14.
2. Addison, J.A., and Parkinson, D. 1978. Influence of collembolan feeding activities on soil metabolism at a high arctic site. Oikos 30: 529-538.
3. Alexander, I.J. 1983. The significance of ectomycorrhizas in the nitrogen cycle. Pages 69-73 in J.A. Lee, S. McNeill and I.H. Rorison, editors. Nitrogen as an ecological factor. Blackwell, Oxford, UK.
4. Anderson, J.M. 1975. Succession, diversity and trophic relationships of some soil animals in decomposing leaf litter. J. Anim. Ecol. 44: 475-495.

5.   Anderson, J.M. and Ineson, P. 1982. A soil microcosm
     system and its application to measurements of
     respiration and nutrient leaching. Soil Biol.
     Biochem. 14: 415-416.
6.   Anderson, J.M., and Ineson, P. 1983. Interactions
     between soil arthropods and microorganisms in carbon,
     nitrogen and mineral element fluxes in decomposing
     leaf litter. Pages 413-432 in J.A. Lee, S. McNeill
     and I.H. Rorison, editors. Nitrogen as an ecological
     factor. Blackwell, Oxford, UK.
7.   Anderson, J.M. and Swift, M.J. 1983. Decomposition in
     tropical forests. Pages 287-309 in S.L. Sutton, T.C.
     Whitmore and A.C. Chadwick, editors. Tropical rain
     forest: ecology and management. Blackwell, Oxford,
     UK.
8.   Anderson, R.V., Coleman, D.C., and Cole, C.V. 1981.
     Effects of saprotrophic grazing on net mineralization.
     Pages 201-216 in F.E. Clark and T. Roswall, editors.
     Terrestrial nitrogen cycles. Ecol. Bull. (Stockholm)
     33.
9.   Ausmus, B.S. 1977. Regulation of wood decomposition
     rates by arthropod and annelid populations. Pages
     180-192 in U. Lohm and T. Persson, editors. Soil
     organisms as components of ecosystems. Ecol. Bull.
     (Stockholm) 25.
10. Ausmus, B.S., Edwards, N.T., and Witkamp, M. 1976.
     Microbial immobilization of carbon, nitrogen
     phosphorus and potassium: implications for forest
     ecosystem processes. Pages 397-416 in J.M. Anderson
     and A. Mcfadyen, editors. The role of terrestrial and
     aquatic organisms in decomposition processes.
     Blackwell, Oxford, UK.
11. Baath, E., Lohm, U., Lundgren, B., Rosswall, T.,
     Soderstrom, B., Sohlenius, B., and Wiren, A. 1978.
     The effect of nitrogen and carbon supply on the
     development of soil organism populations and pine
     seedlings: A microcosm experiment. Oikos 31: 153-163.
12. Bache, B.W. 1980. The acidification of soils. Pages
     375-380 in T.C. Hutchinson and M. Havas, editors.
     Effects of acid precipitation on terrestrial
     ecosystems. Plenum, New York, USA.
13. Berg, B. 1984, in press. Decomposition of root litter
     and some factors regulating the process: long-term
     root litter decomposition in a Scots pine forest.
     Soil Biol. Biochem.
14. Berg, B., and Staff, H. 1981. Leaching, accumulation
     and release of nitrogen in decomposing forest litter.
     Pages 163-178 in F.E. Clark and T. Rosswall, editors.
     Terrestrial nitrogen cycles. Ecol. Bull. (Stockholm)
     33.
15. Blackburn, T.R. 1973. Information and the ecology of
     scholars. Science 181: 1141-1146.
16. Block, W. 1980. Survival strategies in polar
     terrestrial arthropods. Biol. J. Linn. Soc. 14:
     29-38.

17. Borman, F.H., and Likens, G.E. 1979. Pattern and process in a forested ecosystem: disturbance, development and the steady state. Springer-Verlag, New York, USA.

18. Bosatta, E. 1981. A qualitative analysis of the root-microorganism soil system. II. Combined effect of several factors. Ecol. Mod. 13: 237-245.

19. Bosatta, E. and Staff, H. 1982. The control of nitrogen turnover in forest litter. Oikos 39: 143-151.

20. Bosatta, E., Bringmark, L., and Staaf, H. 1980. Nitrogen formation in a Scots pine forest - model analysis of mineralization, uptake by roots and leaching. Pages 565-589 in T. Persson, editor. Structure and function of northern coniferous forests - an ecosystem study. Ecol. Bull. (Stockholm) 32.

21. Breznak, J.A. 1982. Intestinal microbiota of termites and other xylophagous insects. Ann. Rev. Microbiol. 36: 323-343.

22. Bringmark, L. 1980. Ion leaching through a podsol in a Scots pine stand. Pages 341-361 in T. Persson, editor. Structure and function of northern coniferous forests - an ecosystem study. Ecol. Bull. (Stockholm) 32: 25.

23. Brown, J., Miller, P.C., Tieszen, L.L., and Bunnell, F.L. 1980. An Arctic ecosystem: The coastal tundra at Barrow, Alaska. Dowden, Hutchinson and Ross, Stroudsburg, USA.

24. Bunnell, F.L., and Scouller, K.A. 1981. Between-site comparisons of carbon flux in tundra using simulation models. Pages 685-715 in L.C. Bliss, O.W. Heal and J.J. Moore, editors. Tundra ecosystems: a comparative analysis. Cambridge University Press, Cambridge, UK.

25. Caporali, P., Nannjipieri, P., and Pedrazzini, F. 1981. Nitrogen content of streams draining an agricultural and a forested watershed in central Italy. J. Environ. Qual. 10: 72-76.

26. Carlisle, A., Brown, A.H.F., and White, E.J. 1966. Litter fall, leaf production and the effects of defoliation by Tortrix viridana in a sessile oak (Quercus petraea) woodland. J. Ecol. 54: 65-85.

27. Chapin, F.S., Miller, P.C., Billings, W.D. and Coyne, P.E. 1980. Carbon and nitrogen budgets and their control in coastal tundra. Pages 458-489 in J. Brown, P.C. Miller, L.L. Tieszen and F.L. Bunnell, editors. An Arctic ecosystem: The coastal tundra at Barrow, Alaska. Dowden, Hutchinson and Ross, Stroudsburg, USA.

28. Chapman, S.B., and Webb, N.R. 1978. The productivity of a Calluna heathland in southern England. Pages 247-262 in O.W. Heal and D.F. Perkins, editors. Production ecology of British moors and montane grasslands. Springer-Verlag, Berlin, Germany.

29.   Christensen, M. 1981. Species diversity and dominance
      in fungal communities. Pages 201-232 in D.T. Wicklow
      and G.C. Carroll, editors. The fungal community, its
      organization and role in the ecosystem. Marcel
      Dekker, New York, USA.
30.   Clarholm, M. 1981. Protozoan grazing of bacteria in
      soil-impact and importance. Microb. Ecol. 7: 343-350.
31.   Clarholm, M. 1983. Dynamics of soil bacteria in
      relation to plants, protozoa and inorganic nitrogen.
      Institute of microbiology, Report 17. Swedish Univ.
      Agjric. Sci. Uppsala.
32.   Clarholm, M., and Rosswall, T. 1980. Biomass and
      turnover of bacteria in a forest soil and a peat.
      Soil Biol. Biochem. 12: 49-51.
33.   Clarke, F.E., and Rosswall, T. 1981. Terrestrial
      nitrogen cycles. Ecol. Bull. (Stockholm) 33.
34.   Clements, F.E. 1916. Plant succession: An analysis of
      the developments of vegetation. Carneigie Inst. Wash.
      Publ. No. 242.
35.   Clymo, R.S. 1978. A model of peat bog growth. Pages
      187-223 in O.W. Heal and D.F. Perkins, editors.
      Production ecology of British moors and montane
      grasslands. Springer-Verlag, Berlin, Germany.
36.   Cole, D.W. 1981. Nitrogen uptake and translocation by
      forest ecosystems. Pages 219-232 in F.E. Clark and T.
      Rosswall, editors. Terrestrial nitrogen cycles.
      Ecol. Bull. (Stockholm) 33.
37.   Coleman et al. 1977. An analysis of rhizosphere-
      saprophage interactions in terrestrial ecosystems.
      Pages 299-309 in U. Lohm, editor. Soil organisms as
      components of ecosystems. Ecol. Bull. (Stockholm) 25.
38.   Collins, M. 1982. The importance of being a
      bugga-bug. New Scientist 94: 834-837.
39.   Cousins, S.H. 1980. A trophic continuum derived from
      plant structure, animal size and a detritus cascade.
      J. theor. Biol. 82: 607-618.
40.   Cromack, J. 1981. Below-ground processes in forest
      succession. Pages 361-373 in D.A. West, H.H. Shugart
      and D.B. Botkin, editors. Forest succession: concepts
      and applications. Springer-Verlag, New York, USA.
41.   Curry, J.P. 1969. The decomposition of organic
      matter in soil Part II. The fauna of decaying
      grassland herbage. Soil Biol. Biochem. 1: 259-266.
42.   De Angelis, D.L. 1980. Energy flow, nutrient cycling
      and ecosystem resiliance. Ecology, 61: 764-771.
43.   De Selm, H.R., and Shanks, R.E. 1963. Accumulation
      and cycling of organic matter and chemical
      constituents during early vegetational succcession on
      a radioactive waste disposal area. Pages 83-96 in V.
      Schultz and A.W. Klement, editors. Radioecology.
      Reinhold, New York, USA.
44.   Dickson, B.A., and Crocker, R.L. 1953. A
      chronosequence of soils and vegetation near Mt.
      Shasta, California. II. The development of the forest
      floors and the carbon and nitrogen profiles of the
      soils. J. Soil Sci. 4: 142-154.

64

45. Dickinson, C.H., and Pugh, G.J.F. 1974. Biology of plant litter decomposition. Academic Press, London, UK.

46. Dighton, J. 1978. Effects of synthetic lime aphid honeydew on populations of soil organisms. Soil Biol. Biochem. $\underline{10}$: 369-376.

47. Dunger, Von, W. 1969. Fragen der naturlichen und experimentallen besiedlung kulturfeindlicher boden durch lumbriciden. Pedobiol. $\underline{9}$: 146-151.

48. Edwards, C.A., and Lofty, J.R. 1977. Biology of earthworms. Chapman and Hall, London, UK.

49. Elkins, N.Z. and Whitford, W.G. 1982. The role of microarthropods and nematodes in decomposition in a semi-arid ecosystem. Oecologia (Berl) $\underline{55}$: 303-310.

50. Esser, G., Aselmann, I., and Leith, H. 1982. Modelling the carbon reservoir in the system compartment "Litter". Mitt. Geol-Palaont. Inst. Univ. Hamburg $\underline{52}$: 39-58.

51. Flanagan, P.W., and Bunnell, F.L. 1980. Microflora activities and decomposition. Pages 291-334 in J. Brown, P.C. Miller, L.L. Tieszen and F.L. Bunnell, editors. An Arctic ecosystem: The coastal tundra at Barrow, Alaska, USA. Dowden, Hutchinson and Ross, Stroudsburg, USA.

52. Floate, M.J.S. 1970. Mineralization of nitrogen and phosphorus from organic materials of plant and animal origin and its significance in the nutrient cycle of grazed upland and hill soils. J. Br. Grass. Soc. $\underline{25}$: 295-302.

53. Foster, J.R., and Lang, G.E. 1982. Decomposition of red spruce and balsam fir poles in the white mountains of New Hampshire. Can. J. For. Res. $\underline{12}$: 617-626.

54. Frankland, J.C. 1981. Mechanisms in fungal successions. Pages 403-426 in D.T. Wicklow and G.C. Carrol, editors. The fungal cummunity, its organization and role in the ecosystem. Marcel Dekker, New York, USA.

55. Frankland, J.C. 1982. Biomass and nutrient cycling by decomposer basidiomycetes. Pages 241-261 in J.C. Frankland, J.N. Hedger and M.J. Swift, Editors. Decomposer basidiomycetes: their biology and ecology. Cambridge University Press, Cambridge, UK.

56. Frissel, M.J., and Penders, R. 1983. Models for the accumulation of migration of ^{90}Sr, ^{137}Cs, ^{239,240}Pu and ^{241}Am in the upper layer of soil. Pages 63-73 in P.J. Coughtrey, editor. Ecological aspects of radionuclide release. Blackwell, Oxford, UK.

57. Gerson, U., and Chet, I. 1981. Are allochthonous and autochthonous soil microorganisms r- and K- selected? Rev. Ecol. Biol. Sol $\underline{18}$: 285-289.

58. Gorham, E., Vitousek, P.M., and Reiners, W.A. 1979. The regulation of chemical budgets over the course of terrestrial ecosystem succession. Ann. Rev. Ecol. Syst. $\underline{10}$: 53-84.

59. Gosz, J.R. 1981. Nitrogen cycling in coniferous ecosystems. Pages 405-426 in F.E. Clarke and T. Rosswall, editors. Terrestrial nitrogen cycles. Ecol. Bull. (Stockholm) 32.

60. Gray, T.R.G., Hissett, R., and Duxbury, T. 1974. Bacterial populations of litter and soil in a deciduous woodland. II. Numbers, biomass and growth rates. Rev. Ecol. Biol. Sol 11: 15-26.

61. Greenslade, P.J.M. 1982. Selection processes in arid Australia. Pages 125-130 in W.R. Barker and P.J.M. Greenslade, editors. Evolution of the flora and fauna of arid Australia. Peacock, South Australia.

62. Grime, J.P. 1979. Plant strategies and vegetation processes. Wiley, New York, USA.

63. Hagvar, S., and Kjondal, B.R. 1981. Succession, diversity and feeding habits of microarthropods in decomposing birch leaves. Pedobiol. 22: 385-408.

64. Hanlon, R.D.G. and Anderson, J.M. 1979. The effects of Collembola grazing on microbial activity in decomposing leaf litter. Oecologia (Berl.) 38: 93-99.

65. Harding, D.J.L., and Stuttard, R.A. 1974. Microarthropods. Pages 489-532 in C.H. Dickinson and G.J.F. Pugh, editors. Biology of plant litter decomposition. Academic Press, London, UK.

66. Harris, W.F., Santantonio, D. and McGinty, D. 1980. The dynamic belowground ecosystem. Pages 119-129 in R.H. Waring, editor. Forests: fresh perspectives from ecosystem analysis. Oregon State Univ. Press, Oregon, USA.

67. Heal, O.W. 1979. Decomposition and nutrient release in even-aged plantations. Pages 257-291 in E.D. Ford, D.C. Malcolm and J. Atterson, editors. The ecology of even-aged forest plantations. Inst. Terr. Ecol., Cambridge, UK.

68. Heal, O.W. and Felton, J.M. 1970. Soil amoebae: their food and their reaction to microflora exudates. Pages 145-162 in A. Watson, editor. Animal populations in relation to their food resources. Blackwell, Oxford, UK.

69. Heal, O.W. and Ineson, P. 1984 in press. Carbon and energy flow in terrestrial ecosystems - relevance to microflora. In M.J. Klug et al., editors. Current perspectives in microbial ecology. American Society for Microbiology, Washington, USA.

70. Heal, O.W. and MacLean, S.F. 1975. Comparative Productivity in Ecosystems - Secondary Productivity. Pages 89-108 in W.H. van Dobben, and R.H. Lowe-McConnell, editors. Unifying Concepts in ecology. Junk, Hague, Netherlands.

71. Heal, O.W., Latter, P.M. and Howson, G. 1978. A study of the rates of decomposition of organic matter. Pages 136-159 in O.W. Heal and D.F. Perkins, editors. Production ecology of British moors and montane grasslands. Springer-Verlag, Berlin, Germany.

72. Heal, O.W., Swift, M.J. and Anderson, J.M. 1982. Nitrogen cycling in United Kingdom Forests: the relevance of basic ecological research. Phil. Trans. R. Soc. Lond. B, 296:427-444.

73. Heal, O.W., Flanagan, P.W., French, D.D. and MacLean, S.F. 1981. Decomposition and accumulation of organic matter. Pages 587-633 in L.C. Bliss, O.W. Heal and J.J. Moore, editors. Tundra ecosystems : a comparative analysis. Cambridge Univ. Press, Cambridge, UK.

74. Heath, G.W. and Arnold, M.K. 1966. Studies in leaf-litter breakdown. II Breakdown rate of 'sun' and 'shade' leaves. Pedobiol. 6: 238-243.

75. Hoogerkamp, M., Rogaar, H. and Eijsackers, H.J.P. 1983. Effect of earthworms on grassland on recently reclaimed polder soils in the Netherlands. Pages 85-105 in J.E. Satchell, editor. Earthworm ecology, from Darwin to vermiculture. Chapman and Hall, London, UK.

76. Horn, H.S. 1981. Succession. Pages 253-271 in R.M. May, editor. Theoretical ecology. Blackwell, Oxford, UK.

77. Hughes, R.D. and Walker, J. 1970. The role of food in the population dynamics of the Australian Bush Fly. Pages 225-270 in A. Watson, editor. Animal populations in relation to their food resources. Blackwell, Oxford, UK.

78. Huhta, V., Ikonen, E. and Vilkamaa, P. 1979. Succession of invertebrate populations in artificial soil made of sewage sludge and crushed bark. Ann. Zool. Fennici 16: 223-370.

79. Humphreys, W.F. 1977. Production and respiration in animal populations. J. Anim. Ecol. 48: 427-453.

80. Jarvis, P.G. and Leverenz, J.W. 1983. Productivity of temperate, deciduous and evergreen forests. Pages 233-280 in O.L. Lange, P.S. Nobel, C.B. Osmond and H. Ziegler, editors. Physiological plant ecology IV. Ecosystem processes: mineral cycling, productivity and man's influence. Springer-Verlag, Berlin, Germany.

81. Jenkinson, D.S. 1971. The accumulation of organic matter in soil left uncultivated. Rothamsted Exp. Stat. Rep. 1970. 113-137.

82. Jenny, H. 1980. The soil resource; origin and behavior. Springer-Verlag, New York, USA.

83. Kaarik, A.A. 1974. Decomposition of wood. Pages 129-174 in C.H. Dickinson and G.J.F. Pugh, editors. Biology of Plant Litter Decomposition. Academic Press, London, UK.

84. Kallio, P. and Lehtonen, J. 1973. Birch forest damage caused by Oporinia autumnata (Bkh.) in 1965-66 in Utsjoki, N. Finland. Rep. Kevo Subarctic Res. Stn. 10: 55-69.

85. Kirkwood, R.S.M. and Lawton, J.H. 1981. Efficiency of biomass transfer and the stability of model food-webs. J. theor. Biol. 93: 225-237.

86. Kjøller, A. and Struwe, S.   1982. Microfungi in ecosystems: fungal occurrence and activity in litter and soil. Oikos 39: 391-418.

87. Kozlovskaja, L.S.   1969. Der einfluss der exkremente von regenwurmern auf die aktivierung der mikrobiellen prozesse in torfboden. Pedobiol. 9: 1580-164.

88. Lange, O.L., Nobel, P.S., Osmond, C.B. and Ziegler, H. 1983. Physiological plant ecology IV.   Ecosystem processes: mineral cycling, productivity and man's influence. Springer-Verlag, Berlin, Germany.

89. Larsson, S. and Tenow, O.   1980.   Needle-eating insects and grazing dynamics in a mature Scots pine forest in Central Sweden.   Pages 269-306 in T. Persson, editor.   Structure and function of northern coniferous forests - an ecosystem study. Ecol. Bull. (Stockholm) 32.

90. Latter, P.M.   1977.   Decomposition of a moorland litter, in relation to Marasmius androsaceus and soil fauna. Pedobiol. 17: 418-427.

91. Leith, H. 1975. Modelling the primary production of the World.   Pages 237-263 in H. Leith and R.H. Whittaker, editors.   Primary productivity of the biosphere. Springer-Verlag, Berlin, Germany.

92. Lindeman, R.L.   1942. The trophic-dynamic aspect of ecology. Ecology 23: 438-450.

93. Llewellyn, M.J.   1972. The effects of the lime aphid, Eucallipterus tiliae L. (Aphididae) on the growth of the lime Tilia x vulgaris Hayne. I. Energy requirements of the aphid population. J. appl. Ecol. 9: 261-282.

94. Lousier, J.D. and Parkinson, D. 1979. Organic matter and chemical element dynamics in an aspen woodland soil. Can. J. For. Res. 9: 449-463.

95. Luxton, M. 1972. Studies on the oribatid mites of a Danish beechwood soil. Pedobiol. 12: 434-463.

96. MacArthur, R.H. and Wilson, E.D.   1967. The theory of island biogeography.   Princeton Univ. Press, Princeton, USA.

97. McBrayer, J.F. 1977. Contributions of the Cryptozoa to forest nutrient cycles.   Pages 70-77 in J. Matteson, editor.   The Role of Arthropods in Forest Ecosystems. Springer-Verlag, New York, USA.

98. McClaugherty, C.A., Aber, J.D. and Melillo, J.M. 1982. The role of fine roots in the organic matter and nitrogen budgets of two forested ecosystems. Ecology 63: 1481-1490.

99. McGill, W.B., Hunt, H.W., Woodmansee, R.G. and Reuss, J.O. 1981. Phoenix, a model of the dynamics of carbon and nitrogen in grassland soils. Pages 49-115 in F.E. Clark and T. Rosswall, editors. Terrestrial nitrogen cycles. Ecol. Bull. (Stockholm) 33.

100. McIlveen, W.D. and Cole, H.Jr. 1976. Spore dispersal of Endogonaceae by worms, ants, wasps and birds. Can. J. Bot. 54: 1486-1489.

101. MacMahon, J.A. 1981. Successional processes: comparisons among biomes with special reference to probable roles of and influences on animals. Pages 277-304 in D.A. West, H.H. Shugart and D.B. Botkin, editors. Forest Succession: concepts and applications. Springer-Verlag, New York, USA.

102. McNeill, S. and Lawton, J.H. 1970. Animal production and respiration in animal populations. Nature, London 225: 472-474.

103. Major, J. 1969. Historical development of the ecosystem concept. Pages 9-22 in G.M. van Dyne, editor. The ecosystem concept in natural resource management. Academic Press, New York, USA.

104. Marrs, R.H., Roberts, R.D., Skeffington, R.A. and Bradshaw, A.D. 1983. Nitrogen and the development of ecosystems. Pages 113-136 in J.A. Lee, S. McNeill and I.H. Rorison. Nitrogen as an ecological factor. Blackwell, Oxford, UK.

105. Maser, C., Trappe, J.M. and Nussbaum, R.A. 1978. Fungal - small mammal interrelationships with emphasis on Oregon coniferous forests. Ecology 59: 799-809.

106. May, R.M. 1981. Theoretical Ecology. Blackwell, Oxford, UK.

107. Meentemeyer, V. 1978., Macroclimate and lignin control of litter decomposition rates. Ecology 59: 465-472.

108. Mignolet, R. 1972. Etat actuel des connaissances sur les relations entre la microfaune et la microflore edaphiques. Rev. Ecol. Biol. Sol 9: 655-670.

109. Miles, J. and Young, W.F. 1980. The effects of heathland and moorland soils in Scotland and northern England following colonization by birch (Betula spp). Bull. d'Ecol. 11: 233-242.

110. Miller, H.G. 1981. Forest fertilization: some guiding concepts. Forestry 54: 158-167.

111. Moran, V.C. and Southwood, T.R.E. 1982. The guild composition of arthropod communities in trees. J. Anim. Ecol. 51: 289-306.

112. Newell, K. 1984a in press. Interactions between two decomposer basidiomycetes and a collembolan under Sitka spruce: grazing and its potential effects on fungal distribution and litter decomposition. Soil Biol. Biochem.

113. Newell, K. 1984b in press. Interactions between two decomposer basidiomycetes and a collembolan under Sitka spruce: distribution, abundance and selective grazing. Soil Biol. Biochem.

114. Nilsson, I.S., Miller, H.G. and Miller, J.D. 1982. Forest growth as a possible cause of soil and water acidification: an examination of the concepts. Oikos 39: 40-49.

115. Noble, I. and Slatyer, R.O. 1981. Concepts and models of succession in vascular plant communities subject to recurrent fire. Pages 311-335 in A.M. Gill, R.H. Groves and I.R. Noble, editors. Fire and the Australian Biota. Australian Academy of Sciences, Canberra, Australia.

116. Odum, E.P. 1969. The strategy of ecosystem development. Science 164: 262-270.

117. Odum, E.P. 1972. Ecosystem theory in relation to man. Pages 11-24 in J.A. Wiens, editor. Ecosystem structure and function. Oregon State Univ. Press, Corvallis, USA.

118. O'Neill, V., Harris, W.F., Ausmus, B.S. and Reichle, D.E. 1975. A theoretical basis for ecosystem analysis with particular reference to element cycling. Pages 28-40 in F.G. Howell, J.B. Gentry and M.H. Smith, editors. Mineral cycling in southeastern ecosystems. U.S. Energy Res. & Dev. Admin. Symposium Series.

119. Owen, D.F. 1978. Why do aphids synthesize melezitose? Oikos 31: 264-267.

120. Owen, D.F. 1980. How plants may benefit from the animals that eat them. Oikos 35: 230-235.

121. Owen, D.F. and Weigert, R.G. 1976. Do consumers maximize plant fitness? Oikos 27: 488-492.

122. Parkinson, D., Visser, S. and Whittaker, J.B. 1979. Effects of collembolan grazing on fungal colonization of leaf litter. Soil Biol. Biochem. 11: 529-535.

123. Petelle, M. 1980. Aphids and melezitose: A test of Owen's 1978 hypothesis. Oikos 35: 127-128.

124. Petersen, H. and Luxton, M. 1982. A comparative analysis of soil fauna populations and their role in decomposition processes. Oikos 39: 287-388.

125. Phillipson, J. 1973. The biological efficiency of protein production by grazing and other land-based systems. Pages 217-235 in J.G.W. Jones, editor. The biological efficiency of protein production. Cambridge Univ. Press, Cambridge, UK.

126. Pianka, E.R. 1981. Competition and niche theory. Pages 167-196 in R.M. May, editor. Theoretical ecology. Blackwell, Oxford, UK.

127. Pugh, G.J.F. 1980. Strategies in fungal ecology. Trans. Br. mycol. Soc. 75: 1-14.

128. Rafes, P.M. 1971. Pests and the damage which they cause to forests. Page 357-367 in P. Duvigneaud, editor. Productivity of forest ecosystems. UNESCO, Paris, France.

129. Rai, B. and Srivastava, A.K. 1983. Decomposition and competitive colonisation of leaf litter by fungi. Soil Biol. Biochem. 15: 115-117.

130. Raison, R.J. 1979. Modification of the soil environment by vegetation fires, with particular reference to nitrogen transformation: a review. Plant Soil 5: 73-108.

131. Rayner, A.D.M.   1978.,   Interactions between fungi
      colonising hardwood stumps and their possible role in
      determining patterns of colonisation and succession.
      Ann. Appl. Biol. 89: 131-134.
132. Rayner, A.D.M. and Todd, N.D. 1979. Population and
      community structure and dynamics of fungi in decaying
      wood. Adv. Bot. Res. 7: 333-420.
133. Reichle, E.D., McBrayer, J.F. and Ausmus, S. 1973.
      Ecological energetics of decomposer invertebrates in a
      deciduous forest and total respiration budget. Pages
      283-392 in J. Vanek, editor.   Progress in soil
      zoology. Academia, Prague, Czechoslovakia.
134. Reid, C.P.P., Kidd, F.A. and Ekwebelam, S.A. 1983.
      Nitrogen   nutrition,   photosynthesis   and   carbon
      allocation in ectomycorrhizal pine.   Plant Soil 7:
      415-432.
135. Reiners, W.A. 1981. Nitrogen cycling in relation to
      ecosystem succession. Pages 507-528 in F.E. Clark and
      T. Rosswall, editors.   Terrestrial nitrogen cycles.
      Ecol. Bull. (Stockholm) 33.
136. Remacle, J.   1971.   Succession in the oak litter
      microflora in forests at Mesnil-Eglise (Ferage),
      Belgium. Oikos 22: 411-413.
137. Riffle, J.W.   1971.   Effect of nematodes on root-
      inhibiting fungi.   Pages 97-113 in E. Haeskaylo,
      editor.   Mycorrhizae. USDA For. Serv. Mis. Publn.
      1189.
138. Rigler, F.H.   1975. The concept of energy flow and
      nutrient flow between trophic levels. Pages 15-26 in
      W.H. van Dobben and R.H. Lowe-McConnell, editors.
      Unifying concepts in ecology.   Junk, Hague,
      Netherlands.
139. Rorison, I.H., Peterkin, J.H. and Clarkson, D.T.
      1983.   Nitrogen source, temperature and the growth of
      herbaceous plants. Pages 189-209 in J.A. Lee, S.
      McNeill and I.H. Rorison, editors. Nitrogen as an
      ecological factor. Blackwell, Oxford, UK.
140. Rosswall,   T.   and   Kvillner,   E.   1978.
      Principal-components and factor analysis for the
      description of microbial populations. Adv. Microb.
      Ecol. 2: 1-8.
141. Rusek,   J.   1978.   Pedozootische sukzessionen
      wahrendder entwicklung von okosystemen. Pedobiol. 18:
      426-433.
142. Sanchez, P.A., Gichuru, M.P. and Katz, L.B. 1982.
      Organic matter in major soils of the tropical and
      temperate regions.   99-114.   11th Int. Cong. Soil
      Sci., New Delhi, India.
143. Satchell, J.E.   1980. r-worms and K-worms: A basis
      for classifying lumbricid earthworm strategies. Pages
      848-864 in D.L. Dindal, editor.   Soil biology as
      related to land use practices. EPA. Washington, USA.
144. Satchell, J.E. and Lowe, D.G. 1967. Selection of
      leaf litter by Lumbricus terrestris. Pages 102-119 in
      O. Graff and J.E. Satchell, editors. Progress in soil
      biology. Vieweg, Braunschweig, Germany.

145. Seastedt, T.R. and Crossley, D.A. Jr. 1980. Effects of microarthropods on the seasonal dynamics of nutrients in forest litter. Soil Biol. Biochem. 12: 337-342.

146. Shafer, S.R., Rhodes, L.H. and Reidel, R.M. 1981. In-vitro parasitism of endomycorrhizal fungi of ericaceous plants by the mycophagous nematode Aphelenchoides bicaudatus. Mycologia 73: 141-149.

147. Slatyer, R.O. 1977. Dynamic changes in terrestrial ecosystems: patterns of change, techniques for study and applications to management. UNESCO, Paris, France.

148. Soderstrom, B.E. 1979. Seasonal fluctuations of active fungal biomass in horizons of a podzolised pine-forest soil in central Sweden. Soil Biol. Biochem. 11: 149-154.

149. Southwood, T.R.E. 1977. Habitat, the templet for ecological strategies? J. Anim. Ecol. 46: 337-365.

150. Springett, J.A. 1976. The effect of prescribed burning on the soil fauna and litter decomposition in Western Australian forests. Aust. J. Ecol. 1: 77-82

151. Sundman, V. 1970. Four bacterial soil populations characterised and compared by a factor analytical method. Can. J. Micro. 16: 455-464.

152. Swank, W.T. and Waide, J.B. 1980. Interpretation of nutrient cycling research in a management context: evaluating potential effects of alternative management strategies on site productivity. Pages 137-158 in R. Wareing, editor. Forests: fresh perspectives from ecosystem analysis. Oregon State Univ. Press., Corvallis, USA.

153. Swift, M.J. 1976. Species diversity and the structure of microbial communities in terrestrial habitats. Pages 185-222 in J.M. Anderson, and A. Macfadyen, editors. The role of terrestrial and aquatic organisms in decomposition process. Blackwell, Oxford, UK.

154. Swift, M.J. 1977. The roles of fungi and animals in the immobilisation and release of nutrient elements from decomposing branch wood. Pages 193-202 in U. Lohm and T. Persson, editors. Soil organisms as components of ecosystems. Ecol. Bull (Stockholm) 25.

155. Swift, J.M., Heal, O.W. and Anderson, J.M. 1979. Decomposition in terrestrial ecosystems. Blackwell, Oxford, UK.

156. Swift, M.J., Healey, I.N., Hibberd, J.K., Sykes, J.M., Bampoe, V. and Nesbitt, M.E. 1976. The decomposition of branch-wood in the canopy and floor of a mixed deciduous woodland. Oecologia (Berlin) 26: 139-149.

157. Syers, J.K. and Springett, J.A. 1983. Earthworm ecology in grassland soils. Page 67-83 in J.E. Satchell, editor. Earthworm ecology, from Darwin to vermiculture. Chapman and Hall, London, UK.

158. Thompson, W. and Boddy, L. 1983. Decomposition of suppressed oak trees in even-aged plantations II. Colonization of tree roots by cord-and rhizomorph-producing basidiomycetes. New Phytol. <u>93</u>: 277-291.

159. Twinn, D.C. 1974. Nematodes. Pages 421-465 <u>in</u> C.H. Dickinson and G.J.F. Pugh, editors. Biology of Plant Litter Decomposition. Academic Press, London, UK.

160. Ulrich, B. 1980. Production and consumption of hydrogen ions in the ecosphere. Pages 255-282 <u>in</u> T.C. Hutchinson and M. Havas, editors. Effects of acid precipitation on terrestrial ecosystems. Plenum, New York, USA.

161. Usher, M.B. and Parr, T.W. 1977. Are there successional changes in arthropod decomposer communities? J. Environ. Manag. <u>5</u>: 151-160.

162. Van Cleve, K. and Viereck, L.A. 1981. Forest succession in relation to nutrient cycling in the boreal forest of Alaska. Pages 185-210 <u>in</u> D.A. West, H.H. Shugart and D.B. Botkin, editors. Forest succession: concepts and applications. Springer-Verlag, New York, USA.

163. Van Cleve, K., Dyrness, E. and Viereck, L. 1980. Nutrient cycling in interior Alaska flood plains and its relationship to regeneration and subsequent forest development. Pages 11-18 <u>in</u> M. Murray and R. Van Veldhuizen, editors. Forest regeneration at high latitudes. Gen. Tech. Rep. PNW-107.

164. van der Drift, J. and Witkamp, M. 1959. The significance of the breakdown of oak litter by <u>Enoicyla</u> <u>pusilla</u> Burm. Arch. neerl. Zool. <u>13</u>: 486-492.

165. van Rhee, J.A. 1969. Inoculation of earthworms in a newly drained polder. Pedobiol. <u>9</u>:128-132.

166. Van Veen, J.A, McGill, W.B., Hunt, H.W., Frissel, M.J. and Cole, D.V. 1981. Simulation models of the terrestrial nitrogen cycle. Pages 25-48 <u>in</u> F.E. Clark and T. Rosswall, editors. Terrestrial nitrogen cycles. Ecol. Bull. (Stockholm) <u>33</u>.

167. Vitousek, P.M. 1981. Clear-cutting and the nitrogen cycle. Pages 631-642 <u>in</u> F.E. Clark and T. Rosswall, editors. Terrestrial nitrogen cycles. Ecol. Bull. (Stockholm) <u>33</u>.

168. Vitousek, P.M. and Reiners, W.A. 1975. Ecosystem succession and nutrient retention: a hypothesis. Bio. Sci. Am. <u>25</u>: 376-381.

169. Walker, J., Thompson, C.H., Fergus, I.F. and Tunstall, B.R. 1981. Plant succession and soil development in coastal sand dunes of subtropical eastern Australia. Pages 107-131 <u>in</u> D.A. West, H.H. Shugart and D.B. Botkin, editors. Forest succession, concepts and application. Springer-Verlag, New York, USA.

170. Walker, T.W. and Syers, J.K. 1974. The fate of phosphorus during pedogenesis. Geoderma <u>15</u>: 1-19.

171. Wallwork, J.A. 1976. The distribution and diversity of soil fauna. Academic Press, London, UK.

172. Wallwork, J.A. 1982. Desert soil fauna. Praeger, New York, USA.

173. Warnock,A.J., Fitter, A.H. and Usher, M.B. 1982. The influence of a springtail Folsomia candida (Insecta Collembola) on the mycorrhizal association of leek Allium porrum and the vesicular-arbuscular mycorrhizal endophyte Glomus fasciculatus. New Phytol. 90: 285-292.

174. Webster, J. 1970. Coprophilous fungi. Trans. Br. mycol. Soc. 54: 161-180.

175. Webster, J.R., Waide, J.B. and Patten, B.C. 1975. Nutrient recycling and the stability of ecosystems. Pages 1-27 in F.G. Howell, J.B. Gentry and M.H. Smith, editors. Mineral cycling in southeastern ecosystems. U.S. Energy Res. & Dev. Admin. Symp. Series.

176. Whitehead, D.C. 1970. The role of nitrogen in grassland productivity: a review of information for temperate regions. Comm. Bur. Past. Field Crops Bull. 48. Commonwealth Agric. Bureaux, Farnham Royal, UK.

177. Whittaker, R.H. and Woodwell, G.M. 1972. Evolution of natural communities. Pages 137-159 in J.A. Wiens, editor. Ecosystem structure and function. Oregon State Univ., Corvallis,USA.

178. Winogradsky, S. 1924. Sur la microflora autochthone de la terre arable. Compt. Rend. Acad. Sci. (Paris), 178: 1236-1239.

179. Woodmansee, R. G. and Wallach, L.S. 1981. Effects of fire regimes in biogeochemical cycles. Pages 649-669 in F.E. Clark and T. Rosswall, editors. Terrestrial nitrogen cycles. Ecol. Bull (Stockholm) 33.

180. Woods, L., Cole, C.V., Elliott, E.T., Anderson, R.V. and Coleman, D.C. 1982. Nitrogen transformation in soil as affected by bacterial-microfaunal interactions. Soil Biol. Biochem. 14: 93-98.

DECOMPOSITION AND NUTRIENT CYCLING IN AGRO-ECOSYSTEMS

Noorallah G. Juma and William B. McGill[1]

INTRODUCTION

Agriculture manipulates energy fluxes, nutrient dynamics and hydrologic cycles. Descriptions of nutrient cycling and decomposition in agro-ecosystems are site-specific and thus generalizations are of limited validity unless processes and regulatory mechanisms are considered. Consequently, this chapter discusses organic matter status of agricultural soils, some of the mechanisms regulating processes and recently developed ideas on nutrient dynamics in manipulated soil-plant systems.

Unmanaged ecosystems frequently exhibit nutrient cycling characteristics closer to steady-state conditions than those of managed ecosystems, especially agro-ecosystems. Frequent perturbations are characteristic of agro-ecosystems, so that mechanisms affecting responses to perturbations become paramount.

Nutrient cycling and soil development are intimately linked. Soil development processes can be classified into four major groups: additions, removals, transformations, and transfers. A similar approach to nutrient cycling is used to compare unmanaged ecosystems with managed systems in this chapter.

---

[1] Assistant Professor and Professor, Department of Soil Science, University of Alberta, Edmonton, Canada T6G 2E3

Additions

The atmosphere constitutes a major source of C and N
inputs to the soil. Managed soil-plant systems differ from
unmanaged systems in that root production and litter inputs
of C are frequently reduced and inputs of N are increased by
management. For example, C additions to native grassland in
Saskatchewan were estimated at 310 g m^{-2} $year^{-1}$ or about 1.6
times that of associated land continuously cropped to cereals
(Voroney et al. 1981). Nitrogen inputs to native grasslands
averaged only about 0.4 to 0.6 g m^{-2} $year^{-1}$ (Clark et al.
1980) whereas in Alberta cultivated grassland soils received
N at 1.0 to 10 g m^{-2} $year^{-1}$. The reduced C input of roots
and litter reduces energy input to soil organisms and together
with an increased N input, alters the N cycle to emphasize
the mineral N pool and those biological processes affecting
it. Many of these processes are dealt with in other chapters
(Smith and Rice, this volume). Mineralization - immobilization
relations will be emphasized in this chapter.

Transformations

Transformations of elements in soils include oxidation,
reduction, biosynthesis, dissolution, and precipitation.
Soil organisms are directly involved in the first three, and
indirectly in the latter two. Management of soil-plant
systems alters oxidation-reduction relations by: 1) altering
aeration status through tillage, irrigation, and drainage;
and 2) reducing inputs of fixed C as an energy source. The
net result is usually a more oxidative environment for most
of the year. This results in the accumulation of oxidized
substrates which can be quickly reduced under anaerobic
conditions. For example, accumulated NO_3^- is lost by denitrification when the soil becomes saturated. Clark et al.
(1980) concluded that denitrification rarely occurs in
unmanaged grassland, yet it is reported in fertilized
grasslands. Gaseous losses of N (which included oxidative
N_2O generation during nitrification) have been reported to
be 0.31 g/m^2 in a fertilized cropped field and 0.99 g/m^2 in

an adjacent bare fallow field during the May to November
period (Aulakh et al. 1982). About 67% of fall-applied
$^{15}NO_3^-$-N fertilizer (11.2 g N/m^2) was lost due to denitrification in the early spring (Malhi and Nyborg 1983). The
pulsed nature of managed systems further emphasizes the need
for mechanistic and process control information since annual
and monthly fluxes or values for state variables from one
managed system cannot be extrapolated to another.

Transfers

The major transfer agents in ecosystems are plants,
animals and water. In addition, tillage exerts a major
influence in managed systems in two ways: 1) comminution and
incorporation of organic residues into the A_p horizon or
plow layer; and 2) potential disruption of aggregates and/or
relocation of substrates and decomposers. The net results of
such transfers include increased rates of decomposition of
crop residues (Brown and Dickey 1970, Shields and Paul 1973,
Sain and Broadbent 1977, Dormaar and Pittman 1980) and
greater access to organic materials that may otherwise be
physically inaccessible to soil organisms (Toogood and Lynch
1959, Siddoway 1963, Van Veen and Paul 1981).

The tillage role played by soil animals in unmanaged
systems is replaced in part by machinery in agriculture.
However, increased interest in minimum tillage will reduce
the role of machinery and place a greater reliance upon soil
animals (Coleman, this volume).

Losses

Two types of losses must be distinguished: 1) flow
through the system under steady-state conditions; and 2)
nonsteady-state loss. Managed systems have greater inputs of
N than unmanaged systems, and consequently greater losses of
N occur under steady-state conditions. Harvesting accounts
for 30-50% while leaching and volatilization account for 25-35%
of the applied N. The amount of N lost in a managed system is
a function of time and rate of application, cropping system,

and moisture regime (Allison 1966, Campbell and Paul 1978, Cameron et al. 1978, Burford et al. 1981). Under conditions of good management, losses of added N may be as little as 1 to 5%. However, quantities of N lost may reach 50 to 60% in a year (Malhi and Nyborg 1983) although losses of this magnitude are not frequently reported in managed or unmanaged systems.

Another major loss observed in managed soil-plant systems is the loss of soil organic matter through a combination of reduced C inputs, and accelerated decomposition and erosion rates. These losses continue until a new steady-state is reached. The new steady-state on the North American Great Plains is between 40 to 60% of original soil organic matter content and appears to be reached within a century (McGill et al. 1981a).

AMOUNTS OF SOIL ORGANIC MATTER IN VARIOUS AGRO-ECOSYSTEMS

Effects of cultivation on soil organic matter

The organic matter content of a soil reflects the balance between additions and removals. At steady-state, the mass of organic matter in a soil or in a soil organic fraction is calculated by dividing the addition rate, R (g m^{-2} $year^{-1}$) by the decay rate, k (/year). The steady-state organic matter content of uncultivated soil exhibits marked regional variation (Jenny 1980) because both R and k are sensitive to environmental conditions.

Changes in the soil environment, such as cultivating virgin soil or altering a system of agriculture, cause changes both in the rate of addition and the rate of decomposition. Ultimately a new steady-state is reached (Jenkinson 1966, Lathwell and Bouldin 1981). The direction of change depends upon the previous organic matter level as well as the agricultural system (McGill 1983, Robertson 1983). The productivity of cropping systems directly influences the level of organic matter at steady-state (McGill et al. 1981a).

Soil organic matter levels have been compared between native, uncultivated and cultivated sites at many locations. The concentration of organic C and horizon thickness are generally measured at paired sites. However, observed changes in concentration cannot be directly attributed to biological activity because spatial variability of organic matter concentration and horizon thickness are not generally measured. In addition, losses of soil organic matter due to wind and water erosion must also be considered when C levels in cultivated and uncultivated systems are compared (Lucas et al. 1977) because deposition of fine particles has a significant effect on pedon characteristics and organic matter levels. An attempt has been made to summarize data from the literature and assess the trends in organic matter dynamics (Tables 1 to 6).

Comparison of cultivated with native or uncultivated sites from many locations around the world showed that the concentration of organic C decreased by 13 to 60% in the surface horizon depending upon the soil type, management and duration of cultivation (Table 1). Soil organic C concentration decreased in subsurface horizons of Mollisols (Tiessen et al. 1982). In contrast, soil organic C concentration increased in subsurface horizons of two coarse textured Inceptisols receiving corn residues for 35 years (Coote and Ramsey 1983). This trend was not observed in fine textured Inceptisols (Table 1).

Bulk density must be determined when changes in mass of organic matter over time are calculated. Bulk density increases as organic C content decreases (Voroney et al. 1981, Tiessen et al. 1982, Coote and Ramsey 1983). Conversely, an increase in organic matter content causes a decrease in bulk density (Davidson et al. 1967). Changes in mass of organic C at various locations showed that the mass of organic C in the pedon or to the bottom of the B horizon generally declined due to erosion and cultivation time (Table 2). However, there may be a net increase in the mass of C in the subsurface horizons due to incorporation of plant residues and mechanical

Table 1. Changes in concentration of organic carbon due to cultivation.

Location	Soil taxonomy	Years under cultivation	Depth (cm)	Concentration of organic carbon (%)		Change in concentration (%)	Reference
				Uncultivated	Cultivated		
Portage la Prairie, Manitoba	Udic Haploboroll	20	0-20	11.3	8.6	-24	Shutt (1925)
Indian Head, Saskatchewan	Udic Haploboroll	22	0-20	7.4	6.2	-17	Shutt (1925)
Swift Current, Saskatchewan	Aridic Haploboroll	14	0-30	1.99	1.46	-27	Doughty et al. (1954)
	Aridic Haploboroll	14	0-30	2.15	1.64	-24	
	Aridic Haploboroll	12	0-30	2.91	2.41	-17	
Great Plains, U.S.A.[1]	—	37	0-15	1.91	1.10	-42	Haas et al. (1957)
North Dakota[1]	—	43	0-15	3.86	2.25	-41	Haas et al. (1957)
			15-30	1.77	1.17	-34	
			30-45	1.11	0.83	-25	
Ile aux Coudes, Que.	Typic Cryaquept	30	0-38	3.2	2.2	-31	Martel and Deschenes (1976)
Charlevoix, Que.	Typic Cryorthod	30	0-30	5.6	2.7	-52	
Cantons de l'Est, Que.	Typic Cryorthod	30	0-30	3.7	1.7	-54	
Southern Alberta	Aridic Haploboroll	20	0-15	2.79	1.41	-49	Dormaar (1979)
	Typic Haploboroll	65	0-15	3.63	1.44	-60	
	Udic Haploboroll	16	0-15	5.58	3.26	-42	
Saskatchewan	Aridic Haploboroll[2]	65-100	0-15	1.94	1.13	-41	Campbell and Souster (1982)
	Typic Haploboroll	65-100	0-15	3.58	1.90	-46	
	Udic Haploboroll	65-100	0-15	3.69	1.91	-50	
	Typic Cryoboralf	40- 50	0-15	2.53	1.19	-53	
Saskatchewan	Entic Haploboroll	70	0-16	3.77	2.37	-37	Tiessen et al. (1982)
			16-32	2.25	1.96	-13	
	Typic Haploboroll	65	0-15	3.22	1.74	-46	
			15-40	1.26	1.08	-14	
	Udic Haploboroll	90	0-10	4.79	2.00	-58	
			10-30	1.55	1.13	-27	

Table 1. Changes in concentration of organic carbon due to cultivation (continued).

Location	Soil taxonomy	Years under cultivation	Depth (cm)	Concentration of organic carbon (%)		Change in concentration (%)	Reference
				Uncultivated	Cultivated		
Ottawa, Ontario	Aquic Eutrochrept	35	0-10	3.12	2.56	-18	Coote and Ramsey (1983)
			10-20	1.80	2.53	+41	
			20-30	1.22	1.92	+57	
	Alfic Eutrochrept	35	0-10	2.61	1.89	-28	
			10-20	1.71	1.89	+10	
			20-30	1.40	0.97	-31	
	Humaquept	35	0-10	8.68	6.78	-22	
			10-20	3.62	6.26	-27	
			20-30	8.37	5.22	-38	
	Aquic Eutrochrept	35	0-10	2.37	1.47	-38	
			10-20	1.78	1.48	-17	
			20-30	0.70	0.91	+30	

1 Average of 11 locations.
2 Average of soils of three different textures for each great group.

Table 2. Changes in mass of organic carbon due to cultivation (continued).

Location	Soil taxonomy	Years under cultivation	Depth (cm)	Mass organic C (Mg ha^{-1}) Uncultivated	Cultivated	Change in mass (%)	Reference
Rothamsted, England	Aquic Hapludalf	93	0-23	28.3[1]	25.2	-11	Jenkinson and Johnson (1976)
Ile aux Coudes, Quebec	Typic Cryaquept	30	0-17	160	101	-37	Martel and Deschenes (1976)
			17-38	17	18	+6	
			(Total)[2]	177	119	-33	
Charlevoix, Quebec	Typic Cryorthod	30	4-0	37	-	-47	
			0-15	62	52	+63	
			15-30	19	31	-46	
			(Total)	118	83		
Cantons de l'Est, Quebec	Typic Cryorthod	30	0-7	43	20	-53	
			7-30	47	38	-19	
			(Total)	90	58	-35	
Saskatchewan	Udic Haploboroll[3]	70	0-12	63.6	26.4	-58	Voroney et al. (1981)
			12-25	30.6	30.1	-2	
			(Pedon)[4]	120.2	70.4	-41	
	Udic Haploboroll[3]	70	0-15	94.1	40.2	-57	
			15-39	38.0	29.8	-22	
			(Pedon)	152.2	96.7	-36	
	Udic Argiboroll[3]	70	0-13	83.3	49.2	-41	
			13-24	27.0	19.3	-29	
			24-47	26.0	28.5	+10	
			(Pedon)	162.1	103.9	-35	
Saskatchewan	Entic Haploboroll	70	0-31	109	80	-27	Tiessen et al. (1982)
	Typic Haploboroll	65	0-34	90	73	-19	
	Udic Haploboroll	90	0-30	87	43	-51	

Table 2. Changes in mass of organic carbon due to cultivation (continued).

Location	Soil taxonomy	Years under cultivation	Depth (cm)	Mass organic C (Mg ha⁻¹) Uncultivated	Cultivated	Change in mass (%)	Reference
Ottawa, Ontario	Aquic Entrochrept (loamy sand)	35	0-10	40.6	34.6	-15	Coote and Ramsey (1983)
			10-20	23.9	36.4	+52	
			20-30	17.0	27.4	+61	
			(Total)²	81.5	98.4	+21	
	Alfic Eutrochrept	35	0-10	34.5	27.4	-21	
			10-20	22.4	28.2	+26	
			20-30	18.1	15.2	-16	
			(Total)	75.0	70.8	-6	
	Humaquept	35	0-10	78.1	71.9	-8	
			10-20	87.9	73.9	-16	
			20-30	78.7	61.6	-17	
			(Total)	244.7	207.4	-15	
	Aquic Eutrochrept (clay)	35	0-10	33.9	21.2	-37	
			10-20	22.3	23.7	+6	
			20-30	8.4	11.7	+39	
			(Total)	64.6	56.6	-12	

1 This soil was cultivated in 1852; bulk density was first measured in 1882.

2 Total organic C to the bottom of B horizon.

3 Soil sampled from upper, mid and lower slope positions.

4 Total organic C for the pedon. Data for C horizons not reported here.

mixing of soil horizons during tillage. Therefore, recent studies have measured the changes in mass of organic C of pedons at various slope positions (Voroney et al. 1981) and among soil treatments (Jenkinson and Johnson 1976).

Rotations containing clear summerfallow caused a decline of approximately 20% in organic matter concentration (Pittman 1977) but continuous corn production caused a decline of 29 to 63% (Table 3). Tillage disrupts soil aggregates, exposes new surfaces to microorganisms (Ridley and Hedlin 1968) and thus promotes rapid oxidation of organic C (Rovira and Graecen 1957). The practice of row cropping is more destructive to soil organic matter than cereal cropping (Salter and Green 1933, Johnston et al. 1942, Coote and Ramsey 1983) because it requires frequent tillage to reduce weed competition. Also, in the former case soil is left bare during a longer portion of the growing season due to slow initial growth and is subject to erosion.

Grasses and legumes in rotation decrease losses of organic matter through high below ground primary production (McGill and Hoyt 1977) and help maintain organic matter concentration in soil. Continuous pasture increases organic matter concentration. The rapid accumulation of organic matter when arable land was seeded to grass is due to increased annual return of C and decreased decomposition (Jenkinson 1977). Decomposition under grass may be limited by N availability (Huntjens and Albers 1978).

Long-term additions of manure increase organic matter contents (Table 4). Manure is a source of plant nutrients and improves primary productivity (McGill and Hoyt 1977). Application of plant residues reduces the loss of organic matter compared to unamended controls (Table 4). The rate of addition of residues affects the rate of organic matter depletion or accumulation. Application of inorganic fertilizer reduces organic matter losses by increasing production and return of residues, especially roots (Jenkinson and Johnson 1976).

Table 3. Effect of crop rotations on concentration of organic carbon in soils.

Location	Soil taxonomy	Depth (cm)	Rotation[1]	Years under rotation	Concentration of organic C (%) Start	Concentration of organic C (%) End	Change in concentration (%)	Reference
Lethbridge, Alberta	Aridic Haploboroll	0-15	A(6)-RC(1)-CS(3)	11	4.62	4.87	+ 5	Shutt (1925)
		15-30			3.55	4.03	+ 14	
Wooster, Ohio	Typic Fragiudalf	0-17	C-O-W-Cl-T	32	1.02	0.77	- 24	Salter and Green
		0-17	Continuous oats	32	1.02	0.65	- 37	(1933)
		0-17	Continuous corn	32	1.02	0.37	- 63	
Clarinda, Iowa	Typic Hapludoll	0-15	Alfalfa	12	1.97	2.17	+ 10	Van Bavel and
		0-15	Bluegrass	12	1.97	1.95	- 1	Schaller (1950)
		0-15	C-O-Cl	19	1.96	1.87	- 4	
		0-15	Continuous corn	19	1.90	1.35	- 29	
Swan Coastal Plain, Western Australia	Not available[2]	0-13	Continuous pasture	33	0.77	2.03	+160	Barrow (1969)
		0-13	Continuous pasture	40	0.84	2.20	+162	
Lethbridge, Alberta	Typic Haploboroll	0-15	Continuous wheat	20	1.69	1.49	- 12	Pittman (1977)
		15-30			1.51	1.11	- 26	
		0-15	F-W	20	1.63	1.35	- 17	
		15-30			1.48	1.11	- 25	
		0-15	F-W-W	20	1.68	1.36	- 19	
		15-30			1.46	1.09	- 25	
		0-15	F(manured)-W-W	20	1.71	1.49	- 13	
		15-30			1.51	1.14	- 24	
		0-15	F-WW-WW(B)	20	1.62	1.39	- 15	
		15-30			1.46	1.08	- 26	
		0-15	F-W(2)-CW+A(3)	20	1.71	1.50	- 12	
		15-30			1.50	1.20	- 19	
		0-15	F-W(2)-CW(3)	20	1.69	1.47	- 13	
		15-30			1.47	1.14	- 22	
Breton, Alberta	Typic Cryoboralf	0-18	F-W	51	1.54	1.28	- 17	Cannon et al.
		0-18	W-O-B-A+Cl(2)	51	1.42	1.39	- 2	(1984)

[1] Abbreviations: A = alfalfa; RC = row crops; CS = cereals; C = corn; O = oats; W = wheat; Cl = clover; T = timothy; F = fallow; WW = winter wheat; CW = crested wheatgrass; Figures in parenthesis indicate years of successive cropping.

[2] The soil series was Coolup sand.

Table 4.   Effect of additions of organic residues and inorganic fertilizer on concentration of organic carbon.

Location	Soil taxonomy	Cropping system	Duration (yrs)	Depth (cm)	Type of residue or fertilizer	Concentration of organic C (%) Start	Concentration of organic C (%) End	Reference
Winnipeg, Manitoba	Udic Haploboroll	Continuous wheat	37	0-15	Unmanured	na[8]	4.2	Ridley and Hedlin (1968)
					NP[1]		3.7	
					Manure[2]		4.4	
					Manure + NP		4.3	
		Wheat-fallow	37	0-15	Unmanured	na[8]	2.1	
					NP		2.1	
					Manured		2.4	
					Manure + NP		2.7	
Ottawa, Ontario	Aquic Eutrochrepts	Cr(6)-F(8)-Cr(6)[7]	20	0-15	Control	2.91	2.53	Sowden and Atkinson (1968)
					Rye	2.96	2.49	
					Straw	2.91	2.89	
					Alfalfa	2.88	2.85	
					Leaves	2.97	2.97	
					Manure	2.83	3.47	
	Typic Haplohumod	Cr(6)-F(8)-Cr(6)	20	0-15	Control	1.37	1.23	
					Rye	1.06	1.26	
					Straw	1.33	1.70	
					Alfalfa	1.20	1.32	
					Leaves	1.19	1.74	
					Manure	1.11	1.77	
Culbertson, Montana	Typic Argiboroll	Wheat-fallow	8	0- 8	Control	1.05	1.04	Black (1973)
				8-15		0.80	0.77	
				15-30		0.63	0.64	
				0- 8	Wheat @ 1.68t/ha	1.00	1.15	
				8-15		0.79	0.81	
				15-30		0.68	0.65	
				0- 8	Wheat @ 3.36t/ha	1.00	1.22	
				8-15		0.77	0.87	
				15-30		0.65	0.73	
				0- 8	Wheat @ 6.73t/ha	1.02	1.28	
				8-15		0.83	0.99	
				15-30		0.61	0.77	

Table 4. Effect of additions of organic residues and inorganic fertilizer on concentration of organic carbon (continued).

Location	Soil taxonomy	Cropping system	Duration (yrs)	Depth (cm)	Type of residue or fertilizer	Concentration of organic C (%)		Reference
						Start	End	
Rothamsted, England	Aquic Hapludalf	Barley	93[3]	0-23	Unmanured	0.91	0.91	Jenkinson and Johnson (1976)
					Previously manured	1.96	1.58	
					Manured[4]	2.39	3.38	
					NPK Na Mg[5]	1.08	0.96	
					PK Na Mg[6]	1.12	1.10	

[1] Received 50.4 kg/ha monoammonium phosphate (11-21-0).

[2] Received 8.9 t/ha.

[3] 1882-1975 period.

[4] Received 35 t farmyard manure (3.0 t C/ha) annually since 1852.

[5] Received 33 kg P, 90 kg K, 15 kg Na, and 11 kg Mg per ha annually since 1852. Ammonium sulfate (48 kg N/ha) applied annually between 1852 and 1967.

[6] Received 33 kg P, 90 kg K, 15 kg Na, and 11 kg Mg per ha annually since 1852.

[7] Abbreviations: Cr = Cropped; F = Fallow; Figures in parenthesis indicate years of successive cropping or fallow.

[8] Not available.

Zero till and minimum till mulching reduce the rate of organic matter loss compared to conventional tillage practices (Table 5). Aeration, oxidation and erosion are reduced because little physical disruption of aggregates occurs. Doran (1980) showed that the positive effect of zero tillage might be mainly present in surface horizons. There may be a greater net loss of organic matter in subsurface horizons under zero tillage than under conventional tillage systems because crop residues are not incorporated in the soil at these depths.

Dynamics of organic matter in cultivated systems

Trends in organic matter content with time have been monitored at research plots under different management conditions. Organic matter may decrease or increase depending on management practices and the original organic matter content relative to the new steady-state. The trends of soil organic matter content in research plots existing around the world were evaluated by using the data on the total N content of soils. The following equation was used:

$$dN/dt \ = \ -kN+R \qquad\qquad (1)$$

where dN/dt is the rate of change of N content over time, k (/year) is a first-order rate decay rate constant and R is the amount of N added annually to soil. This expression represents first order loss plus constant annual additions and treats soil organic matter as a kinetically homogeneous unit for the duration of the experiments. The annual turnover rate (k) for soil organic matter varied between 1% and 10% (Table 6). Continuous cropping caused higher turnover rates and higher steady-state values compared to fallow systems. Additions of manure similarly increased turnover rate and steady-state level, but additions of fertilizer did not. Further, examination of the amounts of N at steady-state (R/k) showed that either very rapid additions or considerable internal recycling of N occurred

88

Table 5. Effect of tillage intensity on distribution and concentration of organic carbon in soil.

Location	Soil taxonomy	Cropping system	Duration (yrs)	Depth (cm)	Ratio of organic C in different tillage systems[1]		Reference
					NT/CT	MT/CT	
Cherokee, Oklahoma	Udic Argiustoll	Continuous wheat and alfalfa wheat rotation	11	8-15		1.05	Tanchandrphongs and Davidson (1970)
				15-23		1.14	
				23-30		1.12	
				30-38		1.11	
Lexington, Kentucky	Typic Paleudalf	Continuous corn	9	0-8	1.55***		Doran (1980)
				8-15	0.81**		
				15-30	0.90†		
Waseca, Minnesota	Aquic Hapludoll	Continuous corn	9	0-8	1.07†		
				8-15	1.00		
				15-30	0.89		
Morgantown, W. Virginia	Aquic Hapludult and Fragiudult	Continuous corn	5	0-8	1.20*		
				8-15	0.97		
				15-30	1.02		
Lincoln, Nebraska	Pachic Argiustoll-Abruptic Argiaquoll	Continuous corn	3	0-8	1.10†		
				8-15	1.05		
				15-30	0.97		
Sidney, Nebraska	Pachic Haplustoll	Wheat-fallow	9	0-8	1.50*		
				8-15	1.00		
				15-30	0.98		
Sidney, Nebraska	Aridic Argiustoll	Wheat-fallow	10	0-8	1.25		
				8-15	0.98		
				15-30	0.93		
Pendleton, Oregon	Typic Haploxeroll	Wheat-fallow	9	0-8	1.10		
				8-15	0.88**		
				15-30	0.82*		
Grant County, N. Dakota	Typic Haploboroll	Wheat-fallow	25	0-46	1.44		Bauer and Black (1981)
	Typic Argiboroll	Wheat-fallow	25	0-46	0.98		
	Typic Argiboroll and Pachic Argiboroll	Wheat-fallow	25	0-46	1.13		
Ohio	Mollic Ochraqualf	3 rotations[2]	18	0-23	1.15	1.01	Dick (1983)
	Typic Fragiudalf	3 rotations[2]	19	0-23	1.19	1.03	

[1] Abbreviations: NT = no tillage; CT = conventional tillage; MT = minimum tillage.
[2] The three rotations were: 1) continuous corn; 2) corn and soybeans; and 3) corn, oats and alfalfa meadow in a three-year rotation. Rotation effects were not reported.

* p<0.05.
** p<0.01.
*** p<0.001.
† p<0.10.

Table 6. Decay rates (k) and steady-state levels of N (R/k) for various cropping systems around the world calculated using data from research plots with constant management. The data were analyzed using the integrated form of the equation $dN/dt = -kN+R$: ($N = N_0 e^{-kt} + R/t (1-e^{-kt})$).

Conditions	k (/year) Estimate ± S.D.	R/k
Rothamsted (0-23 cm or equivalent weight)[1]; Hoosfield plots;		
plot 7-1 unmanured since 1852	0.0094 ± 0.0055	2.65 ± 1.04 t N ha^{-1}
plot 7-2 manured	0.0313 ± 0.0047	7.62 ± 0.18 t N ha^{-1}
Hays, Kansas (0-18 cm)[2]		
continuous small grains	0.0597 ± 0.0405	0.130 ± 0.0056 % N
alternate small grain and fallow	0.0396 ± 0.0159	0.117 ± 0.0074 % N
Pasture plots at Kybybolite, South Australia (0-15 cm)[3]; from Fig. 7	0.0300 ± 0.0065	1755 ± 179 lb N acre^{-1}
Sheridan, Wyoming[4]; alternate grain – fallow	0.0179 ± 0.0020	15.8 ± 7.7% of original N
Dark Brown Chernozemic loam (0-15 cm) Lethbridge, Alberta		
continuous wheat[5]	0.0733 ± 0.0029	0.146 ± 0.0005 % N
wheat – fallow[5]	0.0608 ± 0.0270	0.131 ± 0.0045 % N
mean of 6, 9, and 10 year rotation[6]	0.0789 ± 0.0453	0.152 ± 0.007 % N
Borbetta, India[7]; unshaded tea		
control soil (calculated orig. N = 0.117%)[8]	0.0970[8]	0.070[8] % N
120:60:60 of N:P205:K2O lb acre^{-1}	0.0853 ± 0.0099	0.083 ± 0.002 % N

1 Jenkinson and Johnson (1976).
2 Hobbs and Brown (1957).
3 Russell (1960).
4 Haas et al. (1957).
5 Freyman et al. (1982).
6 Hill (1954).
7 Gokhale (1959).
8 There were insufficient data to calculate standard deviation.

at several sites. Substantial internal recycling of N is
supported by the following data and discussion.

It is simplistic to treat soil organic matter as a
kinetically homogeneous unit. However, this approach is most
useful for quantifying overall changes in amounts of organic
matter over a period of decades or centuries. Tracer studies
have shown that organic materials in soil have different
decomposition rates and form a continuum ranging from labile
to recalcitrant. Over the short-term, biological activity,
availability of substrates and physical conditions prevailing,
markedly influence decomposition and nutrient cycling in soil.
The following sections emphasize decomposition of substrates
over a short-term, heterogeneity of soil organic matter and
dynamics of C and N. An attempt to unravel the complexity
of the soil system will be made by simulation modelling.

DECOMPOSITION OF ORGANIC MATERIALS OF BIOLOGICAL ORIGIN IN SOIL

Decomposition involves physical breakdown, biochemical
transformation and biophysical stabilization of organic
material. Heterotrophic organisms in soil utilize C in the
organic material for metabolism. Subsequent decomposition
of these organisms and their products results in the formation
of complex, heterogeneous humic materials.

Decomposition has been studied in almost all ecosystems
and has been extensively reviewed (Hunt 1977, Singh and Gupta
1977, Paul and Van Veen 1978, Swift et al. 1979, McGill et al.
1981b). The major conclusions derived from these studies are:
1. Both microorganisms and soil fauna are important in the
   decomposition process. The role of soil macroorganisms,
   especially annelids and arthropods, has been emphasized
   in the comminution of complex substrates; however,
   microorganisms are the primary decomposers, accounting
   for up to 95% of energy utilization (Richards 1974).
2. Added soluble material is rapidly transformed by organisms
   into complex material which is slowly decomposable. The
   decomposition of complex material such as straw shows

biphasic (Jenkinson 1977, Cheshire et al. 1979) or polyphasic behavior (Lynch 1979). Generally, overall decomposition rate can be described by the double exponential model. Fitting decomposition rates to such a model is justified on the basis of kinetic analyses and structure of the materials present in the substrate (Hunt 1977).

3. It is difficult to distinguish and separate partially decomposed material from soil. Complex materials such as straw are substrates as well as habitats for micro-organisms (McGill et al. 1981b), and are intimately associated with other components of soil. Thus, the common methods of measuring decomposition: litter bags, amount of ^{14}C remaining in soil, or amount of $^{14}CO_2$-C evolved from labelled substrates yield data only on net decomposition.

4. Computer simulation techniques utilizing hypothesized efficiencies of utilization of C by microorganisms can account for microbial growth and thereby yield information on actual decomposition of substrates in soil (Paul and Van Veen 1978, McGill et al. 1981b).

5. Lignins and other phenolic materials may be partially altered by microorganisms and stabilized in soil (Haider et al. 1977).

6. The influence of temperature and moisture (Nyhan 1975, Clark and Gilmour 1983), aeration (Reddy and Patrick 1975), residue particle size (Sims and Frederick 1970), N requirement (Jansson 1958) and other factors are generally known, but the mechanisms by which they regulate decomposition, microbial growth, and stabilization of organic material need further elucidation.

Decomposition and architecture of microbial and plant cells

Decomposition of organic matter requires contact between decomposers and substrate so that enzymatic processes can occur. In some cases, extracellular enzymes may be important to initiate the decomposition (Burns and Martin, this volume).

At the microsite level, either within an aggregate or the
digestive tract of a soil animal, such contact must be at a
cellular level. Architecture of the microenvironment and the
cellular residues being decomposed influence the process.
Cell architecture and cell chemistry influence macromolecular
decomposition in soil. It is unlikely that individual
compounds or polymers decompose independently of the components
with which they are associated. Emphasis on major cell
components (walls versus cytoplasm) is therefore likely to be
a useful link between decomposition processes, organisms and
organic materials.

Living cells are diverse; all have structural components,
most have active cytoplasmic components, and some have surface
adherents. Cell walls provide support for the cells. The
general architecture of cell walls involves a fibrous material
embedded in an amorphous support matrix. The fibrous material
resists tension, while the matrix material resists compression
(Kirkwood 1974). The major polymers present in the gram
positive bacterial cell wall are: 1) peptidoglycan, 2) teichoic
and teichuronic acids, 3) polysaccharides and 4) proteins.
Peptidoglycan is the fibrous material present in bacterial
cell walls and provides rigidity to the cell. The cell wall
of gram positive bacteria is usually amorphous because the
four major components are intimately mixed and covalently
bonded so that it is not possible to peel off layers containing
various polymers (Tipper and Wright 1979). The architecture
of gram negative cell envelopes (Fig. 1) is more complex than
the cell wall of gram positive bacteria. Gram negative bacteria
are enclosed by an inner plasma membrane and by an outer cell
envelope. The outer envelope consists of peptidoglycan, and
an outer membrane containing proteins, lipopolysaccharides,
and phospholipids.

Polysaccharides account for 80 to 90% of the dry matter
of fungal cell walls. The remainder is mainly proteins and
lipids although sometimes substantial amounts of pigments
(melanin), pyrophosphate and inorganic ions may be present
(Bartnicki-Garcia 1968). Physically, chitin ($\beta1\rightarrow4$ GluNAc

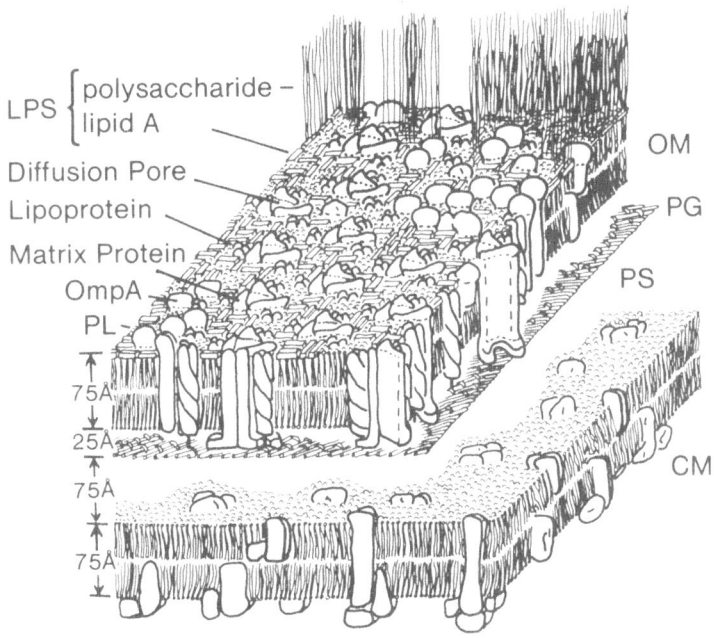

FIGURE 1. A schematic representation illustrating the possible molecular architecture of the E. coli cell envelope. Abbreviations used are: PL, phospholipid; OM, outer membrane; PG, peptidoglycan; PS, periplasmic space; and CM, cytoplasmic membrane. Polysaccharide chains in only some of the LPS molecules are shown (from DiRienzo et al. 1978).

polymer) and cellulose are the most common fibrous materials while proteins and other polysaccharides (glucans, mannans, galactans, heteropolysaccharides) form the matrix and are the cementing agents. Most filamentous fungi in soil have chitosan-chitin or chitin-glucan polysaccharide combinations in their vegetative cells (Bartinicki-Garcia 1968). The overall appearance of fungal cell walls varies with age (Trinci 1978). The primary cell wall is made up of chitin microfibrils and protein. The secondary cell wall consists of chitin fibrils, protein and an amorphous layer of glucan.

Plant cell walls contain cellulose, noncellulosic polysaccharides such as hemicelluloses and pectin, a small amount of protein, bound and free water, appreciable calcium, and other cations. Keegstra et al. (1973) proposed a model of primary cell walls in which the cellulose microfibrils are coated with a layer of xyloglucan (a hemicellulose) one molecule thick. Xyloglucan is glycosidically linked to rhamnogalactouronan (a pectin). The polymers are cross-linked and immobilize cellulose fibers. The role of wall protein is not clearly understood, however, there is some evidence that arabinogalactan may be attached to serine of wall protein and/or arabinosyl tetrasaccharides may be glycosidically attached to hydroxy proline residues (Keegstra et al. 1973). Lignin is deposited as the cell matures and functions to cement and anchor the cellulose fibers together. It stiffens the wall because of its location and prevents chemical and physical damage to the cell (Hall et al. 1974). Lignin also impedes microbial destruction of plant tissue (Zeikus 1981). As the plant matures, cell wall lignification increases resulting in recalcitrant organic material with a high C:N ratio. In contrast to the rigid cell wall, the intercellular material is granular and in a constant flux. These materials are quickly decomposed if released from cells.

The architecture of cells provides a basis to divide cell materials into two broad categories: labile and resistant; characterized by metabolic and structural components, respectively. Cell component chemistry is closely related to cell architecture but individual chemical compounds do not decompose independently of the cellular units into which they are incorporated. In most simulation models plant, animal and microbial residues are divided on the basis of chemical composition. Architecture of the cell wall and cytoplasmic components influences their accessibility to organisms and enzymes, thereby controlling decomposition, and should be considered in descriptions of their decomposition.

Kinetic analysis of decomposition of added organic
materials in soil

A comparison of net decomposition rates of cells, cell
components, microbial biomass in situ, and straw under field
conditions is useful to understand the dynamics of C and N
in soil. Such studies yield additional information on the
transformation and stabilization of organic matter. The
decomposition of carbonaceous materials results in the
formation of new biomass, metabolites and resistant materials.

Kinetic analysis of the decomposition of microbial cells
and their components by the nonlinear least squares method
showed that the data for six microbial species reported by
Nelson et al. (1979) could be fitted to a double exponential
equation:

$$^{14}C \text{ remaining } = Le^{-kt} + Re^{-ht} \quad (2)$$

where R=100-L. The labile component (L) decomposed at a
rapid rate, k (/week) while the resistant component (R)
decomposed at a slower rate, h (/week). The pooled data for
six microbial species showed that 53 to 55% of ^{14}C cells and
cell walls of bacteria, yeast and filamentous fungi (Table 7)
were rapidly decomposed (k=1.5 to 1.6/week). The ^{14}C from
cytoplasm was decomposed faster because a greater proportion
(60%) was in the labile component and the decay rate constant
of that component was also higher (k=2.3/week). Also, the
net decomposition rate constants of the labile component for
the cells, cell walls and cytoplasm were 50 fold higher than
those for resistant components (Table 7). Filamentous fungal
cell walls, especially those of Mucor rouxii and Hendersonula
toruloidea, were more resistant to decomposition because kinetic
analysis showed that the labile component was smaller than in
the bacterial and yeast cell walls. In contrast, the cytoplasmic component of fungal cells was rapidly decomposed.
Kinetic analysis also showed that the decay rate constants
of the cell walls and cytoplasm were higher than intact cells
in some cases (Table 7). This could be attributed to the

Table 7. Kinetic analysis of decomposition of ^{14}C-labelled cells, cell walls, and cytoplasmic fractions of bacteria, yeasts, and filamentous fungi. Raw data obtained from Nelson et al. (1979).

| Material added | Simultaneous estimates ± standard error of estimate[1] | | | RMS[2] |
	L (%)	k (/week)	h ($\times 10^{-3}$ /week)	(%)
Bacillus subtilis				
cells	52.6 + 2.9	1.6 + 0.3	44.1 + 8.5	1.8
cell walls	64.3 + 1.9	1.7 + 0.1	46.5 + 7.3	1.1
cytoplasm	63.5 + 1.4	2.6 + 0.3	37.8 + 5.6	
Pseudomonas aeruginosa				
cells	60.5 + 2.8	1.7 + 0.3	49.7 + 10.0	1.7
cell walls	59.7 + 2.9	1.5 + 0.2	39.9 + 9.5	1.7
cytoplasm	57.2 + 2.0	2.3 + 0.4	34.3 + 6.6	1.5
Hendersonula toruloidea				
cells	60.1 + 1.8	2.2 + 0.3	35.3 + 6.4	1.3
cell walls	48.9 + 2.6	1.9 + 0.4	34.6 + 7.0	1.8
cytoplasm	63.7 + 1.3	3.1 + 0.5	33.5 + 5.4	1.1
Mucor rouxii				
cells	36.2 + 2.4	1.5 + 0.3	36.9 + 5.1	1.5
cell walls	25.3 + 1.5	1.4 + 0.2	34.3 + 2.6	0.9
cytoplasm	60.7 + 1.9	2.2 + 0.3	43.8 + 7.0	1.4
Hansenula holstii				
cells	61.6 + 1.7	1.3 + 0.1	37.3 + 5.6	0.9
cell walls	54.0 + 2.8	1.5 + 0.2	44.3 + 8.1	1.6
cytoplasm	50.3 + 3.1	1.7 + 0.3	42.5 + 8.5	2.0
Candida utilis				
cells	63.5 + 4.3	0.9 + 0.1	42.8 + 13.9	1.6
cell walls	61.6 + 3.2	1.6 + 0.3	46.9 + 11.4	1.9
cytoplasm	61.3 + 2.3	2.1 + 0.3	39.9 + 8.5	1.7
Average[3]				
cells	55.3 + 3.2	1.5 + 0.2	41.6 + 9.6	1.9
cell walls	52.7 + 2.2	1.6 + 0.2	40.5 + 6.4	1.4
cytoplasm	59.5 + 1.4	2.3 + 0.2	38.9 + 4.9	1.0

[1] Equation used: $^{14}C\text{-remaining} = Le^{-kt} + Re^{-ht}$, where R=100-L.

[2] RMS = Root mean square error.

[3] Average of all microbial species studied.

method used to prepare these materials. Sonication destroys the structural integrity of cell components and makes them more vulnerable to microbial attack in soil. The decomposition rate of cytoplasm of soil organisms in situ may be lower due to steric hindrance of cell walls. However, the results obtained from kinetic analysis of decomposition of ^{14}C labelled materials clearly show that rapid decomposition occurred initially. This may be related to cell architecture and composition. The slow rate of decomposition of the resistant component may be due to adsorption of added microbial material and stabilization of metabolites during decomposition of added substrate, and by the slow decay rate of quiescent biomass in soil.

The reaction of microbial components with phenolic materials made them resistant to decomposition. When cell walls and cytoplasm were "mixed" or "complexed" with different preformed model phenolase polymers and allowed to decompose in soil (Nelson et al. 1979), kinetic analysis of those data showed that such complexation reduced both the quantity of the labile component (L) and its decay rate constant (k) by a factor of two to three depending upon the phenolic material used (Table 8). The half life of complexed material ranged from 0.6 to 1.1 years under laboratory conditions.

The decomposition of microbial components in soils were described by two first order reactions even when dealing with cytoplasmic materials added separately from cell wall components. This observation suggests interactions of these materials with soil components created further kinetic heterogeneity. Reaction of cell components with model polymers increased the size of the resistant component and decreased its decay rate constant (Nelson et al. 1979). Together, these observations are consistent with the hypothesis that reaction between soil components and biological structures tends to slow their decomposition. Therefore, even kinetically homogeneous materials appear to be quickly converted to two or more groups of materials within soils leading to formation

98

Table 8.   Decomposition of free, complexed or mixtures of ^{14}C-labelled cell components of bacteria, yeasts and fungi as described by the double exponential model.   Data obtained from Nelson et al. (1979).

Material added	Simultaneous estimates + Standard error of estimate[1]			RMS
	L (%)	k (/week)	h ($\times 10^{-3}$ /week)	(%)
Cell walls[2]	55.3 ± 3.2	1.5 ± 0.2	41.6 ± 9.6	1.9
Cell walls "mixed" with HBA[3]	42.5 ± 1.8	0.9 ± 0.08	31.3 ± 3.7	0.8
Cell walls "complexed" with HBA[3]	32.5 ± 2.5	1.1 ± 0.2	22.9 ± 4.3	1.3
Cell walls "mixed" with HT & P	38.1 ± 1.5	0.7 ± 0.04	26.6 ± 2.6	0.5
Cell walls "complexed" with HT & P	20.1 ± 5.4	0.5 ± 0.2	11.7 ± 6.5	1.4
Cytoplasm[4]	59.5 ± 1.4	2.3 ± 0.2	38.9 ± 4.9	1.0
Cytoplasm "mixed"	47.7 ± 2.6	1.9 ± 0.4	38.8 ± 6.9	1.8
Cytoplasm "complexed"	23.6 ± 1.5	1.2 ± 0.1	17.7 ± 2.3	0.8

[1] Equation used: ^{14}C remaining = $Le^{-kt} + Re^{-ht}$, where R=100-L.

[2] The data used in analyses were averages of decomposition values obtained with five microbial species: B. subtilis, P. aeruginosa, M. rouxii, H. holstii and C. utilis (Nelson et al. 1979).

[3] Mixed, intimately mixed with preformed model phenolase polymers before addition to soil; complexed, reacted with phenol mixtures during phenolase oxidative polymerization.   HBA polymers formed from a mixture of hydrobenzoic acids; HT & P, polymers formed from a mixture of hydrotoluenes plus phenols.

[4] Data obtained with five above mentioned organisms and H. toruloidea.   The HBA and HT & P data were similar for all species and were pooled.

of humads (McGill et al. 1981b) or stabilized organic matter (Paul and Juma 1981).

Dynamics of microbial biomass in soil

Chloroform fumigation (Jenkinson and Powlson 1976) combined with ^{14}C and ^{15}N tracer techniques have facilitated the estimation of microbial biomass decomposition rates in situ in ^{14}C substrate amended or ^{15}N field equilibrated soils. The net decomposition of ^{14}C-labelled biomass was also described by the double exponential model. Kinetic analysis showed that the labile component of the labelled biomass in an Inceptisol and a Mollisol ranged from 34 to 54% of the total microbial biomass compared to 65 to 73% in two Alfisols (Table 9). The decay rate constant for the labile biomass component in the Inceptisol and the Mollisol was almost one-half that in the Alfisols. These rate constants were similar to those obtained for the labile pool when microbial materials were added to soil (Tables 7 and 8). The decay rate constants for the resistant materials were 32 to 65 fold lower than those of the labile component of the biomass (Table 9).

The net decay rate constant of the quiescent biomass measured with ^{15}N for the four soils ranged from 2.0×10^{-2} to 8.8×10^{-3}/week. The net decay rate constants of ^{15}N-labelled biomass for the nonamended Mollisol and Alfisol were almost two fold lower than the net decay rate constant of resistant ^{14}C-labelled biomass (Table 9). The decomposition rate constants are similar for resistant portions of biomass formed in situ and resistant components of labelled microbial cells or cell components added to soil. Consequently, cell wall and cytoplasmic components appear to be meaningful separations of cellular materials and may be useful in describing decomposition in the field.

Table 9. Kinetic analysis of decomposition of labelled microbial biomass in soil as described by single and double exponential models.

Soil	Simultaneous estimates + Standard error of estimate			RMS (%)
	L (%)	k (/week)	h ($\times 10^{-2}$ /week)	
	C-14 glucose amended soils[4] - laboratory incubation			
Aridic Haploboroll, Canada[1]	35.2 + 2.0	3.91 + 0.58	8.50 + 0.10	1.9
Udic Haploboroll, Canada[2]	53.8 + 3.7	1.56 + 0.31	5.05 + 0.80	2.7
Typic Cryoboralf, Canada[2]	65.4 + 2.7	1.62 + 0.20	4.86 + 0.76	2.0
Ochreptic Hapludalf, France[2]	73.3 + 0.6	2.82 + 0.14	4.34 + 0.24	0.5
	N-15 field equilibrated soils[5] - followed by laboratory incubation			
Boralfic Haploboroll, Canada[3]	100		2.59 + 0.13	3.1
Udic Haploboroll, Canada[2]	100		2.25 + 0.10	3.3
Typic Cryoboralf, Canada[2]	100		0.88 + 0.24	8.1
Ochreptic Hapludalf, France[2]	100		2.01 + 0.05	1.9

[1] Paul and Voroney (1980).

[2] Juma, Mary and McGill (unpublished data).

[3] Juma and Paul (1983).

[4] Equation used: Biomass remaining = $Le^{-kt} + Re^{-ht}$, where R=100-L.

[5] Equation used: Biomass remaining = $100e^{-ht}$.

Kinetics of plant residue decomposition under field
conditions

Climatic regimes, chemical composition and architecture
affect the net decomposition rate of plant residues under
field conditions. Single and double exponential equations
best describe the loss of mass of added plant residues in
various terrestrial ecosystems (Wieder and Lang 1982).
Analysis of net decomposition data for ^{14}C-labelled materials
from temperate regions of the world yielded a labile pool
size of 54 to 78% of the total C added with its rate constant
ranging from 4.7 to 1.4/year ($t_{\frac{1}{2}}$ = 0.15 to 0.50 year,
respectively) (Table 10). The decay rate constant for the
resistant material ranged from 0.20 x10^{-2} to 8.1 x10^{-2}/year
($t_{\frac{1}{2}}$ = 3.5 to 8.6 year, respectively). The calculated kinetic
parameters do not explicitly consider the climatic regimes,
straw composition and other factors affecting straw
decomposition.

The quantity of labile material in decomposing ryegrass
was the same in tropical soils in Nigeria as in temperate
regions but its decay rate was 3 to 5 fold higher under
tropical conditions (Jenkinson and Ayanaba 1977). Similarly,
the decomposition rate constant for the labile component
obtained from the composite data of 12 tropical (Costa Rican)
soils was almost 2 to 4 fold higher compared to temperate
soils of Germany (Sauerbeck and Gonzalez 1977). There was a
noticeable flattening of decomposition curves for the Costa
Rican soils which implied no further decomposition of residues
after one year of incubation. This is in contrast to the
trends observed with the tropical soils of Nigeria. The
amount of ^{14}C stabilized in temperate and tropical soils
ranged from 15 to 30% of the added C (Jenkinson and Ayanaba
1977, Sauerbeck and Gonzalez 1977).

The decomposition rate of ^{14}C-medic (Medicago littoralis)
material in four different Australian soils was much faster
than ^{14}C-straw in temperate regions because of favorable
moisture and temperature conditions, and the nature of medic
material. The labile fraction ranged from 61 to 71% of the

Table 10. Kinetic analysis of straw decomposition in field calculated from data reported in the literature.

Substrate	Location	Incubation period (yr)	Soil	Simultaneous estimates[1]			
				L (%)	k (/yr)	($\times 10^{-3}$ /yr)	RMS (%)
14C-ryegrass[3]	England	10	Aquic Hapludalf	70.5 ± 1.2	2.87 ± 0.28	85.1 ± 7.8	0.7
14C-wheat straw[4]	Canada	10	Aridic Haploboroll	78.4 ± 1.9	1.89 ± 0.14	89.0 ± 15.7	1.5
		10	Typic Cryoboralf	71.8 ± 5.0	1.40 ± 0.22	81.3 ± 31.3	2.9
14C-wheat straw[5]	Germany	9	Typic Hapludalf	73.8 ± 3.8	1.47 ± 0.14	89.4 ± 28.5	2.5
		9	Typic Hapludalf	74.7 ± 3.2	1.79 ± 0.16	97.8 ± 27.0	2.5
		4	Typic Hapludalf	62.9 ± 0.6	3.32 ± 0.07	199 ± 6	0.3
		4	Haploboroll	65.0 ± 1.5	3.81 ± 0.27	151 ± 7	0.8
		4	Eutrochrept	63.6 ± 0.9	4.68 ± 0.27	144 ± 10	0.5
		4	Fragiorthod	64.4 ± 2.3	3.62 ± 0.38	155 ± 26	1.2
		4	Humaquept	54.2 ± 0.85	4.51 ± 0.28	156 ± 8	0.5
		4	Calcorthent	53.5 ± 1.40	4.36 ± 0.43	155 ± 13	0.8
14C-ryegrass[6]	Nigeria	2	Oxic Paleustalf	69.6 ± 0.5	9.59 ± 0.28	384 ± 15	0.2
14C-wheat straw[5]	Costa Rica	1	Tropical (12 soils)[8]	71.4 ± 1.5[2]	5.93 ± 0.54[2]	–	5.8
14C medic material[7]	Australia	4	Aridisol	70.5 ± 1.4	26.4 ± 3.2	153 ± 32	1.9
		4	Vertisol	61.0 ± 4.6	21.8 ± 7.2	368 ± 116	5.0
		4	Entisol	69.2 ± 2.8	22.8 ± 5.0	130 ± 59	3.9
		4	Alfisol	70.1 ± 1.3	19.8 ± 1.8	109 ± 27	1.8
15N-medic material[7]	Australia	4	Aridisol	40.6 ± 2.9	6.21 ± 1.11	77.2 ± 22.9	2.7
		4	Vertisol	54.2 ± 7.2[2]	1.28 ± 0.45[2]		8.0
		4	Entisol	33.1 ± 6.1	14.0 ± 9.6	86.8 ± 51.0	7.8
		4	Alfisol	28.0 ± 4.4	10.6 ± 5.2	102 ± 33	5.0

[1] Equation used: 14C or 15N remaining = Le^{-kt} + Re^{-ht}, where R=100-L.

[2] Equation used: 14C or 15N remaining = Le^{-kt} + R, where R=100-L.

[3] Jenkinson (1977).

[4] Voroney (1983).

[5] Sauerbeck and Gonzalez (1977).

[6] Jenkinson and Ayanaba (1977)

[7] Ladd et al. (1981).

[8] Nine soils were Inceptisols, one was an Entisol, one was an Alfisol and one was a Mollisol.

total C added and its rate constant ranged from 26.4 to
19.8/year ($t_{\frac{1}{2}}$ = 0.026 to 0.035 year, respectively). The
decomposition rate constant for the resistant fraction ranged
from 0.37 to 0.11/year ($t_{\frac{1}{2}}$ = 1.88 to 6.35 year, respectively).

The medic-^{14}C remaining as organic residues after four
years ranged from 15 to 20% of the added C. In contrast,
45 to 50% of the ^{15}N organic medic residues remained in soil
which shows that ^{15}N is recycled in soil to a greater extent
than [14]C. The data for decomposition of ^{15}N medic material
in the Vertisol could be fitted only to a single exponential
model which approached the asymtote at 46% (Table 10),
suggesting considerable long-term stabilization.

The decomposition pattern of added residues for a wide
variety of soils shows that the initial decomposition phase
is controlled by the activity of microbes, the availability
of nutrients such as N, and climatic conditions. The resulting
microbial materials and their metabolites undergo stabilization
in soil. Also, only a small portion of added C is stabilized
in soil. Therefore, management practices which return litter
to soil on a continual basis reduce the rate of loss of soil
organic matter. Fertilizer N or organic N applied to soil
undergo cycling in soil such that 30 to 50% of the N is
immediately available to the crop while 30 to 40% is converted
to microbial biomass and its products. The N from the biomass
and labile organic pools is slowly released over a number of
years.

INTERNAL CYCLING OF CARBON AND NITROGEN IN SOIL

Four major approaches have been used to study soil
organic matter over the past 100 years. Acid hydrolysis and
classical fractionation techniques emphasize the chemical
nature of soil organic matter. Studies of organic matter in
particle size fractions have increased our understanding of
the mechanisms of organic matter stabilization. The kinetic
approach combines organic materials by their decomposition
rate and disregards their chemical identity. All these

approaches have increased our knowledge and understanding of soil organic matter. Some of this information has been directly or indirectly used in simulation of C and N dynamics in soil.

Nature and dynamics of C and N using acid hydrolysis

From 30 to 50% of organic C and 80 to 90% of total N is hydrolyzable in boiling 6 M HCl (Sowden 1977, Stevenson 1981). The soluble products are proteins, peptides, amino acids, sugars, uronic acids, pigmented substances, phenolic acids and aldehydes (Schnitzer 1978). Only 40% of total N in soils occurs in the form of protein-like materials. The remainder is in the form of nonprotein and heterocyclic N compounds (Sowden et al. 1977). Complete identification of organic C and N compounds is not possible due to degradation, artefact formation, or adsorption of compounds to insoluble residues during hydrolysis. In addition, some organic materials may be physically protected from acid hydrolysis by soil minerals (Bremner 1965, Freney and Miller 1970, Sowden 1977).

Acid hydrolysis combined with tracer and radiocarbon dating has shown that soil organic matter is composed of labile and recalcitrant materials. Using ^{14}C-labelled soils, a highly labelled C fraction was obtained when the soil was hydrolyzed with dilute acid (Jenkinson 1968, Jansson and Persson 1968), whereas 6 M acid hydrolysis was less selective in removing labelled C from soils.

The radiocarbon ages of soil organic substances range from modern to 3700 years (Paul et al. 1964, Martel and Paul, 1974, Jenkinson and Rayner 1977). Martel and Paul (1974) separated intact and partially decomposed plant residues by flotation and subjected the residue to sequential acid hydrolysis with 0.5 M and 6 M HCl. The light material and acid hydrolysates dated modern. The residue from acid hydrolysis had a radiocarbon age of 1765 ± 65 years compared to 350 ± 65 for the whole soil (Martel and Paul 1974). Jenkinson and Rayner (1977) showed that the radiocarbon age

of decalcified soil was 1450 years while the 6 M acid
hydrolyzate had an age of 515 years and accounted for 65% of
total C. The residue had an age of 2560 years. Thus, 30 to
50% of the soil organic matter had a mean residence time of
1000 years or more. Such material is either chemically
recalcitrant or physically protected from microbial decomposition and may not participate in the short-term cycling
of C and N (Martel and Paul 1974, Paul and Van Veen 1978).

Amino acids generally show greatest changes during N
mineralization and immobilization and in the short-term
cycling of C and N. Thus, up to 50% of labelled N was found
in the amino acid fraction during the initial, rapid net
immobilization periods (Stewart et al. 1963, Broadbent 1968,
Knowles and Chu 1969, Sorensen 1981). This can be attributed
to the formation of microbial proteins during this period
(Ladd and Paul 1973). Some of the amino acid-N is mineralized
on subsequent decay of the microorganisms (Stewart et al.
1963, Knowles and Chu 1969, McGill 1971, Sorensen and Paul
1971, Sorensen 1975). During incubation, the half life of
the amino acid fraction normally increases (Sorensen 1975).
The half life varied from 2.0 to 4.8 years during the 90 to
350 day period and from 3.0 to 8.6 years during the 350 to
730 day period in soils amended with acetate or glucose plus
mineral N (Sorensen and Paul 1971, Sorensen 1975). The half
life of N derived from amino acid was 7.6 years compared to
4.4 years for amino acid C in a clay soil after 90 days
incubation (McGill 1971). The half life of fractions of
labelled C and labelled N in soils amended with ^{14}C-cellulose
and $(^{15}NH_4)_2SO_4$ (Table 11) also showed similar trends
(Sorensen 1981). Some of the amino acids react with phenolic
materials to form humads (McGill et al. 1981b). The longer
half life of amino acid N compared to amino acid C was due to
recycling of N in the soil system. Amino acids can serve as
metabolites and energy sources for microbes; however, amino
acid C is not recycled to the same extent as amino acid N due
to loss of CO_2 from soil. McGill (1971) showed the amino

Table 11.   The half-life of fractions of labelled C and
labelled N in soils calculated for different incubation
periods (from Sorensen 1981).

Soil	Fractions of labelled C and labelled N	Period of incubation in days		
		30-90	90-360	360-1600
		$t_{\frac{1}{2}}$ (year)		
1	Total labelled C	0.2	2	6
	Labelled AA-C[1]	0.4	2	4
	Total labelled organic N	0.6	4	8
	Labelled AA-N[2]	0.5	5	10
4	Total labelled C	0.2	2	7
	Labelled AA-C	0.4	2	6
	Total labelled organic N	0.5	2	11
	Labelled AA-N	0.4	2	11
6	Total labelled C	0.2	2	6
	Labelled AA-C	0.4	2	6
	Total labelled organic N	0.7	5	7
	Labelled AA-N	0.5	2	9
7	Total labelled C	0.5	2	6
	Labelled AA-C	1.2	3	7
	Total labelled organic N	0.4	5	11
	Labelled AA-N	0.8	4	12

[1] AA-C = Amino acid-C.

[2] AA-N = Amino acid-N.

sugar N pool was quickly labelled during the metabolism of
acetate and $(^{15}NH_4)_2SO_4$ in a clay soil. The labelled amino
sugar N pool had a half life of 0.6 years and declined by
70% from 2.0 to 0.6 g/m^2 in 200 days.

Osborne (1977) fractionated six Australian soils before
and after cropping to ryegrass in a greenhouse experiment.
There was no significant relationship among any of the
fractions obtained with hot acid hydrolysis and dry matter

production of ryegrass at three harvest dates. Freney and Simpson (1969) also reported that there was no consistent trend in the mineralization of various N fractions during three to four weeks incubation. Thus, chemical fractionation procedures may not be useful unless combined with techniques to differentiate recently incorporated organic N from older fractions.

It has been suggested that upon immobilization, fertilizer N is first incorporated into amino acids and is then transformed into more stable forms (Broadbent 1968). Allen et al. (1973) compared the chemical distribution of fertilizer N after one growing season and after five years with that of the native humus N under field conditions. Amino acid N accounted for 52% of residual fertilizer N at the end of the first season compared to 36% in native humus N. After five years, amino acids represented 40% of residual fertilizer N. During this period, increases of fertilizer ^{15}N were noted in the NH_4^+ released on hydrolysis and nonexchangeable NH_4^+ fractions (25 versus 13%) and in the acid-insoluble N (26 versus 10%). Smith et al. (1978) studying the distribution of residual fertilizer in a Vertisol and an Alfisol found 61 to 73% of residual fertilizer N was in the amino acid fraction at the end of three cropping seasons. Unlabelled amino acids accounted for 47 to 53% of the total soil N. The redistribution of residual fertilizer N from the amino acid forms as suggested by Allen et al. (1973) had not occurred after three years of cultivation. However, these observations are of limited value because acid hydrolysis is a drastic procedure which releases amino acids from labile and stabilized organo-mineral complexes (Bremner 1965, Swift and Posner 1972).

Nature and dynamics of organic C and N in classical humic fractions

Classical fractionation schemes divide soil organic matter into fulvic acids (FA), humic acids (HA) and humin. The three humic fractions are structurally similar but differ in molecular weight, chemical analysis and functional group

Table 12. $\delta C13$, $\delta C14$ and δ values of soils and δ values of soils and humic fractions (from Campbell et al. 1967).

Sample	$\delta C14$ (0/00)	$\delta C13$ relative to N.B.S. oxalic acid standard (0/00)	δ (0/00)	Equivalent age (years)
Melfort soil	-115.5 + 6.0	-7.3	-102.6 + 6.0	870 + 50
Humic fractions of Melfort soil				
Humin	-146.1 + 6.3	-8.1	-132.3 + 6.3	1140 + 50
Unhydrolyzed humin	-155.0 ∓ 7.1	-7.4	-142 ∓ 7.1	1230 ∓ 60
Humic acids	-154 ∓ 7.3	-7.8	-142.6 ∓ 7.3	1235 ∓ 60
Hydrolyzed humic acids	-13.3 + 6.3	-5.3	-3.0 + 6.3	25 + 60
Unhydrolyzed humic acids	-171.2 + 7.0	-6.8	-159.9 + 7.0	1400 + 60
Fulvic acids	-70.0 ∓11.5	-7.2	-56.6 ∓11.5	470 ∓ 90

content.   The methods of extraction and properties of humic substances have been reviewed by Schnitzer (1978), Bremner (1965) and Oades and Ladd (1977).

The resistant nature of humic materials has been shown with radiocarbon dating techniques.   Campbell et al. (1967) fractionated a Mollisol and subjected the humin and HA to further hydrolysis with 6 M HCl.   The soil and fractions were carbon dated and their equivalent ages were calculated (Table 12).   Results showed that humin, nonhydrolyzable humin, HA and nonhydrolyzable HA had a greater equivalent age than did the hydrolyzed HA and FA.

The dynamics of C and N in humic materials have also been studied using ^{14}C and ^{15}N tracer techniques.   Broadbent (1968) found that the amount of labelled N expressed as percent of total N in HA and FA increased with incubation period in soils amended with straw and $(^{15}NH_4)_2SO_4$ (Table 13). The amount of N as HA was two to three fold higher than FA but the proportion of tagged N to total N in each fraction was similar.   Although immobilized N was present in compounds

Table 13.   Distribution of N in humic and fulvic fractions of three soils treated with labelled ammonium sulfate plus straw (from Broadbent 1968).

Fraction	Days incubated	Total N (g/m^2)	^{15}N (g/m^2)	^{15}N (% of total)
Aiken clay loam				
Humic	1	41.5	0.27	0.6
	60	43.9	2.46	5.6
Fulvic	1	23.5	0.18	0.8
	60	25.4	1.13	4.4
Columbia fine sandy loam				
Humic	1	32.1	0.30	0.9
	60	34.3	2.33	6.8
Fulvic	1	15.3	0.16	1.0
	60	16.8	0.95	5.6
Altamont clay loam				
Humic	1	31.0	0.25	0.8
	60	35.9	2.79	7.8
Fulvic	1	12.8	0.23	1.8
	60	15.0	1.21	8.1

covering a wide range of molecular mass, the bulk of labelled N was present in a high molecular mass fraction (>100,000 daltons).

Paul and McGill (1977) amended two Mollisols with ^{15}N-NO_3^- and unlabelled straw and followed the fate of immobilized fertilizer N for four years under field conditions (Fig. 2). The FA were rapidly labelled by the end of the first year. The ^{15}N from FA was quickly released over the subsequent three years and accounted for 20 to 22% of the ^{15}N remaining at the end of four years. The ^{15}N in HA remained relatively constant over four years and accounted for 8 to 10% of the ^{15}N remaining in soil. The trend for the >0.2 µm

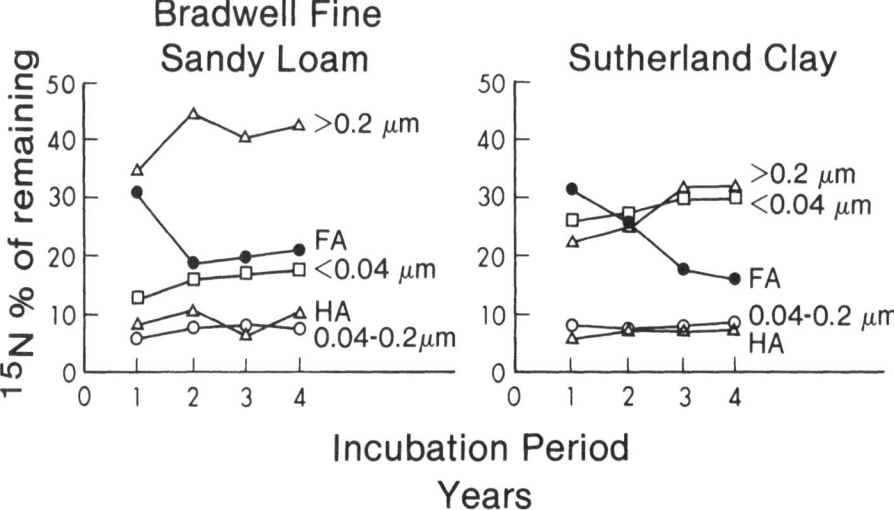

FIGURE 2. Labelled N distribution in Bradwell and Sutherland soils amended with $^{15}N-NO_3$ and unlabelled straw (from Paul and McGill 1977).

fraction was the inverse of that for FA. Fulvic acids exhibited a high turnover rate and the >0.2 μm fraction, which could contain microbial cell wall debris, accumulated N during four years. Similar results were found when the same soils were amended with ^{14}C-acetate and $(^{15}NH_4)_2SO_4$ under laboratory conditions (McGill et al. 1975).

Sochtig and Salfeld (1977) used classical fractionation and acid hydrolysis techniques to compare transformations of organic N during decomposition of straw. They found that α-amino N, hydrolyzable N, FA-N and HA-N increased during immobilization of urea. In contrast, amide N decreased as α-amino N increased.

It is difficult to interpret the results on ^{14}C and ^{15}N transfers because the fractions contain not only the humified carbon and nitrogenous compounds (i.e., compounds bonded as an integral part of humic molecules) but also materials co-extracted from cells, cell debris and other recently synthesized C and N compounds. Thus, fractionation shows

the heterogeneous nature of organic matter. However, the mass
of humic fractions cannot be directly related to state variables
in simulation models describing C and N behavior in soil.

Dynamics of organic N in soil particle size fractions

The C:N ratio of soil particle size fractions decreases
with decreased particle size (Chichester 1969). It ranges
between 50 and 20:1 in sand size fraction and drops to 8.6:1
in fine clay size fraction (Chichester 1970, Amato and Ladd
1980). The properties of C and N in various particle size
fractions have been reviewed by Oades and Ladd (1977).

Chichester (1970) fractionated a ^{15}N-labelled silt loam
soil and incubated the particle size fractions for four weeks
under waterlogged conditions (Table 14). The 53-2 μm fraction
(containing 14% of total N) accounted for 50% while the <2 μm
fraction (containing 59% of total N) accounted for 29% of the
net N mineralized. The atom % excess ^{15}N of N mineralized
from the whole soil and particle size fractions during four
weeks was higher than that of total N (Table 14). The atom
% excess ^{15}N of amino acids present in a pyrophosphate extract
was lower than that of total N in soil and in particle size
fractions except for the <2 to 0.5 μm fraction. Thus, the
immobilized N was neither incorporated into all forms of soil
N to the same degree nor was it uniformly distributed in all
particle size fractions. These data show that techniques to
extract biologically meaningful fractions are needed to study
soil organic matter.

Dynamics of organic C and N in kinetically or biologically defined entities in soil

The organic materials present in soil can also be
described by their relative decomposition rates. Attempts to
study the dynamics of the active and passive fractions of soil
organic matter were first made in the 1950's. Jansson (1958)
noted that the atom % [15]N abundance of NO_3^- formed during
remineralization of recently immobilized fertilizer was higher
than that of total organic N in soils. He postulated that the

Table 14.   Atom percent excess ^{15}N of Chester Silt Loam soil size fractions (adapted from Chichester 1970).

Soil fraction	Total N	Amino acid C	N-mineralized in 4-week incubation
	Atom percent excess ^{15}N		
Whole soil	.470	.404	.897
>53 μm diameter:			
a) Organic			
separate	.950	.884	1.263
b) Remainder	.363	.337	.726
53-2 μm diameter	.311	.275	.717
<2-0.5 μm diameter	.417	.482	.731
<0.5 μm diameter	.366	.543	.858

soil consisted of an active and a passive organic phase. The active organic phase which accounts for 10 to 15% of total N in soil has been linked to microbial biomass.

Further work with ^{14}C-glucose amended soils showed that the active C fraction represented 10 to 20% of total soil C and its C:N ratio was 10 or somewhat less (Jansson 1960). Jenkinson (1966) fumigated ^{14}C-labelled soils and noted a flush of $^{14}CO_2$-C due to the decomposition of killed biomass. Biomass, measured using the fumigation technique, accounted for 2 to 3.5% of the total C in soils (Jenkinson and Powlson 1976) with a half life of 1.69 years (k = 0.41/year) under field conditions in England. Under these conditions, the total labelled carbon remaining in soil had a half life of 3.6 years (k = 0.19/year) (Jenkinson 1965). Amato and Ladd (1980) also noted that ^{14}C- and ^{15}N-labelled biomass was more labile ($t_{\frac{1}{2}}$ = 1.1 year) than labelled C and N remaining in soils ($t_{\frac{1}{2}}$ = 1.7 to 2.3 years). Thus, the active organic phase described by Jansson (1958) corresponds in part to the living biomass, metabolites and/or stabilized microbial residues.

Paul and Juma (1981) combined the techniques of Jansson and Jenkinson and divided soil organic matter into four biologically meaningful phases: the living biomass (by fumigation), the nonbiomass active N (by isotopic dilution calculations in short-term net mineralization experiments with correction for the separate biomass determination) and very old phase (by published ^{14}C dates assuming a C:N ratio for the nonhydrolyzable residue). The remaining soil N made up a fourth phase. They defined this residual fraction as stabilized N and assigned it a half life of about 30 years. This scheme was used to derive the pool sizes of the state variables in the Tramin model (Paul and Juma 1981) which is presented in the next section.

The kinetic approach to study the dynamics of organic fractions has also been used under field conditions. Ladd et al. (1981) measured the formation and decay of isotope-labelled microbial biomass during the decomposition of ^{14}C and ^{15}N labelled medic (Medicago littoralis) material under field conditions. The organic ^{15}N remaining in soil had a longer half life compared to organic ^{14}C (Fig. 3) which is consistant with data reported by Paul and McGill (1977). The labelled material was rapidly incorporated into biomass during the initial, rapid decomposition phase. Biomass ^{14}C and ^{15}N decay showed a biphasic behavior with initial net decomposition rate after maximum growth being faster (^{14}C $t_{\frac{1}{2}}$ = 0.34 year) than the latter phase (^{14}C $t_{\frac{1}{2}}$ = 4.2 year).

O'Brien and Stout (1978) measured the distribution of total organic C, ^{13}C/^{12}C and ^{14}C/^{12}C ratios to a depth of 1 m for 15 years in a soil in New Zealand. Radiocarbon enrichment of the atmosphere, pasture, earthworms and soil were measured following thermonuclear testing which began in 1954 (Fig. 4) and integrated in a model describing the movement and turnover of soil organic matter. The steady-state model indicated that a small fraction of soil organic C (~16%) was very old and uniformly distributed throughout the soil profile. Most of the remainder (other than thermonuclear C) was modern C (less than 100 years old) and

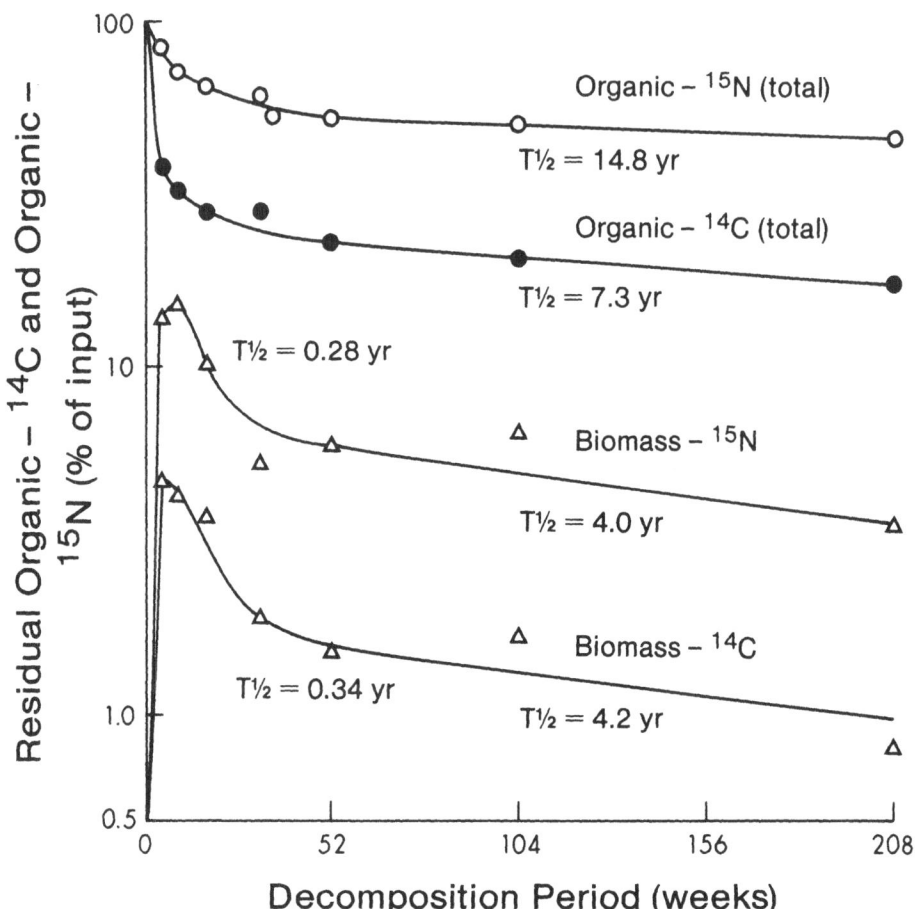

FIGURE 3.   Formation and decay of isotope-labelled
microbial biomass during the decomposition of ^{14}C,
[15]N-labelled M. littoralis in the Roseworthy soil
(from Ladd et al. 1981).

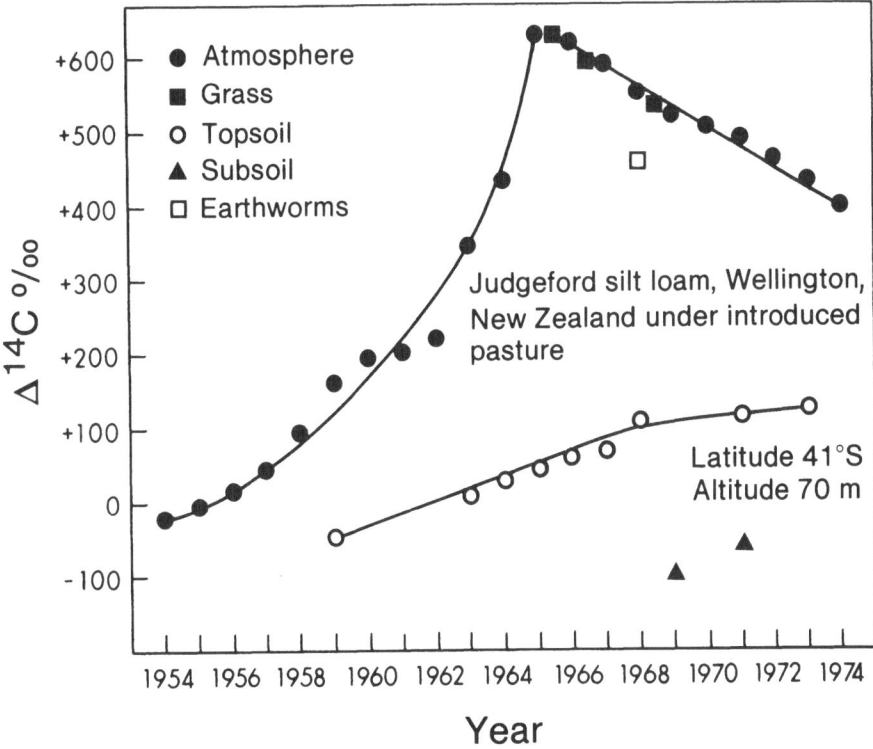

FIGURE 4.   Radiocarbon levels in the atmosphere, the
herbage, the topsoil and subsoil, and the earthworms
in a New Zealand pasture, from 1954 to 1974 (from
O'Brien and Stout 1978).

decreased exponentially with increasing profile depth.   The
model divided soil organic matter by using isotopic ratios
rather than by dating a chemical fraction (Campbell et al.
1967, Jenkinson 1966, Martel and Paul 1974).

Soil organic matter has been studied by a variety of
techniques.   The combination of tracer, chemical, and
biological techniques have aided in evaluating the net
changes occurring within the system.   Integration of these
data in simulation models has provided information about the
dynamic nature of soil organic matter.

MODELLING DECOMPOSITION AND NUTRIENT CYCLING IN SOIL

Mathematical modelling and simulation techniques have been developed for the purpose of providing quantitative description of the behavior of dynamic systems. This has resulted in the integration of available knowledge on decomposition and nutrient cycling in soil and has stimulated interest in mathematical descriptions of different processes. The role of mathematical models in ecosystem research has been reviewed by Hunt and Parton (this volume). Van Veen et al. (1981) compared several mathematical representations of C and N transformations in soils in terms of their biological, technical and physical validity. They noted that only limited data could be used for modelling purposes and more data on control mechanisms were needed. Frissel and Van Veen (1981) have summarized the features of many models used to describe N cycling in the soil-plant system.

Description of decomposition of residues in soil

Early mathematical approaches to decomposition of plant, animal and microbial residues considered their C:N ratios and N content. Microbial growth on the residues was not considered although the N requirement for the microorganisms was calculated by using C:N ratios. Currently, there are two trends used in describing decomposition of residues in soil. The first one uses regression analysis of the decomposition curve and yields data on the number of pools involved and their specific rate constants. The second approach attempts to calculate microbial growth on the residues and yields information on the total decomposition of the residues and formation of microbial materials.

The problem of describing microbial growth on residues is a difficult one. Two approaches have been used. The first one divides the residue into biochemical components which are assigned specific decomposition rates and efficiencies of utilization. The total biomass or a portion of the biomass may be involved (Van Veen et al. 1981).

The first approach ignores the chemical linkage between polymers present in the residues. It also assumes that the sum of decomposition of individual components is equal to decomposition of the residue. This approach has been used in the Tramin model (Juma and Paul 1981). However, it has to be critically examined because association among polymers or adsorption of polymers to clay or humic colloids markedly reduces their decomposition rates.

The second approach divides the residue into labile and resistant fractions based on kinetic analysis of the data or by using assumed ratios of cytoplasmic and cell wall material. McGill et al. (1981b) calculated the structural and metabolic split as follows:

$$F_s = (B_d - B_n)/(B_s - B_n) \qquad (3)$$

where F_s is the fraction of C in the structural component, B_d is the N:C ratio of the dead material (shoot, root or microorganism), B_s is the N:C ratio of the structural component, and B_n is the N:C ratio of the metabolic component. The amount of N in the components is calculated from assigned C:N ratios. The metabolic component is rich in N and is quickly mineralized compared to the structural component. This approach considers the architecture of cells and may gain wider acceptance. Microbial growth on the two components may be described by Michaelis-Menten or first order kinetics. This approach has been used in the Phoenix model (McGill et al. 1981b). The two approaches of describing decomposition of residues have been neither compared nor critically examined.

Modelling internal cycling of C and N in soil

Internal cycling of C and N occurs during microbial growth and death in soil. The size and activity of microbial biomass is explicitly defined in most mechanistic models describing C and N dynamics. The role of soil fauna in comminution, grazing and nutrient cycling may be defined or implied (Hunt and Parton, this volume). The relation of activity of soil fauna either directly to decomposition of residues and soil organic matter, or indirectly to control of internal cycling, has not been well quantified. Therefore, many models do not include faunal activity in the internal cycling of C and N.

Many simulation models have been developed to describe the N cycle in soil (Frissel and Van Veen 1981). The models have been developed for different soils and emphasize certain aspects of the system depending upon the objectives of the research. A systematic analysis of the models based on the mechanisms used and accuracy of description of different processes has not been made. We have compared the mineralization-immobilization submodel of the Phoenix model (McGill et al. 1981b,c) and the Tramin model (Juma and Paul 1981, Paul and Juma 1931) using a common experimental data set (Juma 1981). A comparison of the major features of the models are summarized in Table 15 and the flow charts are presented in Figures 5 and 6.

The experimental data were obtained by incubating a ^{15}N-labelled Boralfic Haploboroll (Weirdale loam) under laboratory conditions. Mineral N, biomass N, total N and ^{15}N abundances, biomass C, total C and CO_2-C evolved were measured over a 12-week incubation period. Biomass C and N were measured by chloroform fumigation, nonbiomass N and metabolite N by isotopic dilution, old N from published radiocarbon dates for humic materials and their C:N ratio, and stabilized N was calculated by difference (Paul and Juma 1981). The pool sizes for the Phoenix model were similarly calculated. The rate constants and detailed description of the processes are given elsewhere (Juma and Paul 1981, McGill et al. 1981b). Minor modifications to rate constants were made to match model outputs to experimental data.

The simulated net changes calculated by both the models were similar to the experimental data (Table 16) but the inputs and outputs from the various pools were markedly different (Table 17). The Phoenix model emphasized biomass

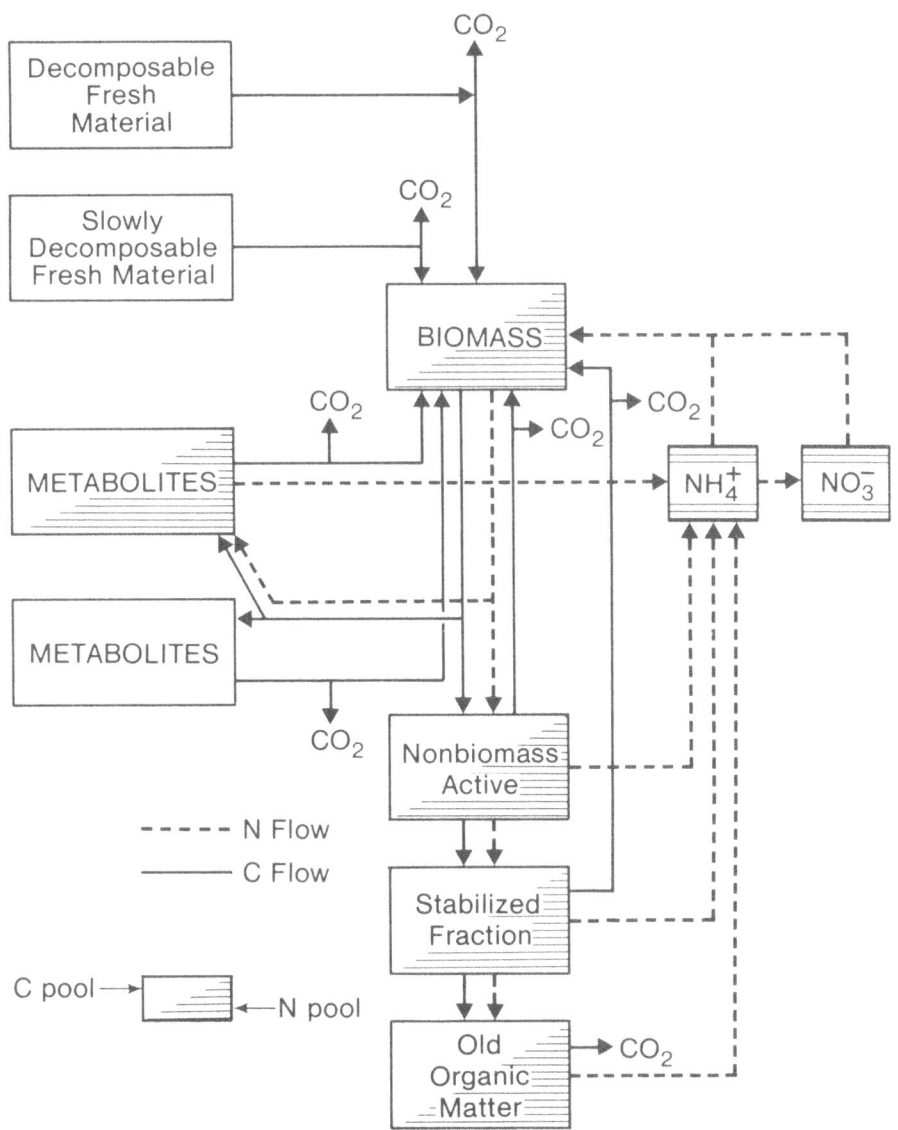

FIGURE 5.   Flow chart for the Tramin model.

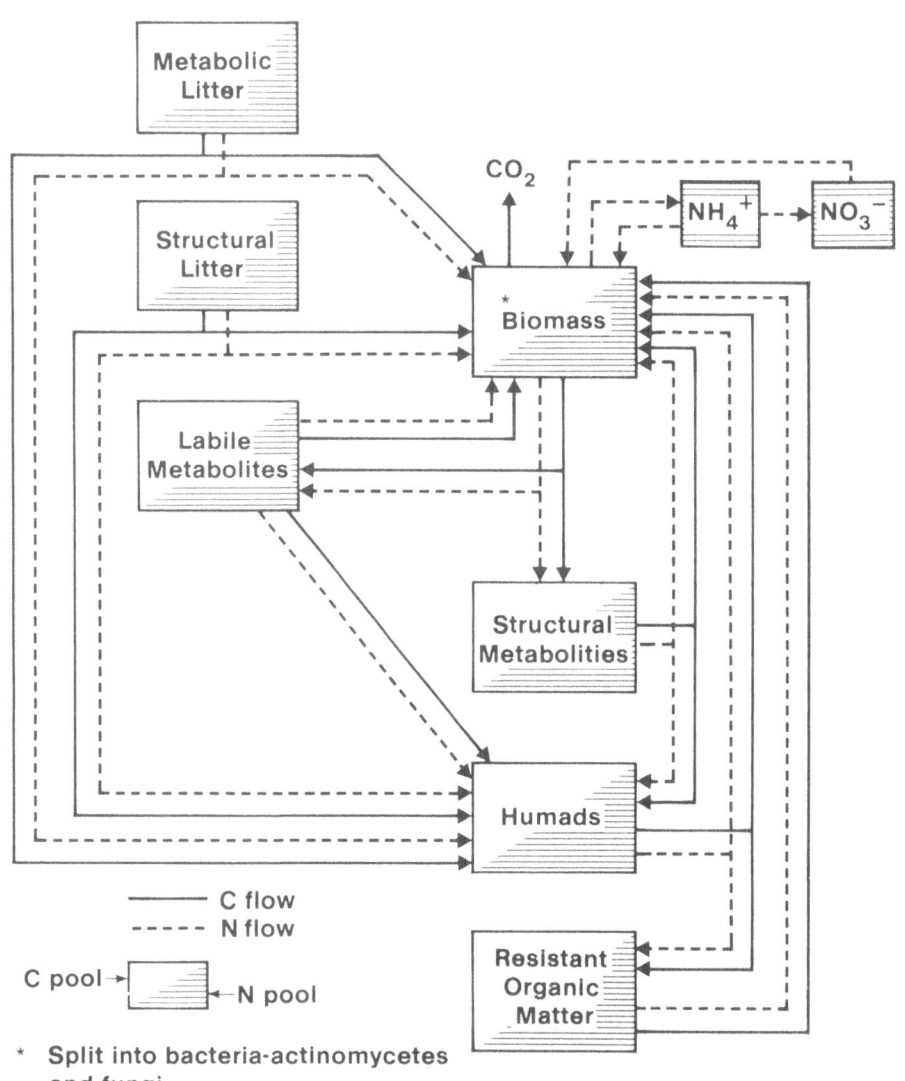

FIGURE 6.   Nitrogen mineralization-immobilization submodel of the Phoenix model.

Table 15. Description of different processes in the Phoenix and Tramin simulation models.[1]

Process	Phoenix	Tramin
Microbial growth	Growth limited by C availability; Bacterial growth faster than fungal growth; Michaelis-Menten kinetics.	Growth limited by C substrate and N availability; Microbial growth has infinite potential; First order kinetics.
Microbial decay	Mortality due to freezing and drying; Plus a density-dependent predation death rate.	First order decay rate.
Uptake of organic C	Uptake from soil solution described by Michaelis-Menten kinetics; Adsorption-desorption reactions explicitly defined.	Implied uptake through solution phase described by first order kinetics.
Uptake of organic N	Uptake of organic molecules from the soil solution by microbes.	Organic N compounds converted to NH_4^+ outside the cell.
NH_4^+ uptake	Dependent on the C:N ratio of bacteria or fungi; Preferential uptake compared to NO_3^-.	Uptake proportional to C used for microbial growth (C:N ratio = 6); Preferential uptake of NH_4^+ compared to NO_3^-.
NO_3^- uptake	Dependent on the C:N ratio of the organisms; Concurrent uptake with NH_4^+.	Uptake of NO_3^- occurs when NH_4^+ is depleted.
NH_4^+ release from biomass	Occurs when the C:N ratio of organisms is low.	Not described.

[1] For detailed descriptions and documentation refer to McGill et al. (1981b,c); Paul and Juma (1981), Tuma and Paul (1981), respectively.

Table 16. Comparison of experimental data with Phoenix and Tramin model outputs at the end of 12 weeks incubation of Weirdale loam soil.

Pool	Experimental	Simulated	
		Phoenix	Tramin
	(mg/m^2)		
Total organic ^{15}N remaining	11.0	11.1	11.3
Biomass ^{15}N	5.2	4.9	5.2
Structural + labile metabolite ^{15}N or Nonbiomass active + metabolite ^{15}N	5.9	6.1	5.7
Mineral ^{15}N	4.4	4.4	4.1
	(g/m^2)		
Mineral ^{14}N	11.5	10.9	12.3
Cumulative CO_2-C evolved	93.8	106.2	85.1
Biomass N	14.6	12.6	13.8

growth because all N was directly incorporated into the bacterial and fungal biomass. In contrast, the Tramin model emphasized NH_4^+ turnover because all N was mineralized outside the cell and only a small portion of NH_4^+ was taken up to meet the needs of the microbial biomass. The rest of NH_4^+ was nitrified. Nitrate uptake calculated by Tramin model was zero because the soil was in net mineralization conditions and enough NH_4^+ was present to meet microbial needs. Biomass decay and net mineralization were similar for both the models. The models showed that net mineralization was

Table 17.   Comparison of various inputs and outputs for
Weirdale soil calculated with Phoenix and Tramin models.

	Phoenix	Tramin
	(g/m^2)	
Biomass growth	28.2	11.9
Bacterial growth	12.6	–
Fungal growth	15.6	–
NH_4^+ uptake	2.4	11.9
NO_3^- uptake	5.0	0
Organic N uptake	20.8	0
Mineral + organic N uptake	28.2	11.9
Biomass decay	16.0	14.7
Net mineralization	8.9	10.3
Gross mineralization	16.3	22.2
Net mineralization/gross mineralization	0.55	0.46

almost one-half of actual mineralization. A simplified
summary of N flux within the various pools is presented
(Fig. 7).

Simulation modelling is a powerful tool for calculating
the inputs and outputs of N in the dynamic soil system.
However, the rate constants used in the simulation model are
different from the net rate constants obtained from
experimental data. For example, net mineralization rate
constants have been measured for a wide variety of soils.
Tracer techniques have yielded net mineralization rate
constants of the recently immobilized N. Simulation models
need actual mineralization and immobilization rate constants
because these processes occur simultaneously in soils.
These rate constants can be determined algebraically for
steady-state systems with three to four pools (Shipley and
Clark 1972). However, the soil is a complex system which is

Phoenix Model

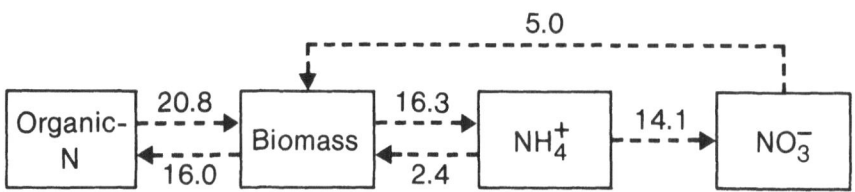

Tramin Model

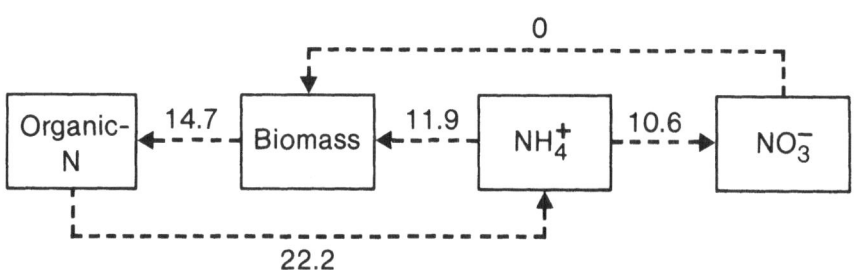

FIGURE 7. Comparisons of N flux (g/m^2) through various pools of Weirdale loam soil calculated with Phoenix and Tramin models.

seldom under steady-state conditions. Therefore, it is not possible to calculate the actual mineralization and immobilization rates from tracer data. Simulation models unravel some of the complexity and produce quantitative data for further evaluation.

The sources of net N mineralized over the 12-week incubation calculated by the models (Table 18) showed that 62 to 77% of N was mineralized from biomass and microbial debris (nonbiomass active fraction or structural and labile metabolites). These components accounted for less than 10% of soil organic matter. The stabilized N contributed the remainder. These data show the dynamic nature of biomass in soil and its role in the internal cycling of N. The flux of N through various pools has been calculated using

Table 18.   Sources of the net N mineralized in Weirdale
loam soil calculated with Phoenix and Tramin models.

Source	Model	
	Phoenix	Tramin
	%	
Biomass	44	28
Structural + labile metabolites	33	-
Nonbiomass active N + metabolites	-	34
Humads or stabilized organic matter	23	38

simulation modelling.   Consequently these data should be
carefully examined as they depend on the concepts used to
develop the models.

The example used in the above simulation exercise is a
simple case because the soil was incubated under laboratory
conditions in the absence of plants.   Hunt and Parton (this
volume) have considered the effects of plant input, fluctuating
moisture and temperature conditions, and many other soil
processes which occur under field conditions.   Simulation
modelling has clearly shown that internal cycling of C and
N is an important feature of the soil system.   The amount of
C and N cycled depends upon the biological availability of
residues and soil substrates, biological activity of the soil,
physical conditions of the soil and other factors such as
tillage, application of pesticides and fertilizers.   Therefore,
controls can be quantified and their interactions treated in
mechanistic simulation models to integrate their combined
effects.

At present, management decisions are based upon average
net amounts of N released from soil to a crop.   Information
about net supply of N is useful, but is not specific for
different types of management.   Also, the full effect of the

change in management may not be known until the new practice has been in effect for several years (Doran et al. 1984, Linn and Doran 1984). It is difficult to experimentally measure the effect of changes in management and other conditions for every soil type; therefore simulation modelling, based on mechanisms controlling different processes in soil, may be a powerful tool to generate information for a specific land area under a particular management. The simulation models should not only provide the correct net results but also be correct in the description of the soil processes. Such software packages, once properly validated, could prove effective tools in managing land for efficient and sustainable agricultural production.

Tracer studies have shown that organic materials of very different chemical composition are included among those which are most readily mineralized. Fractions of soil organic matter which are physically or biologically separable are differentially involved in the turnover of organic matter (Greenland 1971). Clay plays an important physical role in forming organic matter-clay complexes while microorganisms produce polysaccharides which stabilize microaggregates (Greenland 1971, Foster 1981). Cell architecture is also an important biophysical consideration since decomposition of large polymers is not independent of associated materials.

There are two problems in comparing the components of soil organic matter and the magnitudes of their decay rate constants. First, although each component may fall kinetically within a distinct range, it is still chemically heterogeneous. Conversely, chemically similar material (e.g., amino acids) exhibit wide variations in stabilization response. Second, the decay rates of specific components cannot readily be measured directly in the field or the laboratory, unless one can isolate the components and measure isotope dilution rates. Many field scale applications require kinetic homogeneity of groups of materials, but methods to isolate materials from soil are based on chemical characteristics. Ultimately it will be necessary to combine chemical information with micro-site biology and soil physics to fully understand and define changes in parameters due to climate, soil zone, parent material and management. Further, the ability to understand the system and to mathematically describe its dynamics has surpassed techniques to isolate kinetically homogeneous fractions. Therefore, further kinetic approaches to the study of soil organic matter are needed.

RESEARCH NEEDS AND FUTURE OUTLOOK

We have examined the agro-ecosystem at different levels of resolution and have concluded with a comparison of simulation models describing decomposition and internal cycling of C and N. Study of the soil system in detail and integration of information through simulation submodels of various processes can yield valuable information about this dynamic system. Some areas of future research are:

1. The mechanisms of organic to mineral N transformations within or outside the cell need to be clearly defined. The role of extra- and intra-cellular enzymes involved must be elucidated. The mechanisms of regulating NH_4^+ and NO_3^- uptake need further examination.

2. The efficiency of utilization of substrates and their decomposition rate constants cannot be independently measured at the present. These two parameters determine the amount of C used for microbial growth which in turn affects the internal cycling of C and N.

3. The C:N ratios of soil organisms in situ and factors affecting them need further investigation.

4. Biophysical controls, such as fabric reorganization and soil faunal activity, need further investigation because they control spatial relations of substrates and organisms, and nutrient cycling.

5. Soil organic matter must be divided into biologically meaningful and kinetically homogeneous fractions, with emphasis on substrate and site architecture.

6. The mechanisms used to describe different soil processes in simulation models and the flux of nutrients within pools should be examined and compared. Comparison of experimental data with simulation model output should not be an end in itself. More effort should be made to compare and synthesize existing simulation models.

ACKNOWLEDGMENTS

We thank Natural Sciences and Engineering Research Council, Killam Memorial Trusts and Farming for the Future (Agricultural Research Council of Alberta) for financial support. Contribution of the University of Alberta (Alberta Institute of Pedology publication #T-85-1).

REFERENCES

1. Allen, A.L., F.J. Stevenson, and L.T. Kurtz. 1973. Chemical distribution of residual fertilizer nitrogen in soil as revealed by nitrogen-15 studies. J. Environ. Qual., 2:120-124.
2. Allison, F.E. 1966. The fate of nitrogen applied to soils. Adv. Agron., 18:219-258.
3. Amato, M., and J.N. Ladd. 1980. Studies of nitrogen immobilization and mineralization in calcareous soils. V. Formation and distribution of isotope-labelled biomass during the decomposition of ^{14}C- and ^{15}N-labelled plant material. Soil Biol. Biochem., 12:405-411.
4. Aulakh, M.S., D.A. Rennie, and E.A. Paul. 1982. Gaseous nitrogen losses from cropped and summer-fallowed soils. Can. J. Soil Sci., 62:187-195.
5. Barrow, N.J. 1969. The accumulation of soil organic matter under pasture and its effect on soil properties. Aust. J. Exp. Agri. Anim. Husb., 9:437-444.
6. Bartnicki-Garcia, S. 1968. Cell wall chemistry, morphogenesis, and taxonomy of fungi. Ann. Rev. Microbiol., 22:87-108.
7. Bauer, A., and A. Black. 1981. Soil carbon, nitrogen and bulk density comparisons in two cropland tillage systems after 25 years and in virgin grassland. Soil Sci. Soc. Am. J., 45:1166-1170.
8. Black, A.L. 1973. Soil property changes associated with crop residue management in a wheat-fallow rotation. Soil Sci. Soc. Am. Proc., 37:943-946.
9. Bremner, J.M. 1965. Organic nitrogen in soils. Pages 93-149 in W.V. Bartholomew and F.E. Clark, editors. Soil Nitrogen. Amer. Soc. Agron., Madison, Wisconsin.

10. Broadbent, F.E. 1968. Nitrogen immobilization in relation to N-containing fractions of soil organic matter. Pages 131-142 in Isotopes and radiation in soil organic matter studies, IAEA, Vienna.

11. Brown, P.L., and D.D. Dickey. 1970. Losses of wheat straw residue under simulated field conditions. Soil Sci. Soc. Am. Proc., 34:118-121.

12. Burford, J.R., R.J. Dowdell, and R. Crees. 1981. Emission of nitrous oxide to the atmosphere from direct-drilled and ploughed clay soils. J. Sci. Fd. Agric., 32:219-223.

13. Cameron, D.R., C.G. Kowalenko, and K.C. Ivarson. 1978. Nitrogen and chloride distribution and balance in a clay loam soil. Can. J. Soil Sci., 58:77-88.

14. Campbell, C.A., and E.A. Paul. 1978. Effects of fertilizer N and soil moisture on mineralization, N recovery and A-values, under spring wheat grown in small lysimeters. Can. J. Soil Sci., 58:39-51.

15. Campbell, C.A., E.A. Paul, D.A. Rennie, and K.J. McCallum. 1967. Factors affecting the accuracy of the carbon-dating method in soil humus studies. Soil Sci., 104:81-85.

16. Campbell, C.A., and W. Souster. 1982. Loss of organic matter and potentially mineralizable nitrogen from Saskatchewan soils due to cropping. Can. J. Soil Sci., 62:651-656.

17. Cannon, K.R., J.A. Robertson, W.B. McGill, F.D. Cook, and D.S. Chanasyk. 1984. Production optimization on Gray Wooded soils. Farming for the Future project report #79-0132. Department of Soil Science, University of Alberta.

18. Cheshire, M.V., G.P. Sparling, and R.H.E. Inkson. 1979. The decomposition of straw in soil. Pages 65-71 in E. Grossbard, editor. Straw decay and its effect on disposal and utilization. John Wiley, New York.

19. Chichester, F.W. 1969. Nitrogen in soil organo-mineral sedimentation fractions. Soil Sci., 107:356-363.

20. Chichester, F.W. 1970. Transformations of fertilizer nitrogen in soil. II. Total and ^{15}N-labelled nitrogen of soil organo-mineral sedimentation fractions. Plant Soil, 33:437-456.

21. Clark, F.E., C.V. Cole, and R.A. Bowman. 1980. Nutrient cycling. Pages 713-758 in A.I. Bregmeyer and G.M. Van Dyne, editors. Grasslands, systems analysis and man. Cambridge University Press, Cambridge.

22. Clark, M.D., and J.T. Gilmour. 1983. The effect of temperature on decomposition at optimum and saturated soil water contents. Soil Sci. Soc. Am. J., 47:927-929.

23. Coote, D.R., and J.F. Ramsey. 1983. Quantification of the effects of over 35 years of intensive cultivation on four soils. Can. J. Soil Sci., 63:1-14.

24. Davidson, J., F. Gray, and D. Pinson. 1967. Changes in organic matter and bulk density with depth under two cropping systems. Agron. J., 59:375-378.

25. Dick, W.A. 1983. Organic carbon, nitrogen, and phosphorus concentrations and pH in soil profiles as

130

       affected by tillage intensity. Soil Sci. Soc. Am. J.,
       47: 102-107.

26. DiRienzo, J.M., K. Nakamura, and M. Inouye. 1978. The
       outer membrane proteins of gram-negative bacteria:
       Biosynthesis, assembly and functions. Ann. Rev. Biochem.,
       47:481-532.

27. Doran, J.W. 1980. Soil microbial and biochemical
       changes associated with reduced tillage. Soil Sci. Soc.
       Am. J. 44:765-771.

28. Doran, J.W., .W. Wilhelm and J.F. Power. 1984. Crop
       residue removal and soil productivity with no-till corn,
       sorghum and soybean. Soil Sci. Soc. Amer. J., 48:640-645.

29. Dormaar, J.F. 1979. Organic matter characteristics of
       undisturbed and cultivated Chernozemic and Solonetzic A
       horizons. Can. J. Soil Sci., 59:349-356.

30. Dormaar, J.F., and U.J. Pittman. 1980. Decomposition of
       organic residues as affected by various dryland spring
       wheat-fallow rotations. Can. J. Soil Sci., 60:97-106.

31. Doughty, J.L., F.D. Cook, and F.G. Warder. 1954. Effect
       of cultivation on the organic matter and nitrogen of
       Brown soils. Can. J. Agric. Sci., 34:406-411.

32. Foster, R.C. 1981. Polysaccharides in soil fabrics.
       Sci., 214:665-667.

33. Freney, J.R., and R.J. Miller. 1970. Investigation of
       the clay mineral protection theory for non-hydrolysable
       nitrogen in soil. J. Sci. Fd. Agric., 21:57-61.

34. Freney, J.R., and J.R. Simpson. 1969. The mineralization
       of nitrogen from some organic fractions in soil. Soil
       Biol. Biochem., 1:241-251.

35. Freyman, S., C.J. Palmer, E.H. Hobbs, J.F. Dormaar,
       G.B. Schaalje, and J.R. Moyer. 1982. Yield trends on
       long-term dryland wheat rotations at Lethbridge. Can.
       J. Plant Sci., 61:609-619.

36. Frissel, M.J., and J.A. Van Veen. 1981. Simulation of
       nitrogen behaviour of soil-plant systems. Center for
       Agricultural Publishing and Documentation, Wageningen.

37. Gokhale, N.G. 1959. Soil nitrogen status under
       continuous cropping with manuring in the case of unshaded
       tea. Soil Sci., 87:331-333.

38. Greenland, D.J. 1971. Changes in the nitrogen status
       and physical conditions of soils under pastures with
       special reference to the maintenance of the fertility of
       Australian soils used for growing wheat. Soils Fert.,
       34:237-251.

39. Haas, H.J., C.E. Evans, and E.F. Miles. 1957. Nitrogen
       and carbon changes in great plain soils as influenced by
       cropping and soil treatments. US Department of Agriculture,
       Technical Bulletin 1164.

40. Haider, K., J.P. Martin, and E. Rietz. 1977. Decomposition
       in soil of ^{14}C-labeled coumaryl alcohols: Free and linked
       into dehydropolymer and plant lignins and model humic
       acids. Soil Sci. Soc. Am. J., 41:556-562.

41. Hall, J.L., T.J. Flowers, and R.M. Roberts. 1974. Plant
       cell structure and metabolism. Longman, London.

42. Hobbs, J.A., and P.L. Brown. 1957. Nitrogen changes in cultivated dryland soils. Agron. J., 49:257-260.

43. Hunt, H.W. 1977. A simulation model for decomposition in grasslands. Ecol., 58:469-484.

44. Huntjens, J.L.M., and R.A.J.M. Albers. 1978. A model experiment to study the influence of living plants on the accumulation of soil organic matter in pastures. Plant Soil, 50:411-418.

45. Jansson, S.L. 1958. Tracer studies on nitrogen transformations in soil with special attention to mineralization-immobilization relationships. An. R. Agric. Coll. Swed., 24:101-361.

46. Jansson, S.L. 1960. On the establishment and use of tagged microbial tissue in soil organic matter research. Pages 635-641. 7th Int. Cong. Soil Sci., Madison, Wis., USA.

47. Jansson, S.L., and J. Persson. 1968. Co-ordination of humus chemistry and soil organic-matter biology by isotopic techniques. Pages 111-124 in The Use of Isotopes in Soil Organic Matter Studies. IAEA, Vienna.

48. Jenkinson, D.S. 1965. Studies on the decomposition of plant material in soil. I. Losses of carbon from ^{14}C-labelled ryegrass incubated with soil in the field. J. Soil Sci., 16:104-115.

49. Jenkinson, D.S. 1966. Studies of the decomposition of plant material in soil. II. Partial sterilization of soil and the soil biomass. J. Soil Sci., 17:280-302.

50. Jenkinson, D.S. 1968. Chemical tests for potentially available nitrogen in soil. J. Sci. Fd. Agric. 19:160-168.

51. Jenkinson, D.S. 1977. Studies on the decomposition of plant material in soil. V. The effects of plant cover and soil type on the loss of carbon from ^{14}C labelled ryegrass decomposing under field conditions. J. Soil Sci., 28:424-434.

52. Jenkinson, D.S., and A. Ayanaba. 1977. Decomposition of carbon-14 labeled plant material under tropical conditions. Soil Sci. Soc. Amer. J., 41:912-915.

53. Jenkinson, D.S., and A.E. Johnson. 1976. Soil organic matter in the Hoosfield continuous barley experiment. Rothamsted Report for 1976, part 2:87-101.

54. Jenkinson, D.S., and D.S. Powlson. 1976. The effect of biocidal treatments on metabolism in soil. V. A method of measuring soil biomass. Soil Biol. Biochem., 8:209-213.

55. Jenkinson, D.S., and J.H. Rayner. 1977. The turnover of soil organic matter in some of the Rothamsted classical experiments. Soil Sci., 123:298-305.

56. Jenny, H. 1980. The soil resource. Springer-Verlag, New York.

57. Johnston, J., G. Browning, and M. Russell. 1942. The effect of cropping practices of aggregation, organic matter content and loss of soil and water in Marshall silt loam. Soil Sci. Soc. Am. Proc., 7:105-107.

58. Juma, N.G. 1981. Dynamics of soil and fertilizer nitrogen. Ph.D. Thesis, University of Saskatchewan, Saskatoon.

59. Juma, N.G., and E.A. Paul. 1981. Use of tracers and computer simulation techniques to assess mineralization and immobilization of soil nitrogen. Pages 145-154 in M.J. Frissel and J.A. Van Veen, editors. Simulation of Nitrogen Behavior in Soil-Plant Systems. Pudoc, The Netherlands.

60. Juma, N.G., and E.A. Paul. 1983. Effect of a nitrification inhibitor on N immobilization and release of ^{15}N from non-exchangeable ammonium and microbial biomass. Can. J. Soil Sci., 63:167-175.

61. Keegstra, K., K.W. Talmadge, W.D. Bauer, and P. Albersheim. 1973. The structure of plant cell walls. III. A model of the walls of suspension-cultured sycamore cells based on the interconnections of the macro-molecular components. Plant Physiol., 51:188-197.

62. Kirkwood, S. 1974. Unusual polysaccharides. Ann. Rev. Biochem., 43:401-418.

63. Knowles, R., and D.T.H. Chu. 1969. Survival and mineralization and immobilization of ^{15}N-labelled Serratia cells in a boreal forest raw humus. Can. J. Microbiol., 15:223-228.

64. Ladd, J.N., J.M. Oades, and M. Amato. 1981. Microbial biomass formed from ^{14}C, ^{15}N-labelled plant material decomposing in soils in the field. Soil Biol. Biochem., 13:119-126.

65. Ladd, J.N., and E.A. Paul. 1973. Changes in enzymic activity and distribution of acid-soluble, amino acid-nitrogen in soil during nitrogen immobilization and mineralization. Soil Biol. Biochem., 5:825-840.

66. Lathwell, D.J., and D.R. Bouldin. 1981. Soil organic matter and soil nitrogen behavior in cropped soils. Trop. Agric. (Trinidad) 58:341-348.

67. Linn, D.M. and J.W. Doran. 1984. Aerobic and anaerobic microbial populations in no-till and plowed soils. Soil Sci. Soc. Amer. J., 48:1267-1272.

68. Lucas, R.E., J.B. Holtman, and L.J. Connor. 1977. Soil carbon dynamics and cropping practices. Pages 333-351 in W. Lockeretz, editor. Agriculture and Energy. Academic Press, New York.

69. Lynch, J.M. 1979. Straw residues as substrates for growth and product formation by soil micro-organisms. Pages 47-56 in E. Grossbard, editor. Straw decay and its effect on disposal and utilization. John Wiley, New York.

70. Malhi, S.S., and M. Nyborg. 1983. Field study of the fate of fall-applied ^{15}N-labelled fertilizers in three Alberta soils. Agron. J., 75:71-74.

71. Martel, Y.A., and J.M. Deschenes. 1976. Les effets de la mise encultive et de la prairie prolongee sur le carbone, l'azote et la structure de quelques sols du Quebec. Can. J. Soil Sci., 56:373-383.

72. Martel, Y.A., and E.A. Paul. 1974. The use of radiocarbon dating of organic matter in the study of soil genesis. Soil Sci. Soc. Am. Proc., 38:501-506.
73. McGill, W.B. 1971. Turnover of microbial metabolites during nitrogen mineralization and immobilization in soil. Ph.D. Thesis, University of Saskatchewan, Saskatoon.
74. McGill, W.B. 1983. Kinetic effects of quality of soil organic matter. Alberta Soil Sci. Workshop Proc. Pages 1-9.
75. McGill, W.B., and P.B. Hoyt. 1977. Effect of forages and rotation management on soil organic matter. Alberta Soil Sci. Workshop Proc., Alberta Inst. of Pedology, University of Alberta Pub. No. G-77-1.
76. McGill, W.B., C.A. Campbell, J.F. Dormaar, E.A. Paul, and D.W. Anderson. 1981a. Soil organic matter losses. Pages 72-133. Agricultural land: Our disappearing heritage. Alberta Soil Sci. Workshop Proc.
77. McGill, W.B., H.W. Hunt, R.G. Woodmansee, and J.O. Reuss. 1981b. Phoenix - A model of the dynamics of carbon and nitrogen in grassland soils in F.E. Clark and T. Rosswall, editors. Terrestrial Nitrogen Cycles. Ecol. Bull. (Stockholm) 33:49-115.
78. McGill, W.B., H.W. Hunt, R.G. Woodmansee, J.O. Reuss, and K.H. Paustian. 1981c. Formulation, process, controls, parameters and performance of Phoenix: A model of carbon and nitrogen dynamics in grassland soils. Pages 171-191 in M.J. Frissel and J.A. Van Veen, editors. Simulation of nitrogen behaviour of soil-plant systems. Pudoc, Wageningen, The Netherlands.
79. McGill, W.B., J.A. Shields, and E.A. Paul. 1975. Relation between carbon and nitrogen turnover in soil organic fractions of microbial origin. Soil Biol. Biochem., 7:57-63.
80. Nelson, D.W., J.P. Martin, and J.O. Ervin. 1979. Decomposition of microbial cells and components in soil and their stabilization through complexing with model humic acid-type phenolic polymers. Soil Sci. Soc. Am. J., 43:84-88.
81. Nyhan, J.W. 1975. Decomposition of carbon-14 labelled plant materials in a grassland soil under field conditions. Proc. Soil Sci. Soc. Am., 39:643-648.
82. Oades, J.M., and J.N. Ladd. 1977. Biochemical properties: Carbon and nitrogen metabolism. Pages 127-160 in J.S. Russell and E.L. Greacen, editors. Soil factors in crop production in a semi-arid environment. University of Queensland Press, St. Lucia.
83. O'Brien, B.J., and J.D. Stout. 1978. Movement and turnover of soil organic matter as indicated by carbon isotope measurements. Soil Biol. Biochem., 10:309-317.
84. Osborne, G.J. 1977. Chemical fractionation of soil nitrogen in six soils from Southern New South Wales. Aust. J. Soil Res., 15:159-165.
85. Paul, E.A., C.A. Campbell, D.A. Rennie, and R.J. McCallum. 1964. Investigations of the dynamics of soil

134

humus utilizing carbon dating techniques. Trans. 8th Intern. Congr. Soil Sci., Bucharest. Pages 201-208.

86. Paul, E.A., and N.G. Juma. 1981. Mineralization and immobilization of nitrogen by microorganisms. In F.E. Clark and T. Rosswall, editors. Terrestrial Nitrogen Cycles. Ecol. Bull. (Stockholm) 33:179-195.

87. Paul, E.A., and W.B. McGill. 1977. Turnover of microbial biomass, plant residues and soil humic constituents under field conditions. Pages 149-157 in Soil Organic Matter Studies, Vol. 1. IAEA, Vienna.

88. Paul, E.A., and J.A. Van Veen. 1978. The use of tracers to determine the dynamic nature of organic matter. Trans. 11th Intern. Congr. Soil Sci., Edmonton, Alberta. 3:61-102.

89. Paul, E.A., and R.P. Voroney. 1980. Nutrient and energy flows through soil microbial biomass. Pages 215-237 in D.C. Ellwood, J.N. Hedger, M.J. Latham, J.M. Lynch, and J.H. Slater, editors. Contemporary Microbial Ecology, Academic Press, London.

90. Pittman, U. 1977. Crop yields and soil fertility as affected by dryland rotations in southern Alberta. Commun. Soil Sci. Plant Analysis. 8:391-405.

91. Reddy, K.R., and W.H. Patrick. 1975. Effect of alternate aerobic and anaerobic conditions on redox potential, organic matter decomposition and nitrogen loss in a flooded soil. Soil Biol. Biochem., 7:87-94.

92. Richards, B.N. 1974. Introduction to the soil ecosystem. Longman.

93. Ridley, A.D., and R.A. Hedlin. 1968. Soil organic matter and crop yields as influenced by the frequency of summerfallowing. Can. J. Soil Sci. 48:315-322.

94. Robertson, J.A. 1983. Effect of management on soil organic matter. Pages 30-34. Alberta Soil Sci. Workshop Proc.

95. Rovira, A., and E. Graecen. 1957. The effect of aggregate disruption on the activity of microorganisms in the soil. Aust. J. Agric. Res., 8:659-673.

96. Russell, J.S. 1960. Soil fertility changes in long term experimental plots at Kobybolite, South Australia. I. Changes in pH, total nitrogen, organic carbon and bulk density. Aust. J. Agric. Res., 11:902-926.

97. Sain, P., and F.E. Broadbent. 1977. Decomposition of rice straw in soils as affected by some management factors. J. Environ. Qual., 6:96-100.

98. Salter, R.M., and T.C. Green. 1933. Factors affecting the accumulation and loss of nitrogen and organic carbon in cropped soils. Agron. J., 25:622-630.

99. Sauerbeck, D.R., and M.A. Gonzalez. 1977. Field decomposition of carbon-14 labelled plant residues in various soils of the Federal Republic of Germany and Costa Rica. Soil organic matter studies 2:159-170. IAEA, Vienna.

100. Schnitzer, M. 1978. Humic substances: Chemistry and reactions. Pages 1-64 in M. Schnitzer and S.U. Khan, editors. Soil Organic Matter. Elsevier Scientific

Pub. Co., New York.

101. Shields, J.A., and E.A. Paul. 1973. Decomposition of
     ^{14}C-labelled plant material under field conditions.
     Can. J. Soil Sci., 53:297-306.

102. Shipley, R.A., and R.E. Clark. 1972. Tracer methods
     for in vivo kinetics. Academic Press, New York.

103. Shutt, F.T. 1925. Influence of grain growing on the
     nitrogen and organic matter content of the Western
     soils of Canada. Dom. of Canada, Dept. of Agric. Bull.
     N.S. 44.

104. Siddoway, .H. 1963. Effects of cropping and tillage
     methods on dry aggregate soil structure. Soil Sci.
     Soc. Am. Proc., 27:452-454.

105. Sims, J.L., and L.R. Frederick. 1970. Nitrogen
     immobilization and decomposition of corn residue in
     soil and sand as affected by residue particle size.
     Soil Sci., 109:355-361.

106. Singh, J.S., and S.R. Gupta. 1977. Plant decomposition
     and soil respiration in terrestrial ecosystems. Bot.
     Rev., 43:449-528.

107. Smith, S.J., F.W. Chichester, and D.E. Kissel. 1978.
     Residual forms of fertilizer nitrogen in field soils.
     Soil Sci., 125:165-169.

108. Sochtig, H., and J. Salfeld. 1977. Dynamics of organic
     forms of nitrogen in the nitrogen cycle of soil. Soil
     Organic Matter Studies, Vol. 1:285-292. IAEA, Vienna.

109. Sorensen, L.H. 1975. The influence of clay on the rate
     of decay of amino acid metabolites synthesized in soils
     during decomposition of cellulose. Soil Biol. Biochem.,
     7:171-177.

110. Sorensen, L.H. 1981. Carbon-nitrogen relationships
     during the humification of cellulose in soils containing
     different amounts of clay. Soil Biol. Biochem., 13:
     313-321.

111. Sorensen, L.H., and E.A. Paul. 1971. Transformation
     of acetate carbon into carbohydrate and amino acid
     metabolites during decomposition in soil. Soil Biol.
     Biochem., 3:173-180.

112. Sowden, F.J. 1977. Distribution of nitrogen in
     representative Canadian soils. Can. J. Soil Sci.,
     57:445-456.

113. Sowden, F.J., and H.J. Atkinson. 1968. Effect of
     long-term annual additions of various organic amendments
     on the organic matter of a clay and a sand. Can. J.
     Soil Sci., 48:323-330.

114. Sowden, F.J., Y. Chen, and M. Schnitzer. 1977. The
     nitrogen distribution in soils formed under widely
     differing climatic conditions. Geochim. Cosmochim.
     Acta, 41:1524-1526.

115. Stevenson, F.J. 1981. Origin and distribution of
     nitrogen in soil. In F.J. Stevenson, editor. Nitrogen
     in agricultural soils. Agronomy 22:1-42. Am. Soc. of
     Agron., Madison, Wis.

116. Stewart, B.A., L.K. Porter, and D.D. Johnson. 1963. Immobilization and mineralization of nitrogen in several organic fractions of soil. Soil Sci. Soc. Am. Proc., 27:302-304.

117. Swift, M.J., O.W. Heal, and J.M. Anderson. 1979. Decomposition in terrestrial ecosystems. University of California Press, Berkeley.

118. Swift, R.S., and A.M. Posner. 1972. The distribution and extraction of soil nitrogen fraction as a function of soil particle size. Soil Biol. Biochem., 4:181-186.

119. Tanchandrphongs, S., and J.M. Davidson. 1970. Bulk density, aggregate stability and organic matter content as influenced by two wheatland and soil management practices. Soil Sci. Soc. Am. Proc., 34:302-305.

120. Tiessen, H., J. Stewart, and J. Bettany. 1982. Cultivation effects on the amounts and concentration of carbon, nitrogen and phosphorus in grassland soils. Agron. J., 74:831-835.

121. Tipper, D.J., and A.Wright. 1979. The structure and biosynthesis of bacterial cell walls. Pages 261-426 in J.R. Sokatch and L.N. Ornston, editors. The Bacteria. Vol. VII: Mechanisms of adaptation. Academic Press, New York.

122. Toogood, J.A., and D.L. Lynch. 1959. Effect of cropping systems and fertilizers on mean weight-diameter of aggregates of Breton plot soils. Can. J. Soil Sci., 39:151-156.

123. Trinci, A.P.J. 1978. Wall and hyphal growth. Sci. Prog., Oxf., 65:75-99.

124. Van Bavel, C., and F. Schaller. 1950. Soil aggregates, organic matter and yields in a long-term experiment as affected by crop management. Soil Sci. Soc. Am. Proc., 15:399-404.

125. Van Veen, J.A., W.B. McGill, H.W. Hunt, C.V. Cole, and M.J. Frissel. 1981. Simulation models of the terrestrial nitrogen cycles. In F.E. Clark and T. Rosswall, editors. Terrestrial nitrogen cycles. Ecol. Bull. (Stockholm) 33:25-48.

126. Van Veen, J.A., and E.A. Paul. 1981. Organic carbon dynamics in grassland soils. 1. Background information and computer simulation. Can. J. Soil Sci., 61:185-201.

127. Voroney, R.P. 1983. Decomposition of plant residues. Ph.D. Thesis. University of Saskatchewan, Saskatoon.

128. Voroney, R.P., J.A. Van Veen, and E.A. Paul. 1981. Organic carbon dynamics in grassland soils. 2. Model validation and simulation of the long-term effects of cultivation and rainfall erosion. Can. J. Soil Sci., 61:211-224.

129. Wieder, R.K., and G.E. Lang. 1982. A critique of the analytical methods used in examining decomposition data obtained from litter bags. Ecol., 63:1636-1642.

130. Zeikus, J.G. 1981. Lignin metabolism and the carbon cycle. Adv. Microbial Ecol., 5:211-243.

BIODEGRADATION OF ORGANIC RESIDUES IN SOIL

R.G. BURNS[*] and J.P. MARTIN[+]

INTRODUCTION

The biodegradation of plant, animal and microbial debris
is a complex and multifacited process involving an enormous
number and variety of soil organisms.   For the purpose of
this chapter, biodegradation is seen as involving two distinct
yet inter-related phenomena: (1) the breakdown of organic
debris and its macromolecular components by extracellular
enzymes as a prelude to assimilation and mineralization of
soluble low molecular weight products;   and (2) the formation
and fate of polyaromatic substances and polysaccharides (i.e.
humus) arising during microbial metabolism.

Vast quantities of organic residues enter the decomposer
cycle.   This input will vary according to local and regional
conditions which, in the former case, will be determined by
factors such as climate, soil type and the indigenous or
cultivated flora and fauna.   Deposition of plant litter may
range from <10 kg N/ha/yr in tundra regions to 170 kg N/ha/yr
in deciduous tropical forests (Staaf and Berg 1981).   Root
sloughings and exhudates may make a significant contribution
in agricultural areas (e.g. 1200 kg C/ha/yr released from
roots in a wheat field - Jenkinson and Rayner 1977) and
figures for soil microbial biomass, calculated by Jenkinson
and Ladd (1981), range from 170 to 2240 kg C/ha.   Soil fauna
make a comparatively small annual contribiton of less than
100 kg/ha.   The actual mass of organic matter in soil, at

* Biological Laboratory, University of Kent,
    Canterbury, Kent CT2 7NJ, England.
+ Department of Soil and Environmental Sciences, University
    of California, Riverside, California 92521, U.S.A.

any one time, will be determined by the rate of its decomposition and it may exceed annual input. Indeed, soil organic matter is the largest C reservoir and it is estimated that between 1.5 and 3 x 10^{15} kg of C are stored in this form (Schlesinger 1977, Degens 1982).

The biochemical diversity of macromolecular substrates indicates that the soil flora must possess a broad spectrum of extracellular enzymes in order to convert potential substrates to assimilatable metabolites. These substrates include linear (cellulose) and branched (starch) chain homoglycans, heteropolysaccharides such as pectin and hemicelluloses, polyphenolic materials such as lignin, acetyl-glucosamine-containing polymers such as chitin, chitosan and microbial peptidoglycans, and a plethora of lipids, proteins and peptides. Soil properties, such as texture, clay type, pH, organic matter, water tension and aeration, all have an important influence on biodegradation.

Humus is a mixture of numerous organic substances of which humic acid and polysaccharides constitute up to 80% or more of the total mass (Swincer et al. 1969, Martin and Focht 1977). The major components are the humic acid-type molecules which appear to be complex polymers of hydroxy-phenols, hydroxybenzoic acids and other aromatic structures linked to peptides, amino sugars, fatty acids, microbial cell wall and protoplasmic fractions and possibly other constituents (Schnitzer and Khan 1972; Flaig et al. 1975; Hayes and Swift 1978; Saiz-Jimenez et al. 1979). Plant lignins and microbial melanins are important sources of the aromatic units (Martin et al. 1980). Some of the polymeric constituents, such as proteins and uronic acid- and amino sugar-containing polysaccharides, may be protected from rapid biodegradation through strong adsorption to existing humic acids and clays or through the formation of metal ion-clay or metal ion-humic acid complexes (Greenland 1971, Zunino et al. 1982).

Although humus generally constitutes less than 10% of the soil mass, it has a significant affect on soil fertility. It is a slow release N fertilizer, has a high exchange capacity for nutrient cations, exerts a beneficial action on the physical properties of the soil, and is involved in chelation reactions which aid in the micronutrient nutrition of plants. Humus also buffers the soil against rapid changes in pH and may reduce the toxicity of natural and anthropogenic toxins. Finally, humus supports a greater and more varied microbial population (which in turn, promotes mineralization and may even favour the biological control of some plant root pathogens) whilst certain components directly promote plant growth (Flaig et al. 1975, Martin and Focht 1977).

ORGANIC DEBRIS AND MACROMOLECULAR SUBSTRATES

Introduction

The physical and chemical complexity of organic detritus militates against a biologically-simple degradative process. For example, plant cell walls are composed of an intricate network of cellulose fibers together with heteropolymeric pectin, hemicellulose and, in mature tissues, lignin. Furthermore, plant tissues often possess a protective cuticle of gums and waxes and may even contain compounds which are broadly anti-microbial or will inhibit the action of certain degradative enzymes. As far as the living plant is concerned this complex association of organic polymers has essential structural and protective functions and although active resistance mechanisms to microbial attack may disappear on the death of the plant, a variety of formidable barriers have to be overcome before microbial decomposition and assimilation can occur.

The exposure of polysaccharides within plant tissue and their subsequent biodegradation to soluble molecules suitable for cell uptake requires either an individual microbial species possessing a wide range of enzymic capabilities, or a mixed species community. The decomposition of exogenous

plant material is depicted in Fig. 1 and is envisaged as
involving at least four distinct groups of microorganisms:
cellulolytic, hemicellulolytic, pectinolytic and ligninolytic. Of course some species produce a range of depoly-
merases and a few may even be able to degrade polysaccharides
as well as polyuronides and polyphenolics. However, it is
generally true that the decomposition of a complex substrate
such as a plant leaf, dead microbial tissues or an insect
exoskeleton, proceeds more rapidly in the presence of a
mixed community than in a monoculture.

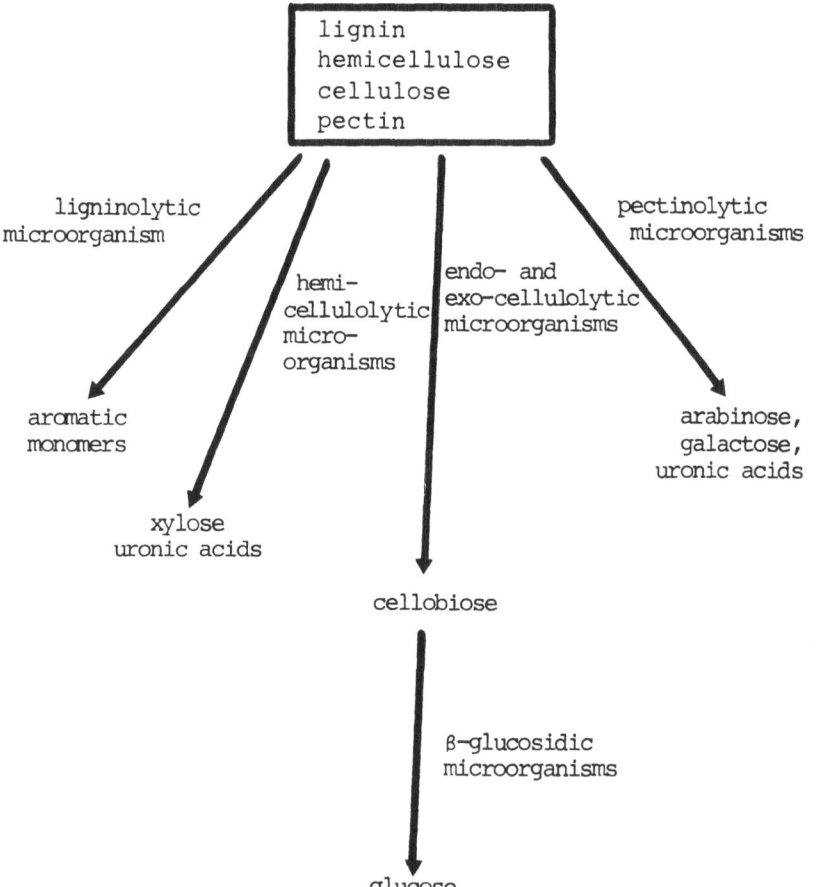

FIGURE 1. Possible communal relationships between microorganisms during the decomposition of plant organic matter.
Lignin, hemicellulose and pectin degradation each involves
a variety of microbial species and enzymes.

While this review places emphasis on microbial aspects
of biodegradation, there is no doubt that the soil fauna,
especially nematodes, annelids and arthropods, play an
important part in the early stages of organic matter breakdown (Schaller 1968, Edwards and Lofty 1972, Neuhauser,
Hartenstein and Connors 1978). This they do either
indirectly by the burrowing activities (increase of surface
area, aeration, spread of inoculum) or directly (comminution, incorporation into soil, digestion). For details of
the complete spectrum of decomposer organisms the reader is
referred to Swift et al. (1979) and other chapters in this
volume.

Components of organic residues

A large proportion of the organic C residues which the
microflora utilizes consist of simple sugars (hexoses and
pentoses), their derivatives (e.g. amino sugars, uronic
acids, sugar alcohols, methylated sugars, deoxysugars) or
their appropriate polymers. Carbon also occurs as a component of lignin, urea, alcohols, fatty acids, purine and
pyrimidine bases and lipids as well as agrochemicals and
xenobiotic compounds. Between 50% and 80% of the dry mass
of plants is carbohydrate, while microbes contain up to
60% organic C. The actual amount and proportion of C in
a cell, tissue or entire organism is determined by species
and age. For instance, although water-soluble carbohydrates
may predominate in young plants, mature plants have a higher
percentage of cellulose, hemicellulose and lignin.

The principal sources of organic N in soil are nucleic
acid bases (purine and pyrimidine), urea, amino compounds
(proteins, peptides, amino acids), chitin (ca. 7% N),
phospholipids, vitamins and teichoic acids. The percent
N in organic debris ranges from <2% in most plants to
>15% in some bacteria (McGill et al. 1981). Clearly C/N
ratios (and to a lesser extent C/P and C/S) have a controlling influence on the rate of substrate degradation.
Phosphorus-containing organic macromolecules include inositol

phosphates, nucleic acids and phospholipids (Stewart and McKercher 1982). Organic S is discussed in Chapter 7.

Plant organic matter. Cellulose is the most abundant polymeric constituent of plant material. Mature wood contains between 40 and 60% (dry mass) cellulose, leaves 10%, stems 30-40% and cotton fibres >90%. Since cellulose is a major component of municipal, industrial and agricultural waste, it is not surprising that great efforts have been made to understand the physical, chemical and microbiological factors which influence its decay. Indeed, cellulosic waste is now regarded as a renewable resource (Tsao 1978, Ghose and Ghosh 1979) and recent studies have emphasized the potential of this polysaccharide not only as a source of methane and fermentable sugar but also as a substrate for microbial biomass or single cell protein production.

Cellulose is an insoluble linear homopolymer composed of β-(1-4)- linked D-glucopyranosyl residues. In the plant cell wall individual cellulose molecules are clumped together to form microfibrils which, in turn, aggregate to produce fibers. The microfibrils are stabilized by the formation of hydrogen bonds between the hydroxyl groups of adjacent cellulose chains. Some areas of the fibril are highly ordered and are known as crystallites while more loosely associated zones have an amorphous structure. This distinction between physically different regions of the cellulose is important when consider-ing the accessibility of the substrate to cellulases. Cellulose structure is discussed in detail by Cowling and Brown (1979) and Fan et al. (1980).

Hemicelluloses are a diverse group of alkali-soluble polysaccharides, intimately associated with cellulose, pectins and lignins in plant cell walls and ranking next to cellulose as the second most abundant plant-derived substrate in the biosphere. Based on mass, mature wood may contain greater than 30% hemicellulose, while wheat straw contains about 20%. Hydrolysis of hemicellulose yields a mixture of hexose sugars (e.g. D-galactose, D-mannose), pentose sugars (D-xylose, L-arabinose) and uronic acids (glucuronic acid, galacturonic acid). Xylose polymers (xylans) are the dominant hemicelluloses and may account for up to 35% of plant mass, for example, wheat straw hemicellulose is 90% xylan. The biochemistry of the major hemicelluloses (i.e. xylans, glucomannans, galactoglucomannans, xyloglucans and β-D-glucans) has been reviewed by Aspinall (1980).

Pectic substances are structural polysaccharides which form a minor component of the cell walls of higher plants and rarely contribute more than 1% of the total mass. In the primary cell wall, cellulose microfibrils have a rigidity which is largely due to a hemicellulose-pectin network. Pectins are heteropolysaccharides composed of α-(1-4)-linked galacturonic acid (pectic acid) in which the carboxyl groups may be esterified to various degrees with methyl groups. Depending on the plant source galacturonan may have rhamnose inserted in it or have side-chain sugars (i.e. rhamnose, galactose, arabinose, xylose). Pectic substance and pectinases have been reviewed by Rombouts and Pilnik (1980).

Starch is a glucan serving as an energy reserve for plants and is stored in roots (tubers), stems (corns) and swollen leaf structures (bulbs). It is also found in high concentrations in cereal grains (e.g. barley, maize, wheat, oats). Starch is composed of an essentially linear struc-ture (in which the glucose dimer, maltose, is linked in the α-(1-4) position) as well as a branched structure containing α-(1-4) and α-(1-6) linkages. The former is known as amylose while the latter, amylopectin, is the major component of most starches. Starch and starch degrading enzymes have been discussed recently by Manners (1979).

Lignin is an important carbon-containing constituent of vascular plants and 15 to 34% of wood mass. Predictably, lignin composition and quantity vary with the age and species of the plant: mature plants have more lignified tissue than young plants; conifers generally contain more lignin than do hardwood species and their respective compositions differ.

Lignins are polymers of coniferyl, sinapyl and p-coumaryl
alcohols joined by a variety of intermonomer linkages.
Ether bonds are the dominant linkages but alkyl-alkyl,
alkyl-aryl and aryl-aryl also exist. Lignin is also found
in intimate association with cellulose.

Animal organic constituents. Animal tissues generally
contain a smaller proportion of carbohydrate than do those
of plants and microorganisms. Chitin, however, is a major
organic component of arthropod exoskeletons as discussed
later. A second important animal carbohydrate is glycogen.
This substance is similar in both structure and function to
the starch of plant cells. Thus, it is a polymer composed
of α-(1-4) and α-(1-6) linked glucose units and acts as a
medium for energy storage.

A large number of animal carbohydrates occur as
glycoproteins in which carbohydrates are linked through
glycosidic bonds. The carbohydrate moiety varies in size
from mono-to polysaccharide. Glycoproteins are common
constituents of animals but are also found in some plants
and microorganisms. They are also extremely diverse in
structure and function; glycoproteins have structural
(collagen, hyaluronic acid, chondroitin sulphates, heparin),
protective (immunoglobulins), hormonal (thyroglobulin), and
enzymic (acetylcholin-esterase) functions. They also serve
as food reserves (casein, ovalbumin).

Microbial organic constituents. It is often stated
that chitin is second only to cellulose in abundance and it
has been estimated that cellulose and chitin together
account for the deposition of >1 x 10^{15} kg each year of new
carbohydrate. Exoskeletons of soil arthropods (insects,
arachnids) may contain up to 80% chitin (Jeuniaux, 1963).
Chitin is also an essential structural component of fungi,
the cell walls of which frequently contain greater than 10%
by mass of this polymer. A few protozoa also contain small
amounts of chitin. Chitin is a long chain polymer of N-
acetylglucosamine linked by β-(1-4) bonds and is thus

chemically related to cellulose.   Like cellulose, chitin
also forms crystalline structures which are arranged side by
side and linked through hydrogen bonds.   A detailed description of chitin and its properties can be found in Muzzarelli
(1977).

Chitosan occurs in the hyphal walls of zygomycete fungi,
such as the genera Mucor, Mortierella and Rhizopus, and may
contribute as much as 33% to the dry mass of the wall
(Bartnicki-Garcia 1968).   Chitosan is a close relative of
chitin in two ways: it is physically-associated with chitin
and is probably derived from chitin by deacetylation.

The major structural polymer of virtually all bacterial
walls is peptidoglycan which contributes from 5 to 10%
(Escherichia coli) to 60 to 70% (Micrococcus luteus) of the
dry mass of the cell wall.   Peptidoglycan is similar to
chitin since it is a linear chain made up of β-(1-4) linked
acetyl amino sugars, but differs due to the presence of
N-acetylmuramic acid.   Short peptides link the parallel
chains of amino sugars to form a rigid network.   Details of
the structure of peptide glycans and other microbial macromolecules can be found in Rogers et al. (1980).

Lipopolysaccharides form part of the outer membrane of
Gram-negative bacteria and comprise 20 to 30% of the dry
mass of the cell wall.   As their name suggests they are
composed of a polysaccharide and a lipid, yet are extremely
varied and complex macromolecules.   The lipid fraction
consists of β-(1-4) or β-(1-6) linked glucosamine carrying
long chain fatty acids (e.g. myristic acid).   The lipid is
linked, through an eight carbon sugar to the polysaccharide
which contains, in addition to a number of common sugars
(glucose, galactose), L-rhamnose and heptoses.

Teichoic and teichuronic acids are major cell wall
components of Gram-positive bacteria and may constitute 30%
to 50% of the dry mass of the wall or about 10% of the total
cell.   There are several types of teichoic acid but all
contain polyglycerol phosphate or polyribitol phosphate.
In many bacterial species, the polyphosphate chain contains

sugar and amino sugar substituents. Teichuronic acids are polymers made up of alternating units of uronic acid and a hexose or hexosamine.

Several species of bacteria are capable of producing an extracellular layer of polysaccharide. The possible functions of this extracellular gum or slime layer have been discussed by Dudman (1977) although the conditions favouring its production in vitro (high nutrients, aerobiosis) may be rarely encountered in soil. Microbial exopolysaccharides produced by soil microorganisms include: levans (Pseudomonas, Xanthomonas, Bacillus), xanthan (Xanthomonas), curdlan (Alcaligenes, Agrobacterium), alginate (Azotobacter, Pseudomonas) and a large variety derived from Rhizobium and Arthrobacter species. Pullulan is produced by the fungus Aureobasidium. Powell (1979) has recently described many of these exopolysaccharides.

The cytoplasm and cytoplasmic membrane of microorganisms are an important source of substrates after their death and lysis. Phospholipids constitute 30 to 40% of the cytoplasmic membrane of all microorganisms and some yeasts and fungi contain polymers of manose and other sugars. All Gram-positive bacterial membranes contain lipoteichoic acid The cytoplasmic contents of prokaryotes include a plethora of enzymes, substrates and metabolites too numerous to be mentioned here. However, carbohydrate storage material, notably starch and glycogen, occur in many fungi, yeasts, protozoa and algae, as well as in species of Clostridium.

The biodegradation of three major organic matter components, namely cellulose, lignin and proteins, is considered in the following section. These important examples illustrate many of the problems facing the degradative microflora and how these problems have been overcome.

Cellulose.

Enzymology of cellulose breakdown. Cellulolysis has been reviewed extensively in recent years (e.g. Gong and Tsao 1979, Goksøyr and Eriksen 1980, Eriksson and Johnsrud 1982) and only a brief description is presented here.

The hydrolysis of cellulose to glucose is effected by the cooperative activities of three enzymes.

1. Cellulase (EC 3.2.1.4), also known as endocellulase, endo-1, 4-β-glucanase, carboxymethylcellulase and C_x.

2. Exo-cellobiohydrolase (EC 3.2.1.91), also known as cellobiohydrolase, 1,4-β-D-glucan cellobiohydrolase, exocellulase, Avicelase and C_1.

3. β-D-glucosidase (EC 3.2.1.21), also known as cellobiase.

Endocellulase, cellobiohydrolase and cellobiase will be used in the following description of cellulolysis and cellulase(s) as an imprecise or as a collective noun.

Endocellulase is produced by all cellulolytic microorganisms and is assayed most frequently by measuring changes in the viscosity of carboxymethylcellulose. Endocellulase randomly attacks internal 1-4 bonds rendering cellulose more accessible to other cellulases. It, therefore, enhances the rate of degradation by cellobiohydrolase by releasing many short chain oligosaccharides. Furthermore, there is a synergistic relationship between endocellulase and cellobiohydrolase in cellulose biodegradation and this may arise from the formation of an enzyme complex which prevents the re-association of internal breaks because cellobiohydrolase is already in position to effect further hydrolysis.

Endocellulase from different species (or even within the same species) may contain a number of active sub-units or possibly different enzymes. The multiplicity of endocellulases (i.e. molecular mass from 5,000 to 52,000 daltons) may be a useful property in the hydrolysis of inaccessible native cellulose fibers with their intrinsic variability (amorphous and crystalline zones) and their close association with other polymers.

Cellobiohydrolase attacks cellulose through terminal cleavage releasing the dimer, cellobiose. When incubated with odd-numbered cellodextrins glucose will also be released.   Although multiple forms of cellobiohydrolase have been reported they are not as common as the endocellulases.

Cellobiase hydrolyzes cellobiose to glucose and as many as five different types have been identified in Sporotrichum pulverulentum (Deshpande et al. 1978).   The molecular size of the different cellobiases may be determined by associated sugar moieties (most are glycoproteins) or by a process of cleavage by proteolytic enzymes.

Truly cellulolytic microorganisms (e.g. Trichoderma reesei (= viride), T. koningii, Sporotrichum pulverulentum) produce all three enzymes of the cellulase complex.   Other fungi and most bacterial species produce only endocellulases and must therefore depend upon other microorganisms in order to use cellulose as a carbon source.   Cellobiase itself is synthesised by a vast array of microbial species and organisms possessing this property are not strictly described as cellulolytic.

Native cellulose breakdown in soil.   Microorganisms which utilize exogenous high molecular mass materials are confronted by three types of problems: the intrinsic chemical and physical complexity of naturally-occurring polymers; the inaccessibility of potential substrates in cell remains due to their association with other molecules; and the influence of soil components on microorganisms - substrate interactions.   These constraints to the rapid assimilation of organic residues are well-illustrated with reference to cellulose.

1.   As stated earlier cellulose (and chitin) fibrils contain compact crystalline regions and loosely-associated amorphous zones.   The former structure is much less accessible than the latter and consequently the ratio of crystalline to amorphous cellulose will influence the extent of extracellular cellulase pene- tration and thus the rate of depolymerization.   In this regard lower molecular weight cellulases are more effective than their higher molecular mass equivalents.

The size of cellulose polymers varies widely and ranges from <2000 glucose units (ca. 3.5×10^5 daltons) in mature wood to as many as 10,000 (ca. 1.5×10^6 daltons) in cotton fibers. Filter paper (a frequently used substrate in experimental studies) has a low degree of polymerization (500-2000 glucose units). Chain length may have an important bearing on cellulolysis because it is believed that the specificity of endo- and exo-cellulases is due to their requirement to bind to subsites on the polysaccharide during β-(1-4) bond cleavage. For example, cellobiohydrolase acts readily as chain length increases suggesting that the enzyme needs to be bound to groups of glucose units in order to achieve maximum activity. Obviously chain length will determine solubility and the number of bonds that need to be broken prior to the formation of mono- and di-saccharides.

The hydration state of the polymer will have a direct effect on hydrolytic reactions but, in addition, swollen cellulose presents a greater surface area (and therefore more exposed glucosidic bonds) for enzyme contact compared with dehydrated forms.

2. Cellulose in plant cell walls is physically and, in some instances chemically linked to other polymers, notably pectins, hemicelluloses and lignins. Thus cellulolytic organisms depend upon other hydrolases in order to gain access to these substrates. Sometimes a variety of hydrolases is produced by an individual species (e.g. cellulolytic microorganisms may also be ligninolytic) but very often a community of different microbial species is involved directly or indirectly in the degradation of cellulose (Burns 1982a). Cell wall polymers are usually overlaid with a protective cuticle of gums and resins. These

complex materials may require a specialized microflora for their degradation. Filamentous fungi and actino-mycetes, which are capable of vigorous and sustained growth, probably have an important role in the pene-tration of cell wall structures.

The presence of certain polymers, such as hemicellu-loses, may actually accelerate cellulose degradation by stimulating an enzymatically-varied microflora or by acting as a co-substrate. On the other hand, lignin and phenolic substances in general, retard cellulose breakdown.

3. The biodegradation of all substrates, including pure cellulose and native cellulose, is a very differ-ent process in the presence of soil. This is because soil contains particulate and colloidal fractions which strongly influence microbe-substrate interactions. These fractions, clay minerals and humus, have a high unit surface area (>500 m^2/g) and carry a cation exchange capacity of between 600 and 7500 $\mu eq/g$. The surface area available for biological activity is affected by clay type and level of hydration and the cation exchange capacity is influenced to a lesser (clays) or greater (humus) extent by pH. As a result of these properties, together with the ionic state of microbial cells, their extracellular enzymes and the variety of substrates and metabolites, microbial activity predominates at soil surfaces and is influenced by properties of those surfaces. The characteristics of soil interfaces have been discussed in a number of recent reviews (Burns 1980, 1983, Marshall 1980, Stotzky and Burns 1982).

Cellulose, in contrast to proteins (discussed later) and polyuronic acids, is non-ionic and not adsorbed to any significant extent by soil. However, carbon-containing metabolites (especially amino acids) produced during cellu-lolysis may become bound to soil surfaces and thus soils

containing highly adsorptive clays may retard the process of
mineralization (Sørensen 1975, 1981).   Interestingly
Sørensen (1981) proposed that the presence of clay also
increased the efficiency of cellulose assimilation, and
calculated that, in the soil with the highest clay content
C conversion efficiency (55%) approached that of bacteria
grown under optimum conditions in vitro.

Hydrolase activity in soil.   The presence of active
extracellular enzymes in soil is essential for the depolymerization of cellulose and the many other high molecular
weight substrates.   However, enzymes secreted into the
aqueous phase of the soil are unstable due to a combination
of adsorption, denaturation and degradation.   We know, for
instance, that narrow bands of sterile smectite clay,
inserted into cellulose agar, retard or prevent altogether
the diffusion of cellulases or the radial growth of cellulolytic microorganisms (Hope and Burns 1983).   The adsorption of enzyme-protein on or within clay minerals generally
results in a decline or total loss of activity (for exceptions see Stotzky 1974, Haska 1981) while destruction due to
physicochemical factors (e.g. pH, ionic strength) is rapid
as is the assimilation of soluble enzymes by proteolytic
soil microorganisms.   If the general unsuitability of soil
to free extracellular enzymes is considered in conjunction
with the temporal and spatial heterogeneity of suitable
substrates, it is apparent that microbial activity which
depends upon constitutive enzyme production does not offer
competitive advantage to soil microorganisms.   Thus,
continuous production of exoenzymes such as cellulases (even
at low levels) represents an energetic deficit to the microbial species.   Regulation by de-repression is equally
ineffective because there may be no substrate in the vicinity
and induction requires the presence of a soluble trigger
molecule which can be taken up by the cell.   Nutritional
strategies available to the soil microflora have been
discussed in detail elsewhere (Burns 1983).

Nonetheless, it is apparent that many enzymes do survive after secretion but that they do so by becoming complexed with the soil humic fraction.   These immobilized or accumulated enzymes have been the focus of much research in recent years (Kiss et al. 1975, Skujins 1976, Burns 1977, 1978a) and are believed to play an important part in the early stages of substrate breakdown (Burns 1978b) as well as having a complex ecological function in which they serve as stable extracellular detectors which pass information (in the form of enzyme inducers or chemo-attractants) to microbes located in the vicinity (Burns 1982b).   A large number of immobilized soil enzymes habe been characterised (Table 1) and it is clear that soils have an indigenous and stable enzymatic capacity that is independent of current microbial activity and that is, amongst other things, extremely resistant to proteinases, dehydration and exposure to elevated temperatures.   It is suggested that many of the enzymes have become immobilized following the death and lysis of microbial cells and their release from intracellular locations.   Certainly an enzyme such as β-glucosidase, which has a soluble low molecular mass substrate, would fall into this category.   Other enzymes, such as endocellulase and cellobiohydrolase are secreted during normal cell growth. The mechanism(s) of immobilization are not known but entrapment within, co-valent bonding to, and co-polymerization with the humic molecule have all been suggested (see Burns 1978b).   It is also possible that some enzymes may be incorporated into highly-recalcitrant melanins prior to externalization.

It is difficult to distinguish activity due to immobilized enzymes from activities associated with other soil fractions (e.g. proliferating cells, cell debris).   However, it is believed that in at least some instances immobilized enzymes play a significant role in the turnover of substrate. For instance, it has been variously estimated (e.g. Paulson and Kurtz 1969, Dalal 1975) that urease not associated with proliferating cells accounts for <10% to >90% of urea

hydrolysis and a significant proportion of this activity
will be due to immobilized enzyme. Although it appears
unlikely that native insoluble cellulose is hydrolyzed by
immobilized endocellulase and cellobiohydrolase (however,
the depolymerisation of poorly soluble laminarin is achieved
by humic-1,3-β-glucanase complexes [Lethbridge, Bull and
Burns 1978]), the often rate-limiting enzyme, β-glucosidase,
does form stable complexes with humic matter (Hope, Alexander
and Burns 1980).

TABLE 1.   Hydrolases immobilized by soil clays and/or humates

EC number[*]	Recommended name	Substrate
3.1.1.1	Carboxylesterase	Hydroxy-methylcou-marin butyrate
3.1.1.2	Arylesterase	Phenyl acetate
3.1.1.3	Triacylglycerol lipase	4-Methyl umbellifer-one nonanoate
3.1.3.1	Alkaline phosphatase	p-Nitrophenyl phosphate
3.1.3.2	Acid phosphatase	p-Nitrophenyl phosphate
3.1.6.1	Arylsulphatase	p-Nitrophenyl sulphate
3.2.1.1	α-Amylase	Starch
3.2.1.2	β-Amylase	Starch
3.2.1.4	Cellulase	Cellulose, carboxy-methylcellulose
3.2.1.6	Endo-1,3(4)β-D glucanase	Laminarin
3.2.1.8	Xylanase	Xylan
3.2.1.21	β-Glucosidase	p-Nitrophenyl β-D glucoside, cellobiose
3.4.11-17	Peptidases	N-benzoyl L-arginine amide, benzyloxycarbonyl phenylalanyl leucine
3.4.21-24	Proteinases	Casein, gelatine
3.5.1.1	Asparaginase	Asparagine
3.5.1.2	Glutaminase	Glutamine
3.5.1.4.	Amidase	Formamide, acetamide
3.5.1.5	Urease	Urea

* Enzyme commission number authorized by the International
  Union of Biochemistry.

Lignin. It is clear that the presence of lignin in virtually all mature plant materials restricts the activity of many polysaccharase-producing microorganisms. Not only does lignin form a mechanical barrier against microbial and enzymatic penetration of organic residues but some of its aromatic constituents may be bacteriostatic and fungistatic while others may selectively inhibit certain carbohydrate depolymerizing enzymes. For instance, Varadi (1972) demonstrated that a number of phenolic substances (e.g. p-hydroxybenzyl alcohol, vanillin, syringaldehyde, p-coumaric acid, cinnamic acid) repressed the production of cellulase and xylanase in the fungus Schizophyllum commune and recently Vohra et al. (1980) reported that ferulic acid reduced β-glucosidase activity to zero while partially inhibiting endo- and exocellulase activity. Thus it may be realistic to see lignin degradation as the primary rate-limiting step in polysaccharide mineralization. Certainly pre-treatment of organic matter to remove lignin accelerates the subsequent decay of residual polysaccharides (Chahal et al. 1979).

Most studies of the biodegradation of lignin have involved white-rot Basidiomycetes such as Coriolus versicolor, Phanerochaete chrysosporium and Pleurotus ostreatus, all of which are able to totally degrade the substrate to carbon dioxide. Unfortunately the structure of lignin creates problems in understanding its breakdown because, in contrast to the polysaccharides, it does not contain repeating units nor are the linking bonds easily hydrolyzed. Indeed, hydrolases are probably of little importance in lignin depolymerization. In addition, lignin is chemically heterogenous and large quantities of a standardised lignin substrate are not easily available. Traditionally, heterogenous substrates, such as milled wood lignin or lignin derived from pulping operations have been used, but in recent years {^{14}C}-labelled plant and model lignins have become available and, as a result, considerable progress has been made in describing degradation processes (Kirk et al. 1977).

The initial and necessarily extracellular attack on the lignin polymer may involve the oxidation of the Cα - Cβ bond of the propyl side chains to yield aromatic residues, or demethylation followed by oxidative cleavage of aromatic rings still attached to the polymer to yield aliphatic carboxyl groups. The low molecular mass aromatic structures produced as a result of these activities (e.g. vanillin, vanillate, syringaldehyde, guaicylglycerol, p-hydroxybenzoate, coniferaldehyde, p-coumarate) are suitable for uptake and metabolism by a large number of fungal and bacterial species (Cain, 1980), but in soil may become attached to clay and humic colloids or be repolymerized to form novel humic structures.

Lignin breakdown is an aerobic process (Hackett et al. 1977) even though the metabolites are substrates for both aerobes and anaerobes. As mentioned previously, the intermonomer linkages of lignin are not attacked by hydrolases but require oxygenases (Dagley 1978). Two further modes of attack, which are discussed by Zeikus (1980) and Shimada (1980) are conversion of ether-containing linkages to esters before hydrolysis and the requirement for superoxide radicals; the latter may even contribute to the chemical degradation of lignins.

The enzymes participating in lignin biodegradation have yet to be unequivocally identified and described. No doubt part of the problem is the difficulty in obtaining specific ligninolytic activities in cell-free preparations and it is possible that ligninases are bound to cell walls and unstable when released, and that direct contact between hyphae and substrate is required for degradation (Rosenberg 1978). One group of enzymes, the phenol oxidases (e.g. laccase, o-demethylase) are frequently implicated in lignin degradation (Freudenberg and Neish 1968) and may have a variety of functions including demethyoxylation, α-carbinol oxidation and side-chain elimination as well as polymerization. In some cases, peroxidase will also cause substantial demeth-oxylation. Ander and Eriksson (1976) have suggested that

phenol oxidases have a secondary and regulatory role in
cellulose degradation since they destroy phenols which are
potential inhibitors of cellulases.   Cellobiose: quinone
oxidoreductase is an extracellular enzyme which has been
found in culture filtrates of Polyporus versicolor and
Phanerochaete chrysosporium (Westermark and Eriksson 1974,
1975)   and Sporotrichum (Chrysosporium) thermophile
(Canevascini and Meier 1978) and is also believed to have a
dual role in lignin and cellulose decay.   This enzyme reduces
quinones (produced by the activity of phenol oxidases) to
catechols and simultaneously oxidizes cellobiose to cellobiono-
σ-lactone (cellobionic acid).

There has been some effort directed towards determining
the role of microorganisms other than the white-rot Basidiomycetes in lignin biodegradation.   Many soil bacteria are
able to metabolize low molecular mass (<500 daltons) lignin-
related aromatic monomers and dimers but studies of bacterial
degradation of the lignin macromolecule have not been conclus-
ive and even if it occurs the rate is slow compared to that
achieved by the white-rot fungi.   Yeasts can grow on di-
lignols as can a range of common soil Fungi Imperfecti
(Fusarium, Aspergillus, Penicillium) yet no degradation of
native lignin has been reported while the ligninolytic
activity of the brown-rot fungi is limited to demethylation
and some oxidation reactions (Kirk 1971).   Due to the
intimate relationship between lignin and glucans most naturally-
occurring plant-derived substrates are referred to as ligno-
celluloses.

The study of the breakdown of these mixed substrates has
been stimulated by the ability to preferentially label the
various components such that ^{14}C-{lignin}-lignocelluloses
and ^{14}C-{glucan}-lignocelluloses can be produced.   Consequently, it is possible to differentiate between those microorganisms degrading the glucan fraction, those degrading the
lignin, and those degrading both.   For example Crawford and
Crawford (1976) showed that Thermomonospora fusca was
primarily utilising the cellulose and not the lignin whereas

the white-rot fungus, <u>Polyporus versicolor</u>, degraded both components of the substrate. Crawford and Sutherland (1979) discovered that while some <u>Streptomyces</u> species would decompose both principal components of lignocellulose and that others would only attack cellulose, none would solely degrade lignin. Indeed many of the studies of lignocellulose breakdown (e.g. Ander <u>et al</u>. 1980, Hall <u>et al</u>. 1980) suggest that microbial growth on the lignin component requires the presence of the co-substrate cellulose so that although the presence of lignin is likely to depress cellulose hydrolysis, the presence of cellulose is essential for lignin decay to occur at all.

Probably the greatest and as yet unfaced challenge to those investigating lignocellulose decay is to understand the combined activities of mixed populations. There is little doubt that the overall mineralization of lignocellulose in natural environments proceeds slowly by the combined activities of a variety of microbial and faunal species.

<u>Proteins and peptides</u>. There is a vast array of proteins in plant, animal and microbial debris ranging from the easily degraded non-structural proteins (e.g. enzymes) to the somewhat more recalcitrant structural proteins and protein-containing macromolecules (e.g. keratin, lipoproteins, peptidoglycans). This variety of substrates explains the enormous diversity of proteinases (proteases) many of which are produced by bacteria or fungi (see Payne 1980 for details). It is probably true that there are virtually no proteins that are entirely protected from microbial hydrolysis. The classification of peptide hydrolases is somewhat unsatis- factory mainly due to similarity of reaction and the lack of strict substrate specificity. However, two principal groups of enzymes acting on peptide bonds are recognised: peptidases and proteinases or proteases. Peptidases are exohydrolases which remove terminal amino acids or dipeptide units from proteins and peptides: proteinases are endohydrolases which solubilise proteins by cleaving internal peptide bonds. Extracellular proteolytic enzymes, like other groups of

microbial extracellular enzymes, share few common chemical
or physical features.   They have no general pattern of amino
acid composition although a low incidence of cysteine
residues is reported.   Only a few are glycoproteins.
However, proteinases and peptidases in general have relatively low molecular mass (20,000 - 40,000 daltons), often require
divalent metal ions such as Zn^{2+} or Mn^{2+} and may exhibit
enhanced stability in the presence of Ca^{2+}.   Collectively,
proteolytic enzymes are active from pH 2 to 10.

The utilization of exogenous proteins by microorganisms
depends upon the hydrolysis of these high molecular mass
substrates to peptides and amino acids prior to cellular
absorption.   The direct uptake of proteins by microbial
cells is rare although Cohen (1980) has discussed the possible
occurrence of protein permeases in fungi and alternative uptake
mechanisms such as micropinocytosis.   The maximum size for
oligopeptide uptake by bacteria and filamentous fungi is
probably about five amino acid residues (600-700 daltons)
(Payne and Gilvarg 1978);   a restriction indicative of cell
wall porosity as well as the availability of suitable peptide
transport systems.   Those cells lacking peptide transport
systems yet having a requirement for peptides must secrete
peptidases (or be associated with species that do) and possess
amino acid transport proteins.   Peptidases in many bacteria
and fungi are predominantly, but not exclusively, either
cytoplasmic or periplasmic (Wolfinbarger 1980).   However,
some bacteria, such as Clostridium histolyticum and Bacillus
subtilis, produce both extracellular proteinases and peptidases.

In soil, proteins, peptides and amino acids are adsorbed
to clay surfaces and associated in various ways with humic
materials.   The factors which influence the rate of binding
and strength of retention (and ultimately the rate of biodegradation) include: type of colloid, concentration and
valency of ions occupying the adsorption sites, bulk phase
and colloid surface pH values, and the molecular size, pI and
concentration of the adsorbate.   Most proteins will inter-
calate with expanding lattice clays (catalase and pepsin are

exceptions) and this will reduce the rate of their biodegradation - obviously by preventing direct contact by proteolytic
microorganisms, less obviously by masking sites usually
exposed to extracellular proteinase attack

Stotzky (1980) has summarized a series of studies
concerning the availability of clay-protein mxitures to
microorganisms as C and/or N sources. The complexity of
microbe clay-protein interactions is illustrated by one
series of observations: proline bound to Ca^{2+} - or H^{+}-
saturated montmorillonite was used only as a N source;
arginine bound to H^{+}-montmorillonite was available as both
a C and an N source, whereas arginine bound to Al^{3+}-mont-
morillonite served either as a N source (in medium containing
dextrose) or as a C source (in medium containing NH_3NO_3)
but not both; aspartic acid and cysteine were unavailable
under any circumstances when adsorbed to clays.

Recently Marshman and Marshall (1981) suggested that
proteins are bound to two different sites on clay minerals.
One site was susceptible to attack by bacterial proteinases
and the other not. The difference between the sites was
dependent upon the nature of the bacterial proteinases,
together with the properties of the adsorbant (clay) and
the protein. They also reported that growth was diauxic
with protein hydrolysis occurring during the second phase
of growth, and that the rate of utilization of protein, was
influenced by the protein/clay ratio. At low protein/clay
ratio no substrate was available; at intermediate ratios
growth rate but not yield was affected; and at high protein/
clay ratios the clays had no effect on microbial activity.

Humic-protein interactions are discussed later in this
chapter.

HUMUS

Humic acid polymers. Humic acid, the major component
of humus, appears to consist of complex polymers of hydroxy-
phenols, hydroxybenzoic acids, and other aromatic structures
with linked peptides, amino sugar compounds, fatty acids,

microbial cell wall and protoplasmic fractions, and possibly other constituents (Flaig et al. 1975, Hayes and Swift 1978, Kononova 1972, Nelson et al. 1979, Saiz-Jimenez et al. 1979, Schnitzer and Kahn 1972, Stevenson 1982). The ability of numerous phenolic compounds to undergo enzymatic and autoxidative polymerization reactions is probably of major importance in the formation of humic acid molecules. During the microbial metabolism of these compounds, transformations such as β-oxidation of side chains with the release of acetic

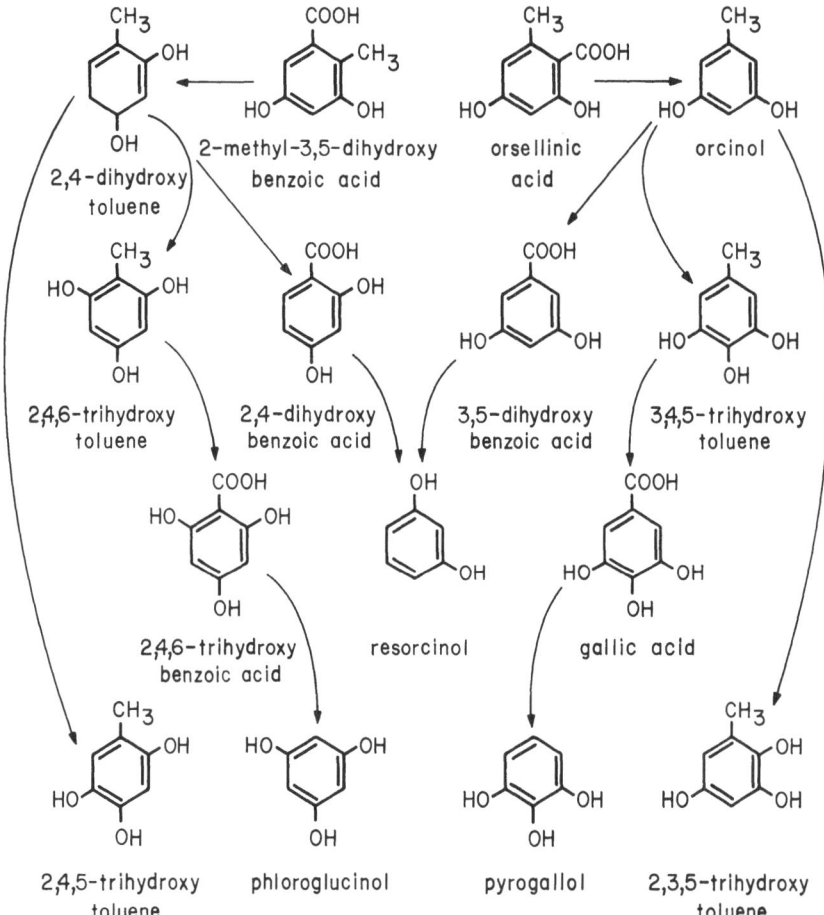

FIGURE 2. Synthesis and transformation of phenols
by Epicoccum nigrum.

acid, decarboxylation, demethoxylation, and the formation of
additional hydroxyl groups, give rise to a large variety of
phenolic substances (Fig.2).   Many of these, such as ortho-
dihydroxy and trihydroxy phenols, readily autoxidize to form
polymers at pH 6 and above (Haider et al. 1974).   Also these
and less reactive phenols such as ferulic acid, coniferyl
alcohol, vanillic acid and orcinol, are readily oxidized by
microbial phenolases and peroxidases that are either

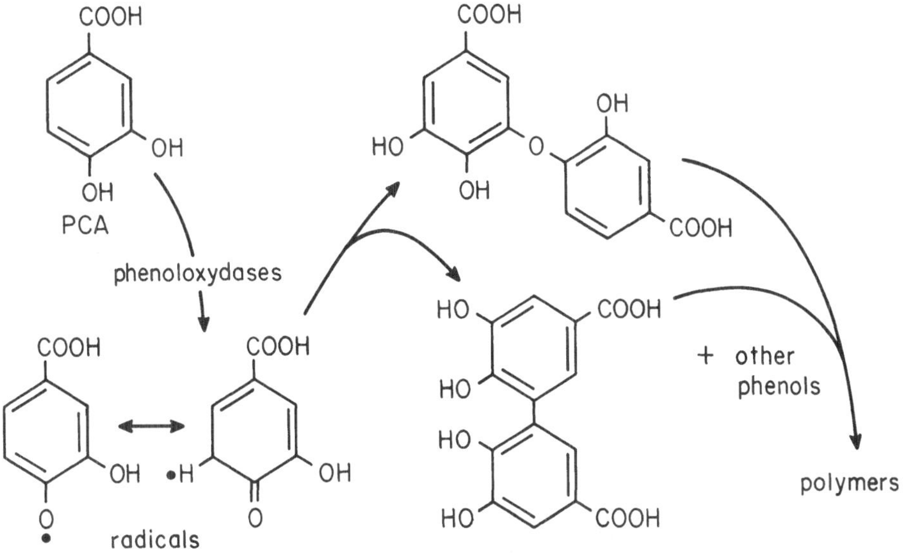

FIGURE 3.   Possible polymerization reactions of phenols
involving radicals.

associated with the soil humic colloids or are synthesized
de novo by soil organisms (Martin et al. 1972, Ladd and
Butler 1975, Burns 1978a, Suflita and Bollag 1981).   Other
phenols, such as p-hydroxybenzoic and 2,5-dihydroxybenzoic
acids, may be linked into the polymers (Martin and Haider
1980a).

The phenol oxidation process produces radicals which are stabilized by forming dimers or on further oxidation form quinones which link other phenols or substances with free amino groups (such as peptides, amino sugar polysaccharides and microbial cell wall structures) through nucleophylic addition reactions (Flaig et al. 1975, Martin and Focht 1977). Condensed ring aromatic structures and parts of lignin and melanin molecules may also be linked into the humus polymer molecules (Figs. 3 and 4).

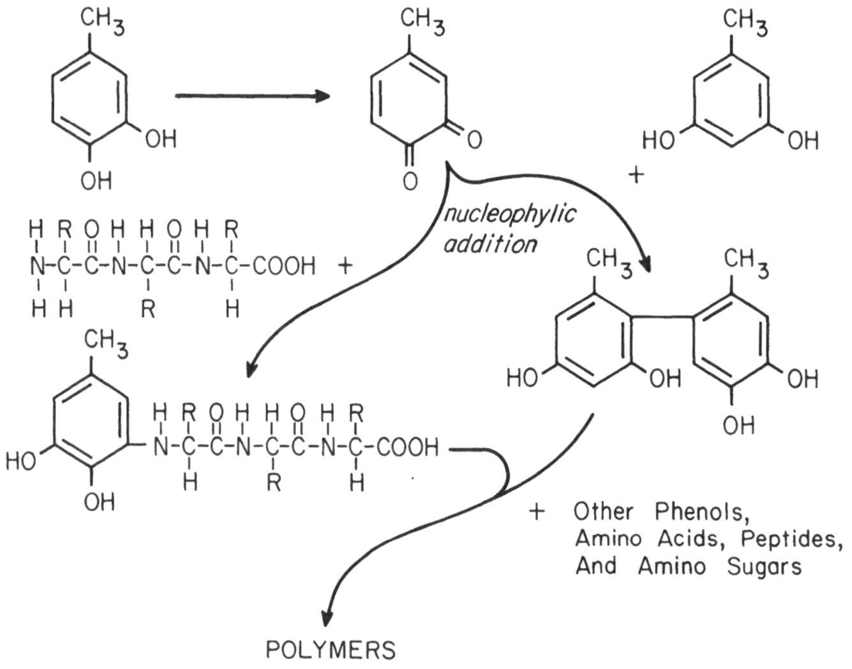

FIGURE 4.   Possible polymerization reactions of phenols involving quinone formation.

Considerable information about the structure and content of humic materials has been obtained from various types of comparative degradative analyses.   These analyses include: hydrolysis, oxidative degradation, reductive degradation, pyrolysis and many others (Schnitzer and Kahn 1972).

In interpreting the results one must be careful not to
conclude that the partial breakdown products which are recovered represent original structural units since the products may
be altered by the reaction or be released in a different
chemical arrangement than the starting units (Martin et al.
1974, Maximov et al. 1977, Hayes and Swift 1978, Mathur and
Schnitzer 1978).

Simple acid hydrolysis will release amino acids, peptides
and sugars from polysaccharide units associated with humic acids.
The humic acid polymers contain about 1 to 6% nitrogen (Kononova
1966). Up to 50% or more of this nitrogen can be recovered as
amino acid-N upon acid hydrolysis and the amino acids include
essentially all those found in plant or microbial proteins
(Bremner 1967, Sowden et al. 1976). Actually these values
are low because some of the amino acids are destroyed by
hydrolysis and the terminal amino acid of a peptide chain
which is linked to a phenol unit is not readily cleaved
(Haider et al. 1965). Therefore, it is evident that peptides
or proteins represent the major form of N in humic acid
polymers and if a humic acid contains 4% N, then 16% or more
of the polymer probably consists of peptide structures.

Mild acid hydrolysis procedures are used to estimate the
quantities of polysaccharides associated with humic acid
polymers (Greenland and Oades, 1975). When soils are
extracted with alkali and the extract acidified to precipitate
the humic acid fraction most of the polysaccharides remain in
the supernatant liquid as fulvic acid and in the extracted
soil as humin. Only small quantities (1 to 5%) are associated with humic acid (Iverson and Sowden 1962, Flaig et al.
1975, Linhares and Martin 1979b).

Methylation and oxidation of humic acids with permanganate or cupric oxide, forms a series of di-, tri-, tetra-,
penta- and hexabenzene carboxylic acids, phenols and fatty
acids (Schnitzer 1977). It has been suggested that many of
these fragments represent original structural units but it
is more probable that most are formed by oxidative splitting

of covalent linkages and cleavage of susceptible rings with
the oxidation of side chains originally present or linked to
carboxyl groups (Maximov et al. 1977, Hayes and Swift 1978).
Similar structures are formed upon the oxidation of model
humic acid polymers synthesized from simple phenolic
substances (Mathur and Schnitzer 1978).

Table 2.   Some phenolic substances identified upon
            sodium amalgam reductive degradation of
            soil humic acids and fungal melanins.

Hydroxy acids	Hydroxy phenols and toluenes
2,4-Dihydroxybenzoic	2,6-Dihydroxytoluene
2,3-Dihydroxybenzoic	3-Hydroxytoluene
2,3,4-Trihydroxybenzoic	4-Hydroxytoluene
p-Hydroxybenzoic	2,4-Dihydroxytoluene
Protocatechuic	Orcinol
Gallic	2,3,5-Trihydroxytoluene
m-Hydroxybenzoic	2,4,5-Trihydroxytoluene
p-Hydroxycinnamic	Phloroglucinol
2,6-Dihydroxybenzoic	Methylphloroglucinol
3,5-Dihydroxybenzoic	Pyrogallol
3,4,5-Trihydroxybenzoic	Resorcinol
Vanillic	p-Cresol
Ferulic	Dimethyl resorcinol
Syringic	

Sodium-amalgam degradation of humic acids yields about 15 to 35% ether soluble material and a large variety of simple phenolic substances (Hayes and Swift 1978, Martin et al. 1974, Piper and Posner 1972). The phenols include both lignin and microbial derived compounds (Table 2). Piper and Posner (1972) reported yields of simple phenolic substances equivalent to 12-32% of the starting material. Martin et al. (1974) obtained yields of 3 to 6% while model humic acids and fungal melanins yielded 4 to 32% simple phenolic substances. Tests with pure aromatic compounds indicated that many compounds (especially the orthodi- and trihydroxy phenols and naphthalenic substances) are partially destroyed by the reduction process. It is therefore probable that the actual quantities of phenolic substances in the polymers are much higher than estimated by this method.

Studies with model humic acids have shown that the simple phenols released by the sodium-amalgam degradation are the same as those used to synthesize the polymer or slightly altered compounds formed by decarboxylation or migration of an hydroxyl group to the other benzene ring of a ring to ring ether linkage (Martin et al. 1974).

The application of pyrolysis-mass spectrometry techniques to the characterization of organic polymers has been of great value in recent years. Pyrolysis of soil humic acids yields numerous fragments of which the most prominent are S-and N-containing fragments related to proteins, benzenes and phenols. Also numerous somewhat less prominent moieties are related to polysaccharides and aliphatic structures. Typical lignin-type peaks are very small or absent altogether (Martin et al. 1977, Menzelaar et al. 1977). Pyrolysis of whole soils yields spectra with more dominant peaks related to polysaccharides (Saiz-Jimenez et al. 1979).

Recently some attention has been given to nuclear magnetic resonance examination of humic polymers and lignins and complex but reasonably well differentiated spectra have been obtained (Gonzalez-Vila et al. 1976, Hatcher et al. 1980b). Most spectra show peaks or shoulders in the areas of resonance for

aliphatic, protein, aromatic, polysaccharide and carboxylic
acid structures. Some aquatic humic acids show weaker
signals in the aromatic area (Hatcher et al. 1980a). The
analyses support other methods of analyses since they also
indicate the complex nature of the humic polymers.

   Polysaccharides. Collectively soil polysaccharides
constitute the second most abundant component of soil humus
(Foster 1981, Martin and Haider 1971). Numerous investigators have indicated values for these polymers ranging from 5
to 25% of the soil fraction (Cheshire 1977, Greenland and
Oades 1975). Inasmuch as the method for determining soil
polysaccharides generally gives artificially low values, the
actual percentages may range higher (Gupta and Sowden 1965,
Swinzer et al. 1968). The soil polysaccharide fraction is
active in binding soil particles into aggregates (Harris
et al. 1966, Griffiths and Burns 1972, Martin et al. 1955,
Foster 1981).

   Most plant polysaccharides readily decompose in soil
and it is probable that the major portion of this recalcitrant
fraction originates through microbial synthesis (Cheshire
1977). Also soil polysaccharides contain amino sugar units
which are synthesized by many microbes but are not generally
present in plant polysaccharides (Benzing-Purdie 1981).
Microbial polysaccharides are also subject to decomposition
but the rate varies with specific polymers (Martin 1971).
The soil polysaccharide fraction contains about 10% uronic
acid units (Greenland and Oades 1975). These and units with
cis hydroxyls or phosphoric acid esters form salts or
complexes with di- and trivalent metal ions which may greatly
reduce their susceptibility to degradation (Martin et al.
1966). Polysaccharides may be complexed with clay minerals
through hydrogen bonding and metal ion-clay linkages which
increase their resistance to biodegradation (Harris et al.
1966, Guckert et al. 1977).

   Complex polysaccharides containing amino acid as well
as amino sugar units may be stabilized by linkage into humic
acid polymers through nucleophylic addition to quinones

(see Fig.4) or by strong hydrogen bonding to humic acids
(Bondietti et al. 1972, Martin et al. 1978).   It has also
been suggested that microbial and plant polysaccharides at
all stages of decomposition could be recombined through the
action of soil enzymes which are stabilized in humus or
released during autolysis of cells.   Those structures,
which are resistant to biodegradation or readily form complexes
with clays and metal ions would persist and contribute to the
soil polysaccharide fraction (Martin and Haider 1971, Burns
1978a).

Polysaccharide fractions isolated from soils normally
contain ten or more major sugar units and many others in
smaller amounts (Greenland and Oades 1975).   These include
glucose, mannose, galactose, arabinose, xylose, rhamnose,
fucose, galacturonic acid, glucuronic acid, glucosamine and
galactosamine.   Attempts to isolate significant fractions
with fewer structural units have been unsuccessful.
Generally it has been concluded that it is extremely difficult
to separate a complex mixture of polysaccharides into single
types.   This also suggests, as indicated earlier, that soil
polysaccharides are, indeed, composed of very complex polymers.

When soil humus is extracted with sodium hydroxide
solutions, the major portion of the polysaccharide material is
recovered in the fulvic acid and humin fractions (Swincer
et al. 1969, Linhares and Martin 1979b, Saiz-Jimenez et al.
1979).   Polysaccharides containing amino acid units are
largely retained in the humin fraction (Guckert 1975).   It
appears, therefore, that most of the polysaccharides or most
of the humus molecules containing polysaccharide moieties do
not precipitate upon acidification of the sodium hydroxide
extract or are not soluble under the conditions of extraction.
Some metal complexes of uronic acid-containing polysaccharides
are insoluble in alkaline solution (Martin et al. 1966) and
it is also likely that some clay-metal-polysaccharide
complexes are insoluble.

Contribution of Organic Residue Carbons to Humus

Rapidly degraded substrates. The use of uniformly or specifically ^{14}C-labelled organic substrates has greatly facilitated our understanding of the transformations involved in humification and the distribution of the C in various humus fractions (Jenkinson 1971, Paul and Van Veen 1978, Haider and Martin 1979). Readily available small molecular weight substrates such as sugars, amino acids, aliphatic acids and pyrimidines are metabolized within a few hours or days. Carbon is rapidly evolved as CO_2 but initially about 40 to 60% of the carbon may be transformed into microbial cells and products (Wagner 1975, Paul and Van Veen 1978) which are also subject to further biodegradation and transformations although C is released at a reduced rate. After one to three months about 75 to 85% of the original C will have been evolved as CO_2 and the residual carbon becomes relatively resistant to further biodegradation. About 10 to 40% of the residual C (1.5 to 10% of the original) will be present in biomass and the remainder will be a constituent of the new humus (Kassim et al. 1981). With time the percentage of the residual substrate carbon in biomass declines and the percentage in humus increases. In most soils the biomass constitutes about 2 to 4% of the organic carbon (Jenkinson and Powlson 1976, Anderson and Domsch 1978, Paul and Van Veen 1978). The importance of modelling C dynamics in decomposition models is discussed in Chapter 11.

Small amounts of the residual carbon from readily bio-degradable substrates are found in the aromatic portions of the soil humus (Martin et al. 1974) but the bulk is found in protein and polysaccharides which are solubilized by acid hydrolysis (Jenkinson 1971, Oades and Wagner 1971, Martin et al. 1980). This would be expected as a major portion of the metabolized carbon would be incorporated into microbial protoplasm, cell wall polymers and polysaccharides. Polymers with amino or sulphydryl groups, such as proteins and poly-saccharides could be stabilized by linkage to quinones during enzymatic or autoxidative polymerization of phenols (Mason 1953

and 1955, Flaig et al. 1975, Bondietti et al. 1972) and through
adsorption to or complexing with metal ions and soil colloids.

Polysaccharides, proteins, lipids and crop and microbial
residues.

Sixty per cent or more of most organic residues returned
to the soil consist of cellulose and other polysaccharides.
Some crop residues such as legumes and microbial tissues contain
relatively large (6-25%) percentages of proteins.   Specific
plant and microbial tissues may also contain appreciable amounts
of lipids.   These materials are all readily biodegradable but
especially during the early stages of decomposition many decompose more slowly than the simple sugars and acids (Table 3).
After 3 to 9 months 60 to 75% of C is evolved as CO_2 (Sauerbeck
and Gonzalez 1977, Paul and Van Veen 1978) and about 10 to 30%
may be present in microbial biomass (Kassim et al. 1981). A smaller
loss of protein carbon as CO_2 probably indicates that some
of the original polymers (or more likely partially degraded
units) are stabilized by incorporation into the soil humus
(Verma et al. 1975).

More complex organic materials, such as crop residues and
some microbial cells, are degraded at a still slower rate.
Generally after 3 months about 50 to 60% of the carbon is
evolved as CO_2 and after one year the loss is about 60 to 70%
(Jenkinson 1971, Jenkinson and Ayanaba 1977, Sauerbeck and
Gonzalez 1977, Paul and Van Veen 1978.   Relatively smaller
amounts of residual C are incorporated into biomass (2 to 13%)
while the major portion is stabilized in the new humus.

Small amounts of partially degraded protein and polysaccharide C may be used for synthesis of aromatic polymers
(Haider et al. 1974, Martin et al. 1974) but most of it is
first converted to microbial proteins, polysaccharides and
related cell constituents and products.   During the subsequent
degradation of the cells and products parts of these polymer
molecules are stabilized in soil humus and released upon acid
hydrolyses.

Phenols, hydroxybenzoic acids and other aromatic substrates.
Simple phenolic substances may be present in plant residues

Table 3.   Decomposition of a variety of organic substrates
in a fertile, neutral pH, sandy loam top soil
(Greenfield sandy loam).

Substrate	Decomposition after weeks [*]						
	1	2	4	8	12	20	28

A. Woods, barks, and organic fertilizers

Substrate	1	2	4	8	12	20	28
Prunewood	12	25	32	40	45	52	54
Activated sewage sludge	25	33	38	47	48	50	51
Dairy cow manure (fresh)	18	25	33	39	43	48	50
Chicken manure (fresh)	34	36	40	43	44	45	46
Apricot wood	8	14	20	28	34	41	43
Horse manure	5	6	9	23	31	36	38
Redwood bark	<1	1	2	4	8	32	36
White fir wood	0	<1	3	6	8	18	27
Digested sewage sludge	2	3	4	15	18	21	23
Redwood wood	0	0	0	<1	4	13	19
Canadian peat moss	<1	2	3	4	8	14	18
White fir bark	0	<1	1	4	6	9	15
Worm manure (casts)	2	2	3	8	9	11	11
Douglas fir bark	2	3	4	5	7	9	10
Black peat	<1	<1	2	2	2	3	3
Incense cedar wood	<1	<1	<1	1	1	2	3
Soil humic acid	<1	<1	<1	<1	1	1	2

Table 3. (Continued)

Substrate	Decomposition after weeks[*]						
	1	2	4	8	12	20	28

	B. Plant residues and products						
Glucose	73	80	82	88	89	90	90
Starch	48	58	69	76	81	84	86
Cellulose	27	40	52	71	77	79	84
Lima bean straw	36	47	57	71	75	78	79
Tomato tops	37	44	54	62	65	68	70
Wheat straw	26	37	45	55	59	63	67
Sudan grass	34	45	54	60	64	65	66
Corn stalks	30	38	47	53	59	61	62
Avocado leaves	14	19	23	32	36	39	41
Ponderosa pine needles	12	18	23	26	28	20	32
Corn stalk lignin	5	7	12	18	22	27	30

	C. Proteins, amino acids and pyrimidines						
Glycine	74	78	83	88	89	-	-
Casein	58	67	72	81	84	-	-
Uracil	60	68	72	76	79	-	-
Cysteine	54	64	69	73	75	-	-
Tyrosine	56	61	66	71	73	-	-

	D. Cellular material and products of soil organisms						
Arthrobacter sp. cells	60	72	79	83	85	86	87
Earthworms	59	68	72	78	80	82	84

Table 3. (Continued)

Substrate	Decomposition after weeks[*]						
	1	2	4	8	12	20	28

D. Cellular material and products of soil organisms (continued)

	1	2	4	8	12	20	28
Leuconostoc dextranicus polysaccharide	50	61	70	76	79	81	83
Penicillium vinaceum cells	56	64	72	75	76	78	79
Anabaena flos-aqua cells	38	45	51	60	66	73	75
Polysaccharides from blue-green algae	28	41	50	59	65	72	73
Chromobacterium violaceum cells	45	52	56	61	66	69	71
Hendersonula toruloidea cells	27	34	41	46	48	-	-
Chromobacterium violaceum polysaccharide	12	24	29	32	35	-	-
Azotobacter indicus polysaccharide	10	15	20	26	30	-	-
Anthraquinones from A. glaucus	7	9	12	15	16	-	-
S. atra black melanin	0	1	4	8	10	-	-
Hendersonula toruloidea melanin	3	3	4	5	6	-	-
Eurotium echinulatum melanin	1	2	3	4	4	-	-

Table 3. (Continued)

Substrate	Decomposition after weeks [*]						
	1	2	4	8	12	20	28
E. Organic acids and phenolic substances							
2,4-D side chain carbon	18	68	77	83	86	87	88
Benzoic acid	68	75	78	81	82	83	84
Pyruvic acid	62	69	75	79	81	-	-
Acetic acid	61	68	73	76	79	-	-
Vanillic acid	47	60	64	68	71	73	74
2,4-D ring carbon	6	48	65	69	71	73	75
Ferulic acid	41	59	62	67	68	69	69
Coumaryl alcohol	15	26	31	36	45	48	51
Catechol	11	14	18	20	22	24	26

* Percentage of applied C evolved as CO_2.

and are released during the biodegradation of plant phenolic polymers such as lignins (Flaig et al. 1975) and are formed through microbial synthesis (Haider et al. 1974).   Some common soil fungi synthesize up to 30 or more phenolic compounds (Martin and Haider 1971).   Epicoccum nigrum for example, synthesizes orsellinic, cresorsellinic, 6-methyl-salicyclic and other phenolic acids during the early stages of growth (Haider and Martin 1967).   These are transformed by decarboxylation, oxidation of methyl groups and introduction of additional hydroxyl groups to form a wide variety of

phenolic compounds (Fig. 2).   Microbes also synthesize
p-hydroxycinnamic acid-type phenols or release them from
lignins during decomposition.   These are also transformed
to numerous phenolic compounds by the above process and
through β-oxidation of side chains with release of acetic
acid units.   In pure culture most simple phenolic substances
are readily degraded by many species of microbes (Table 3)
but in soil the C loss is not as great as would be expected
if the aromatic rings were cleaved to form simple aliphatic
acids (Martin and Haider 1976).   Generally, C losses from
aliphatic acids and glucose average over 80% in a 3 month
period while losses from phenols vary from about 10 to 71%.
This indicates stabilization of a portion of the intact
benzene rings in soil humus as concentration of added simple
phenols rapidly decreases in unsterilized soil (Haider and
Martin 1975, Martin et al. 1979).   The more reactive the
compound with respect to radical formation or oxidative
polymerization reactions the greater the stability in the
soil.   It appears likely that a portion of the intact
compounds and early transformation phenols is stabilized
through autoxidative and enzymatic polymerization reactions
which create humus or form linkages into existing humus.

Another stabilization mechanism could be linkage into
fungal phenolic polymers (Martin and Haider 1971).   When
lignin-derived phenols are placed under the colonies of
melanin-forming fungi the added compounds of their transformation products become constituent units of the polymers
(Martin and Haider 1976).   When ^{14}C-labelled ferulic or
vanillic acid, for example, was placed under pads of
Hendersonula toruloidea or Stachybotrys chartarum 70 to 78%
of the ring carbon was recovered in the cell wall and culture
medium melanins.

The incorporation of phenolic carbons into biomass also
depends upon the specific C bonding and the reactivity of the
compound.   In general the greater the reactivity with respect
to oxidation the smaller the percentage of residual C

incorporated into biomass. In one test 1.6, 4.5, 12.5, 18.7 and 19% of the residual ring carbons of catechol, ferulic acid, p-hydroxycinnamic acid, anisic acid and glucose respectively, were present in biomass after 12 weeks incubation in a fertile sandy loam top soil with neutral pH (Kassim et al. 1982). Also, 50% of residual C from ferulic acids was released upon acid hydrolysis of the incubated soil further indicating an incorporation of the intact ring structures into humus.

Hydroxyanthracene and naphtholenic derivatives are synthesized by numerous plants and microorganisms (Robinson 1963). Indeed hydroxyanthraquinones may constitute up to 30% of the mass of some fungal species (Thomson 1971). Eurotium echinulatum and Aspergillus glaucus strains synthesize up to 20 or more anthraquinones which may be separated from the growth media or extracted from the mycelium (Saiz-Jiminez et al. 1975, Linhares and Martin 1979a). Reductive degradation of the melanin polymers of these fungi yields anthraquinones. Such polymers with anthraquinones are highly resistant to microbial degradation (Linhares and Martin 1978).

Anthraquinone and naphthalenic derivatives have also been found in small quantities in soil and have been isolated from soil humus. Mathur (1971) obtained 2-methyl-1,4-naphthoquinone from a podzol fulvic acid to which cultures and cell-free enzymatic preparations of Poria subacida were added. Kumada et al. (1961) obtained anthraquinones from humic acid of volcanic soils from Japan by alkaline degradation with sodium permanganate.

Naphthalenes, anthracenes and other condensed ring aromatic compounds are subject to microbial decomposition but the degradation rate may vary greatly (Dagley 1971). During a 12-week incubation period in a fertile soil 69 and 47%, respectively, of the C of emodin and chrysophanic acid was evolved as CO_2 while only 23% of the carbon of an anthraquinone mixture synthesized by Aspergillus glaucus was lost (Linhares and Martin 1979a).

Lignin. Lignin is the second most abundant organic polymer synthesized by terrestrial plants (Kirk et al. 1977, Crawford 1981). Because it is a phenolic polymer and is relatively resistant to biodegradation it is considered an important source of structural units for soil humus formation (Hurst and Burges 1967, Oglesby et al. 1967, Flaig et al. 1975). During its transformation to humus it undergoes profound alterations and its properties, in many respects, are quite different from those of humic and fulvic acid polymers (Haider et al. 1974). Differences include marked increases in exchange acidity and N such as peptides, and an increase in carboxyl, a decrease in methoxyl groups, and wide differences in the compounds recovered by various degradative procedures and pyrolysis (Meuzelaar et al. 1977, Maximov et al. 1977, Schnitzer 1977, Saiz-Jimenez et al. 1979).

Lignin molecules at all stages of biodegradation, including simple phenols and hydroxy dimers and trimers, could be recombined by enzymatic or autooxidative polymerization reactions with copolymerization of peptides, amino sugar units and various other microbial and plant cell constituents to form new polymers. The partial degradation products could also be linked to existing soil humus molecules (Martin and Haider 1980b).

Recent progress in the labelling of plant and model lignins has greatly facilitated our knowledge of the breakdown of lignin and the transformations that occur during humification in soil (Hackett et al. 1977, Haider et al. 1977, Crawford 1981). During a six month incubation period about 15 to 25% of the ring and 2-side chain C of model and cornstalk lignins was evolved as CO_2. This compares with about a 30 to 38% loss of 1- and 3- side chain and methoxyl C. After 2 years, losses of 2-side chain and ring carbons increase to about 35 to 40% and that of the 1-, 3-, and methoxyl C to about 55% or more (Martin et al. 1980). Drying followed by rewetting, and additions of readily biodegradable organic substrates exerted no effect on the

degradation rate (Haider and Martin 1981). Upon extraction
and fractionation of soils incubated with specifically ^{14}C-
labelled lignins, the major portion of the residual carbons
are recovered in the humic acid fraction (Martin et al. 1980,
Stott et al. 1983).

A relatively small percentage of the residual lignin
C is present in soil humus polymers which are solubilized
upon acid hydrolysis (Martin et al. 1980). Although most
residual carbohydrate and other highly degradable carbons
are found in peptide and polysaccharide polymers (Mayaudon
and Simonart 1959, Wagner and Mutatkar 1968, Jenkinson 1971)
the major portion of the residual lignin C is in aromatic
complexes. If the aromatic lignin rings had been cleaved,
a substantial portion of this C would be utilized by microbes
and thus would be found in the peptide and carbohydrate
fractions of humus. Biomass estimations support this
conclusion since after six to twelve months about 10 to 30%
C from readily-metabolized substrates is present in biomass
versus 1% or less for lignin C (Kassim et al. 1981). The
very small amount of the lignin C incorporation into C
biomass is somewhat less than would be expected on the basis
of C lost as CO_2. This suggests that the lignin may be
partly degraded by cometabolism reactions and much of its
released C is not utilized for biosynthesis.

The studies on lignin biodegradation all indicate that
lignin is a major precursor for soil humus formation.
However, degradative analyses of humus, as previously discussed,
show that lignin undergoes profound changes during humification.

Effect of concentration. Most biodegradation studies
using ^{14}C-labelled or unlabelled plant residues have indicated
that the percentage of amended C evolved as CO_2 and the amount
stabilized in new humus is not significantly influenced by
additions up to 2% if incubation is carried out for at least
three to six months (Pinck and Allison 1951, Sauerbeck 1968,
Jenkinson 1977). Most studies on concentration effects
have not included very minute amounts of the substrate nor

amounts over 2%. In a recent experiment by Martin and Haider (1979a) the proportion of C evolved from substrate concentrations ranging from 1 to 50,000 ppm were the same for cellulose, wheat straw and protein. For simple soluble substrates such as glucose and ferulic acid, however, the C evolution increased proportional to the increase in application rate. About twice the percentage of the added C from the 1 to 10 ppm addition was stabilized in the soil humus and biomass at 12 weeks compared to 10,000 ppm additions. Also about twice the amount of added ferulic and/or p-hydroxy-cinnamic acid carbons were stabilized than for glucose C. Although the major portion of residual C was present in the new humus, proportionately more of the glucose C was in the biomass and more of the phenolic C in humus. These differences associated with concentration are probably related to the microbial populations involved in the decomposition processes. Small amounts of a soluble substrate would be well dispersed in the soil and would be insufficient to stimulate a specific microbial population. Also the autochthonous populations could produce more phenolases which could polymerize some of the ring structures before cleavage in the case of the phenols and could synthesize more aromatic substrates from the glucose. However, small quantities of an insoluble substrate would not be evenly dispersed throughout the soil and would constitute microsites of relatively abundant carbon source. At higher concentrations the zymogenous populations or specific organisms better adapted to compete for the abundant C could convert a larger proportion of substrate C to CO_2 or to more readily degradable metabolites.

Such concentration effects may be pertinent to pesticide degradation tests. Many pesticides which are applied in low concentrations contain aromatic structures. Relatively small C losses may not necessarily mean that pesticides remain in the soil in unaltered form but instead they could be partially degraded and the aromatic portion linked into the humus.

Effect of clay, humus and soil pH. In extensive field studies various investigators have noted that the amount of humus in the soil, the plant cover and the soil pH exert little influence on the amount of organic C stabilized (Sauerbeck and Fuhr 1968, Shields and Paul 1973, Sauerbeck and Johnen 1974, Jenkinson 1977). Shorter incubation periods with a wide variety of organic substrates have given similar results with the exception that the initial degradation rate may be slower in acid soils (Jenkinson 1971, Martin et al. 1974, 1978, Zunino et al. 1982).

The quantity and type of clay, however, may greatly affect the amount of organic carbon stabilized. As noted earlier with cellulose, clays will often increase microbial growth rate especially during the early degradation of readily available substrates but will reduce the total loss of C as CO_2 by increasing the efficiency of C utilization by the microbes and by complexing with decomposition products and humic substances (Greenland 1965, 1971, Stotzky 1967, Bondietti et al. 1972, Filip 1975, Guckert et al. 1977, Turchenek and Oades 1979).

Soils containing the clay, allophane (a mixture of poorly crystalline alumino-silicates with variable SiO_2/Al_2O_3 ratios) are unusually high in humus (Thorpe and Smith 1949, Aomine and Jackson 1959). It is generally believed that the humic colloids are stabilized by complexing with active aluminium groups or sesquioxides associated with the allophane surfaces (Wada and Higashi 1976). Griffith and Schnitzer (1975) found that a fulvic acid fraction was unusually high in silica even after repeated purification procedures. They suggested that a silica-humus complexing may account for some of the resistance to biodegradation.

Zunino et al. (1982) compared the loss of C from a wide variety of readily degradable organic substrates in normal agricultural soils of Chile and California and allophanic soils of Chile. The presence of allophane in soil or additions of allophane to a normal soil reduced C losses from most substrates by 36 to 67%. In a continuing study

(Martin et al. 1982) C losses over a one year period from more resistant polymer substrates such as plant lignins and fungal melanins were reduced to even a greater extent in the allophanic soils.

Effect of readily available carbon source on humus stability - the priming effect. The introduction of isotopic techniques made it possible to determine the influence of one type of organic substrate on the decomposition of another or of the existing soil humus (Jenkinson 1966, Sauerbeck 1966, Jenkinson 1971). In a series of tests it was noted that when two readily available substrates, one ^{14}C-labelled and the other unlabelled, were applied to soil, both decomposed at the same rate as when either was applied alone (Martin et al. 1974, Haider and Martin 1975). Broadbent and Norman (1946) suggested that the addition of a readily degradable organic substrate to soil may greatly enhance the evolution of CO_2 from the soil humus. This priming effect has been investigated by several workers and the results have been summarized by Jenkinson (1971) and Sauerbeck (1966). It was concluded that fresh organic substrates added to the soil may sometimes increase or decrease the evolution of CO_2 from the soil humus, but the effects are very small and transient.

Another approach to resolving the controversy concerning the priming action has been to add a ^{14}C-labelled substrate to the soil, incubate for a period of time and then apply a readily available unlabelled substrate to see if the second addition will increase the evolution of [14]CO_2 (Sauerbeck 1966, Jenkinson 1971). However, the results from such investigations have also been inconclusive. A third approach has been to determine the influence of readily available organic substrates on the decomposition of more resistant substrates or polymers when both are added simultaneously to the soil. Extensive tests using different soils have indicated that the rate of decomposition of reactive phenols, model humic acid polymers, fungal melanins and model and cornstalk lignins under favorable moisture conditions and constant temperature depends upon the substrate and the soil and is not significantly influenced by the addition of readily available organic substrates (Haider and Martin 1981, Martin and Haider 1979b).

Microbial melanins. Many soil organisms, including fungi and bacteria (including actinomycetes) synthesize melanins and either secrete them or retain them within cells or spores (Reisinger and Kilbertus 1974). For many years it has been suggested that these polymers may be similar to soil humic acids (Martin and Haider 1971, Haider et al. 1975, Martin and Focht 1975).

It has been stated that fungal melanins usually have structures based on catechol as opposed to the indole melanins of animal origin (Nicolaus 1968) but numerous recent studies indicate there is a tremendous variation in the number and kinds of structural units in these polymers. Some contain mostly resorcinol type phenols derived from orsellinic acid, some both orsellinic and p-hydroxycinnamic acid derived phenols and others a variety of anthraquinone structures as well as numerous phenols. Hendersonula toruloidea synthesizes over 40 and Eurotium echinulatum over 50 phenolic and aromatic substances most of which become constituent units of the melanins (Martin et al. 1972, Saiz-Jimenez et al. 1975). During the polymerization process through enzymatic or autoxidative reactions, proteins and various cell wall components may be linked into the molecules (Verma and Martin 1976, Nelson et al. 1979).

The melanins from Epicoccum nigrum, Aspergillus sydowi, Hendersonula toruloidea and Eurotium echinulatum are similar to soil humic acids with respect to elemental composition, high exchange acidity, amino acids released upon acid hydrolysis, phenols released upon sodium-amalgam reductive degradation, resistance to microbial degradation, types of structure released upon oxidative degradation and pyrolysis, and low polysaccharide content (Saiz-Jimenez et al. 1975, Schnitzer and Neyroid 1975, Hayes and Swift 1978, Linhares and Martin 1978, 1979a). The melanin from a Stachbotrys

Table 4. Decomposition of [^{14}C]-labelled Stachybotris atra melanins and various cellular fractions in Greenfield sandy loam.

Fraction	N source	Percentage decomposition (weeks)				
		1	2	4	8	12
Isolate producing very dark pigment	Asparagine					
Melanin extracted from cells		<1	2	3	5	6
Melanin recovered from medium		<1	2	5	8	10
Ether extract of cells		5	11	23	31	35
Whole cells		12	19	29	37	41
Extracted cells		29	43	58	64	68
Non-dialysable culture medium polymers		27	42	58	64	67
Isolate producing light gray melanin and dark gray mycelium						
Culture medium melanin	NaNO$_3$	5	16	32	45	50
	Asparagine	4	8	17	33	38
Cells	NaNO$_3$	10	20	27	33	37
	Asparagine	10	19	29	37	42
Non dialysable culture medium polymers	NaNO$_3$	31	37	43	48	51
	Asparagine	26	32	42	52	56

All cultures grown under stationary conditions.

species contains more aliphatic structure and that from
Aspergillus niger spores contains over 50% polysaccharide
(Saiz-Jimenez et al. 1979).   The decomposition of S. atra
melanins compared with other cellular fractions is shown in
Table 4.

The reactions involved in polymer formation from fungal
phenols involve autoxidative and enzymatic oxidative
processes.   Martin and Haider (1968) demonstrated that
phenols such as 2,3,5-, 2,4,5-, 2,3,6- or 3,4-5-trihydroxy-
toluenes react with oxygen even under weakly acid or neutral
conditions to form reactive quinones or radicals.   These
then react with each other or with less reactive phenols and
with amino acids or peptides in the mixture to form polymers.
Some of the reactions involved have been discussed by Mason
(1955), Bremner (1967), Musso (1967) and Haider et al. (1975).

Fairly well resolved ^{13}C resonance spectra have recently
been obtained for humic acids (Hatcher et al. 1980a,b) and
for a few fungal melanins (Gonzalez-Villa et al. 1976).
Eurotium echinulatum melanin gave strong signals in the
aromatic and aliphatic-protein regions and weaker signals in
the polysaccharide area.   The spectra showed similarities
to those from some soil humic acids.   The strongest signals
for Aspergillus niger spore melanins were in the polysaccharide areas as expected, and those for Stachybotrys melanin
in the aliphatic-protein area with weak signals in the
aromatic zone.

In a recent study the ^{13}C NMR spectra for a
number of fungal melanins and soil were compared to
peat humic acid (Ludemann et al. 1982).
Melanins from E.nigrum, A. sydowi, H. toruloidea, E.
echinulatum, and A. glaucus gave complex ^{13}C NMR spectra
with strong peaks, shoulders, or plateaus in the areas of
resonance for aliphatic, peptide, polysaccharide, aromatic
and carboxylic acid structures.   All these melanins and
the soil humic acid showed strong peaks or shoulders at
about 20, 25, 28, 35, 40, 58, 75, 80, 110, 120, 130, 136,
175 and 180 ppm.   There were a few differences in the

strength and specific signals among the melanins and the
humic acid which would be expected as the melanins had never
been in soil.   Also humic acids vary greatly in content of
peptides and other structural units.   Stachybotris atra
melanins gave weak signals in the aromatic ring area.

Model humic polymers.   Various investigators have
prepared model humic acid polymers in an attempt to better
understand the possible formation mechanisms involved and
the nature and properties of the humic acid polymers.
Most studies have involved relatively simple mixtures.
Flaig (1964) summarized investigations on the coupling of
hydroquinone and other phenols under alkaline conditions or
with mild oxidizing reagents.   Haider et al. (1965) studied
the linkage of amino acids and peptides with phenols during
oxidation with fungal phenoloxidases.   When peptides were
incorporated all the amino acids except the terminal acid
linked to the benzene ring were released upon acid hydrolysis.

Ladd and Butler (1966) prepared 23 model phenolic
polymers with or without N as amino acids, peptides or
proteins.   On the basis of molecular weight distribution
and amino nitrogen released upon acid hydrolysis, the
polymers incorporating proteins were most similar to natural
humic acids.   Goh and Stevenson (1971) reported that the
infra-red spectra of natural and model humic acids prepared
from catechol or protocatechuic acid, with or without
glycine or peptides, were similar.

Haider and Martin (1970), Martin et al. (1972) Verma
et al. (1975), Martin and Haider (1976), Verma and Martin
(1976), have prepared numerous simple and more complex model
humic acid-type phenolic polymers in order to determine
their properties, compare these with soil humic acids and
fungal humic acid-type melanins and to study the degradation
in soil of specific constituent carbons using ^{14}C-labelled
structural units.   Some of the model polymers were made by
autooxidation of phenolic mixtures at an elevated pH but
most were prepared at pH 6.5 using basidiomycete phenolase

as the oxidizing agent. With respect to exchangeable acidity
elemental analysis, molecular mass distribution, amino acids
and N released upon acid hydrolysis, or with proteases,
resistance to biodegradation in soil, fragments released upon
pyrolysis and phenolic constituents released upon Na-amalgam
reductive degradation, the polymers prepared from a relatively
large variety of phenolic compounds and peptides or proteins
were the most similar to the soil humic acids and to fungal
humic acid-type melanins (Haider et al. 1974, Martin et al.
1972, Martin et al. 1974, Martin and Haider 1971, Meuzelaar
et al. 1977).

Recently Martin and Haider (1980a) compared the use of
basidiomycete phenolase and peroxidase for synthesizing model
humic acid polymers. Yields of humic acid plus fulvic acid
of the phenolase-catalyzed humic acid-type polymers varied
from 25 to 50% reaction products. Using an adaptation of
the peroxidase - H_2O_2 method for preparing model lignins the
yields of humic polymers were almost doubled and a greater
percentage of each polymer was recovered as humic acid.

Zunino et al. (1982) have suggested that greater
progress in the area of metal binding by humic-type polymers
can be made by using model humic acid-type polymers in which
the binding sites are better defined. Polymers which contain
largely phenolic hydroxyl groups, both hydroxyl and carboxyl
groups, and either of the above types with N- and S-containing amino acids or peptides, can be obtained by using
specific precursors. Mathur and Schnitzer (1978) reported
that many model phenolic polymers are similar to soil humic
acids with respect to fragments released upon permanganate
oxidation and that it is possible to select from a variety
of starting materials the type of polymer most suitable for
the nature of the reaction one wishes to study.

Humic-pesticide complexes. Phenolic substances are
of major importance in the formation of humic acid polymers
(Flaig et al. 1975) and numerous pesticides and other
synthetic organic chemicals contain benzene ring moieties.

As with natural organic substrates, it is necessary to introduce hydroxyl groups into the benzene moiety for ring cleavage to occur (Dagley 1971). At this stage the aromatic compounds could be oxidatively polymerized and linked into humus molecules, or undergo ring cleavage. It would be expected, therefore, that phenolic partial degradation products of man-made organic compounds would be linked into soil humus polymers.

Catechols or chlorocatechols are intermediate degradation products of many pesticides containing benzene ring structures (Hill and Wright 1978) which readily link into humic acid-type polymers. Martin and Stott(1981) and Stott et al. (1983) found 80 to 90% of chlorocatechols present in model polymer reaction mixture with peroxidase were linked into the polymers formed. Sjoblad and Bollag (1981) and Bollag et al. (1977 and 1980) isolated phenol oxidase enzymes from soil organisms and from soil which brought about the oxidative coupling of lignin and pesticide derived phenols. The xenobiotic compounds included several chlorinated phenols, naphthalenic compounds and halogenated anilines. A mixture of the natural and the synthetic phenols resulted in the formation of cross-coupling structures. Bartha (1971) and Bartha and Pramer (1970) have presented evidence that aniline rings released during the biodegradation of pesticides such as acylanilide, phenylcarbamate and nitroaniline fungicides, which contain substituted aniline rings, become covalently linked to humus.

In a study by Wolf and Martin (1976), melanic fungi were shown to incorporate significant amounts of the benzene ring portion of the herbicides 2,4-D and chlorpropham into relatively resistant humic acid type polymers (melanins) that they formed, while the side chain carbons were utilized for synthesis of general cell components which were readily degraded by soil microorganisms.

Recently several investigators have expressed a concern that pesticide residues or toxic partial decomposition products could be stabilized in soil humus and at a later date could be released, be adsorbed by soil organisms and plants, and thereby present a serious environmental contamination problem (Bartha 1980). Such a phenomenon would be extremely unlikely. It should be stressed first that potentially toxic compounds have been linked into humic molecules since humus was first formed on the earth. These compounds include phenols, naphthalenes, anthracene derivitives, aflatoxins and many other toxic substances (Martin and Stott 1981). Soil organisms and plants also synthesize chlorinated hydrocarbons such as the fungal toxins griseofulvin, islanditoxin and ochratoxin (Ciegler et al. 1971). The latter contains a chlorinated phenol component. Large numbers of halogenated hydrocarbons are synthesized by marine plants and animals (Sims et al. 1978). During the linkage of amino acids into humic acid molecules aniline structures may be formed (Flaig et al. 1975). The linkage of toxic compounds into soil humus can more logically be viewed as a detoxification process.

A second consideration is that the original toxic compounds may not be released during the microbial degradation of humic polymers. If ring structures are involved the ring could be cleaved while the compound is still linked to the polymer and the side chains could be degraded by oxidative and decarboxylation reactions. If the compound is released in its original form it is immediately subject to microbial degradation or repolymerization reactions. Natural toxic compounds, such as many phenols and anthraquinones present in humus, have not caused a toxic condition in the soil as they are slowly released or degraded. Also there is no evidence that toxic components of pesticide molecules linked into humus have been released and contaminated the soil environment, or exerted a toxic affect on soil organisms or plants.

REFERENCES

1. Ander, P., and K-E. Eriksson. 1976. The importance of
   phenol oxidase activity in lignin degradation by the
   white-rot fungus Sporotrichum pulverulentum. Archiv.
   Microbiol. 109: 1-8.
2. Ander, P., A. Hatakka, and K-E. Eriksson. 1980.
   Degradation of lignin-related substances by Sporotrichum
   pulverulentum, pages 1-15, in T.K. Kirk, T. Higuchi and
   H-M. Chang, editors. Lignin biodegradation : microbiology, chemistry and potential applications, vol.11.
   CRC Press, Boca Raton, Florida.
3. Anderson, J.P.E., and K.H. Domsch. 1978. Mineralization
   of bacteria and fungi in chloroform-fumigated soils.
   Soil Biol. Biochem. 10: 207-213.
4. Aomine, S., and M.L. Jackson. 1959. Allophane determination in Ando Soils by cation exchange capacity delta
   value. Soil Sci. Soc. Amer. Proc. 23: 210-214.
5. Aspinall, G.O. 1980. Chemistry of cell wall polysaccharides,
   pages 473-500, in J. Preiss, editor. The biochemistry
   of plants, vol. 3. Academic Press, London.
6. Bartha, R. 1971. Fate of herbicide-derived chloroanilines
   in soil. J. Agr. Food Chem. 19: 385-387.
7. Bartha, R. 1980. Pesticide residues in humus. ASM News
   46: 356-360.
8. Bartha, R., and L. Bordeleau. 1969. Cell-free peroxidases
   in soil. Soil Biol. Biochem. 1: 139-143.
9. Bartha, R., and D. Pramer. 1970. Metabolism of acylanilide
   herbicides. Adv. Appl. Microbiol. 13: 317-341.
10. Bartnicki-Garcia, S. 1968. Cell wall chemistry, morpho-
    genesis and taxonomy of fungi. Ann. Rev. Microbiol.
    22: 87-108.
11. Benzing-Purdie, L. 1981. Glucosamine and galactosamine
    distribution in soil as determined by gas liquid
    chromatography of soil hydrolysates: Effect of acid
    strength and cations. Soil Sci. Soc. Am. J. 45: 66-70.
12. Bollag, J-M, S-Y. Liu, and R.P. Minard. 1980. Cross-
    coupling of phenolic humus constituents and 2,4-dichloro-
    phenol. Soil Sci. Soc. Amer. J. 44: 52-56.
13. Bollag, J-M., R.P. Sjoblad, and D.P. Minard. 1977.
    Polymerization of phenolic intermediates of pesticides
    by a fungal enzyme. Experientia 33: 1564-1566.
14. Bondietti, E., J.P. Martin, and K. Haider. 1972.
    Stabilization of amino sugar units in humic-type polymers.
    Soil Sci. Soc. Amer. Proc. 36: 597-602.
15. Bremner, J. 1967. Nitrogenous compounds, pages 32-66,
    in A.D. McLaren and G.H. Peterson editors. Soil
    biochemistry. Marcel Dekker, New York.
16. Broadbent, F.E., and A.G. Norman. 1946. Some factors
    affecting the availability of the organic nitrogen in
    soil - a preliminary report. Soil Sci. Soc. Amer.
    Proc. 11: 264-267.

17. Burges, N.A., H.M. Hurst and S.B. Walkden. 1964. The phenolic constituents of humic acid and their relation to lignin of the plant cover. Geochim. Cosmochim. Acta 28: 1547-1564.
18. Burns, R.G. 1977. Soil enzymology. Sci. Prog. (Oxford) 64: 275-285.
19. Burns, R.G. 1978a. Enzymes in soil : some theoretical and practical considerations, pages 295-339, in R.G. Burns editor. Soil enzymes. Academic Press, London.
20. Burns, R.G. 1978b. Soil enzymes. Academic Press, London.
21. Burns, R.G. 1980. Microbial adhesion to soil surfaces : consquences for growth and enzyme activities, pages 249-269, in R.C.W. Berkeley, J.M. Lynch, J. Melling, P.R. Rutter and B. Vincent editors. Microbial adhesion to surfaces. Ellis Horwood, Chichester.
22. Burns, R.G. 1982a. Carbon mineralization by mixed cultures, pages 473-541, in A.T. Bull and J.H. Slater, editors. Microbial interactions and communities. Academic Press, London.

23. Burns, R.G. 1982b. Enzyme activity in soil : location and a possible role in microbial ecology. Soil Biol. Biochem. 14: 423-427.
24. Burns, R.G. 1983. Extracellular enzyme-substrate interactions in soil, pages 249-298, in J.H. Slater, R. Whittenbury and J.W.T. Wimpenny, Microbes in their natural environment, 34th Symp. Soc. Gen. Microbiol. Cambridge University Press, Cambridge.
25. Cain, R.B. 1980. The uptake and catabolism of lignin-related aromatic compounds and their regulation in microorganisms, pages 21-60 in T.K. Kirk, T. Higuchi and H-M. Chang, editors. Lignin biodegradation : microbiology, chemistry and potential applications. vol. 1. CRC Press, Boca Raton, Florida.
26. Canevascini, G., and H. Meier. 1978. Cellulolytic enzymes of Sporotrichum thermophile. Abst.XII Int. Cong. Microbiol., Munich 1978.
27. Chahal, D.S., M. Moo-Young, and G.S. Dhillon. 1979. Bioconversion of wheat straw and wheat straw components into single-cell protein. Can. J. Microbiol. 25: 793-797.
28. Cheshire, M.V. 1977. Origins and stability of soil polysaccharides. J. Soil Sci. 28: 1-10.
29. Ciegler, A., S. Kadis, and S.J. Ajl. 1971. Microbial toxins vol. 6: Fungal toxins. Academic Press, New York.
30. Cohen, B.L. 1980. Transport and utilization of proteins by fungi, pages 411-430, in J.W. Payne, editor. Microorganisms and nitrogen sources. John Wiley, Chichester.
31. Cowling, E.B., and W. Brown. 1979. Structural features of cellulosic materials in relation to enzymic hydrolysis. Adv. Chem. Ser. 95: 152-187.

32. Crawford, R.L. 1981. Lignin biodegradation and transformation. John Wiley, New York.
33. Crawford, D.L., and R.L. Crawford. 1976. Microbial degradation of lignocellulose : the lignin component. Appl. Environ. Microbiol. 31: 714-717.
34. Crawford, D.L., and J.B. Sutherland. 1979. The role of actinomycetes in the decomposition of lignocellulose. Dev. Ind. Microbiol. 20: 143-151.
35. Crawford, D.L., and J.B. Sutherland. 1980. Isolation and characterization of lignocellulose-decomposing actinomycetes, pages 95-125, in T.K. Kirk, T. Higuchi, and H-M. Chang, editors. Lignin biodegradation : microbiology, chemistry and potential applications. vol. 11. CRC Press, Boca Raton, Florida.
36. Dagley, S. 1971. Catabolism of aromatic compounds by microorganisms. Adv. Microbiol. Physiol. 6, 1-46.
37. Dagley, S. 1978. Microbial catabolism, the carbon cycle and environmental pollution. Naturwiss. 65: 85-95.
38. Dalal, R.C. 1975. Urease activity in some Trinidad soils. Soil Biol. Biochem. 7: 5-8.
39. Degens, E.D. 1982. Tracing man's carbon dioxide. ICSU Newsletter 17: 4-5.
40. Deshpande, V., D-E. Eriksson, and B. Pettersson. 1978. Production, purification and partial characterization of 1,4-β-glucosidase enzymes from Sporitrichum pulverulentum. Eur. J. Biochem. 90: 191-198.
41. Drew, S.W., and K.L. Kadam. 1979. Lignin metabolism by Aspergillus fumigatus and a white-rot fungus. Dev. Ind. Microbiol. 20: 153-161.
42. Dudman, W.F. 1977. The role of surface polysaccharides in natural environments, pages 357-414, in I. W. Sutherland, editor. Surface carbohydrates of the prokaryotic cell. Academic Press, London.
43. Edwards, C.A., and J.P. Lofty. 1972. Biology of earthworms. Chapman and Hall, London.
44. Eriksson, K-E., and S.C. Johnsrud. 1982. Mineralisation of carbon, pages 134-153, in R.G. Burns and J.H. Slater, editors. Experimental microbial ecology. Blackwell Scientific Publications, Oxford.
45. Fan, L.T., Y-H. Lee, and D.H. Beardmore. Mechanism of enzymatic hydrolysis of cellulose : effects of major structural features of cellulose on enzymatic hydrolysis. Biotechnol. Bioeng. 22: 177-199.
46. Filip, Z. 1975. Wechselbeziehungen zwischen Mikroorganismen and Tonmineralen und ihre Auswirkung auf die Bodendynamik. Habilitationsschrift, Univ. Giessen.
47. Flaig, W. 1964. Chemische Untersuchungen an Humusstoffen. Chem. 4: 253-265.
48. Flaig, W., H. Beutelspacher, and E. Rietz. 1975. Chemical composition and physical properties of humic substances, pages 1-211, in J.E. Gieseking, editor. Soil components, vol. 1. Organic components. Springer-Verlag, New York.

49. Foster, R.C. 1981. Polysaccharides in soil fabrics.
    Soil Sci. 214: 665-667.
50. Freudenberg, K., and A.C. Neish. 1968. Constitution
    and biosynthesis of lignin. Springer-Verlag, Berlin.
51. Gascho, G.F., and F.J. Stevenson. 1968. An improved
    method for extracting organic matter from soil.
    Soil Sci. Soc. Amer. Proc. 32: 117-118.
52. Ghose, T.K., and P. Ghosh. 1979. Cellulase production
    and cellulose hydrolysis. Proc. Biochem. Nov: 20-24.
53. Goh, K.M., and F.J. Stevenson. 1971. Comparison of
    infra-red spectra of synthetic and natural humic
    and fulvic acids. Soil Sci. 112: 392-400.
54. Goksøyr, J., and J. Eriksen. 1980. Cellulases, pages
    283-330, in A.H. Rose, editor. Economic microbiology,
    vol. 5. Academic Press, London.
55. Gong, C-S., and G.T. Tsao. 1979. Cellulase and biosyn-
    thesis regulation. Ann. Rep. Ferm. Proc. 3: 111-140.
56. Gonzalez-Vila, F.J., H. Lentz, and H.D. Ludemann. 1976.
    C^{13} nuclear magnetic resonance spectra of natural
    humic substances. Biophys. Biochem. Res. Commun.
    72: 1063-1070.
57. Greenland, D.J. 1965. Interactions between clays and
    organic compounds in soils. Part 11. Adsorption of
    soil organic compounds and its effect on soil properties
    Soils Fert. 28: 415-425.
58. Greenland, D.J. 1971. Interactions between humic and
    fulvic acids and clays. Soil Sci. 111: 34-41.
59. Greenland, D.J., G.R. Lindstrom, and J.P. Quirk. 1961.
    Role of polysaccharides in stabilization of natural
    soil aggregates. Nature (London) 191: 1283-1284.
60. Greenland, D.J., and J.M. Oades. 1975. Saccharides
    pages 213-261, in J.E. Gieseking, editor. Soil
    components, vol. 1. Organic compounds. Springer-
    Verlag, New York.
61. Griffith, S.M., and M. Schnitzer. 1975. The isolation
    and characterization of stable metal-organic
    complexes from tropical volcanic soils. Soil Sci.
    120: 126-127.
62. Griffiths, E., and R.G. Burns. 1972. Interaction
    between phenolic substances and microbial polysaccharides
    in soil aggregation. Pl. Soil 36, 599-612.
63. Guckert, A. 1975. Origine et devinir des polysaccharides
    du sol, pages 116-127, in G. Kilbertus, O. Reisinger,
    A. Mourey and J.A. Camela, editors. Proc. 1st. Sym.
    Biodegradation et Humification, Univ. of Nancy,
    Nancy, France.
64. Guckert, A., H.H. Tok, and F. Jacquin. 1977. Biodegrad-
    ation de polysaccharides bacteriens adorbes sur une
    montmorillonite, pages 403-411, in Soil organic
    matter studies vol. 1. lAEA-FAO Vienna.
65. Gupta, U.C., and F.J. Sowden. 1965. Studies on methods
    for the determination of sugars and uronic adids in
    soils. Can. J. Soil Sci. 45: 237-240.

66. Hackett, W.F., W.J. Conners, K. Kirk, and J.G. Zeikus. 1977. Microbial decomposition of synthetic ^{14}C-labelled lignins in nature : lignin biodegradation in a variety of natural materials. Appl. Environ. Microbiol. 33: 43-51.

67. Haider, K., and J.P. Martin. 1967. Synthesis and transformation of phenolic compounds by Epicoccum nigrum in relation to soil humus formation. Soil Sci. Soc. Amer. Proc. 31: 766-772.

68. Haider, K., and J.P. Martin. 1970. Humic acid-type phenolic polymers from Aspergillus sydowi culture medium, Stachybotrys spp. cells and autoxidized phenol mixture. Soil Biol. Biochem. 2: 145-156.

69. Haider, K., and J.P. Martin. 1975. Decomposition of specifically carbon-14-labelled benzoic and cinnamic acid derivatives in soil. Soil Sci. Soc. Amer. Proc. 39: 657-662.

70. Haider, K., and J.P. Martin. 1979. Abbau and Umwandlung von Pflanzenruckstanden und ihren Inhaltsstoffen durch die Mikroflora des Bodens. S. Pflanzenernaehr. Bodenkund. 142: 456-475.

71. Haider, K., and J.P. Martin. 1981. Decomposition in soil of specifically ^{14}C-labelled model and cornstalk lignins and coniferyl alcohol over two years as influenced by drying, rewetting and additions of an available C substrate. Soil Biol. Biochem. 13: 447-450.

72. Haider, K., L.R. Frederick, and W. Flaig. 1965. Reactions between amino acid compounds and phenols during oxidation. Plant Soil 22: 49-64.

73. Haider, K., J.P. Martin, and Z. Filip. 1974. Humus biochemistry, pages 195-244, in E.A. Paul and A.D. McLaren, editors. Soil biochemistry. vol. 4. Marcel Dekker, New York.

74. Haider, K., J.P. Martin, and E. Rietz. 1977. Decomposition in soil of ^{14}C-labelled coumaryl alcohols, free and linked into dehydropolymer and plant lignins and model humic acids. Soil Sci. Soc. Amer. J. 41: 556-562.

75. Hall, P., W. Glasser, and S. Drew. 1980. Enzymatic transformations of lignin, pages 17-31, in T.K. Kirk, T. Higuchi and H-M. Chang, editors. Lignin biodegradation : microbiology, chemistry and potential applications. vol. 11. CRC Press, Boca Raton, Florida.

76. Harris, R.F., G. Chesters, and O.N. Allen. 1966. Dynamics of soil aggregation. Adv. Agron. 18: 107-169.

77. Haska, G. 1981. Activity of bacteriolytic enzymes adsorbed to clays. Microbial Ecol. 7: 331-334.

78. Hatcher, P.G., R. Rowan, and M.A. Mattingly. 1980a. ^{1}H and ^{13}C NMR of marine humic acids. Org. Geochem. 2: 77-85.

79. Hatcher, P.G., D.L. VanderHart, and W.L. Earl. 1980b. Use of solid-state ^{13}C NMR in structural studies of humic acids and humin from Holocene sediments. Org. Geochem. 2: 87-92.

80. Hayes, M.H.B., and R.S. Swift. 1978. The chemistry of soil organic colloids, pages 179-320 in D.J.Greenland and M.H.B. Hayes, editors. The chemistry of soil constituents. John Wiley and Sons, New York.

81. Hill, I.R., and S.J.L. Wright. 1978. Pesticide microbiology. Academic Press, New York.

82. Hope, C.F.A., and R.G. Burns. 1983. Extracellular cellulase activity in soil. Soc.Gen. Microbiol. Quart. 10: 12.

83. Hope, C.F.A., J.M. Alexander, and R.G. Burns. 1980. β-D-Glucosidase activity in soil. Soc. Gen. Microbiol. Quart. 8: 41.

84. Hurst, H.M., and N.A. Burges. 1967. Lignin and humic acids, pages 260-286, in A.D. McLaren and G. H. Peterson, editors. Soil biochemistry. Marcel Dekker, New York.

85. Iverson, K.C., and F.J. Sowden. 1962. Methods for the analysis of carbohydrate material in soil: 1. Colorimetric determination of uronic acids, hexoses and pentoses. Soil Sci. 94: 245-250.

86. Jenkinson, D.S. 1966. The priming action, pages 199-207, in Report of FAO/IAEA Tech. Meet. on the Use of Isotopes in Soil Organic Matter Studies, Braunschweig, Germany, 1963.

87. Jenkinson, D.S. 1971. Studies on decomposition of ^{14}C-labelled organic matter in soil. Soil Sci. 111: 64-70.

88. Jenkinson, D.S., and A. Syanaba. 1977. Decomposition of carbon-14 labelled plant material under tropical conditions. Soil Sci. Soc. Am. J. 41: 912-915.

89. Jenkinson, D.A., and J.N. Ladd. 1981. Microbial biomass in soil : measurement and turnover, pages 415-471, in E.A. Paul and J.N. Ladd, editors. Soil biochemistry. vol.5. Marcel Dekker, New York.

90. Jenkinson, D.A., and D.S. Powlson. 1976. The effects of biocidal treatments on metabolism in soil. V. A method for measuring soil biomass. Soil Biol. Biochem. 8: 209-213.

91. Jenkinson, D.S., and J.H. Rayner. 1977. The turnover of soil organic matter in some of the Rothamsted classical experiments. Soil Sci. 123: 298-305.

92. Jeuniaux, C. 1963. Chitine et chitinolyse. Mason et Cie, Paris.

93. Kassim, G., J.P. Martin, and K. Haider. 1981. Incorporation of a wide variety of organic substrate carbons into soil biomass as estimated by the fumigation procedure. Soil Sci. Soc. Amer.J. 45: 1106-1112.

94. Kassim, G., D.E. Stott, J.P. Martin, and K. Haider. 1982. Stabilization and incorporation into biomass of phenolic and benzenoid carbons during biodegradation in soil. Soil Sci. Soc. Amer. J. 46: 305-309.

95. Kirk, T.K. 1971. The effects of microorganisms on lignin. Ann. Rev. Phytopath. 9: 185-210.

96. Kirk, T.K., W.J. Connors, and J.G. Zeikus. 1977. Advances in understanding the microbiological degradation of lignin. Rev. Adv. Phytopath. 11: 369-394.
97. Kiss, S., M. Dragan-Bularda, and D. Radulescu. 1975. Biological significance of enzymes in soil. Adv. Agron. 27: 25-87.
98. Kononova, M.M. 1966. Soil organic matter. Pergammon Press, Oxford.
99. Kononova, M.M. 1972. Current problems in the study of soil organic matter. Pochvovedenie 7: 27-36.
100. Kumada, K., A. Suzuki, and K. Aizawa. 1961. Isolation of anthraquinone from humus. Nature (London) 191: 415-416.
101. Ladd, J.N., and H.H.A. Butler. 1966. Comparison of some properties of soil humic acids and synthetic phenolic polymers incorporating amino derivatives. Aust. J. Soil Res. 4: 41-54.
102. Ladd, J.N., and J.H.A. Butler. 1975. Humus-enzyme systems and synthetic organic polymer-enzyme analogs, pages 143-193, in E.A. Paul and A.D. McLaren, editors. Soil biochemistry. vol.4. Marcel Dekker, New York.
103. Lethbridge, G., A.T. Bull, and R.G. Burns. 1978. Assay and properties of 1,3-β-glucanase in soil. Soil Biol. Biochem. 10: 389-391.
104. Linhares, L.F., and J.P. Martin. 1978. Decomposition in soil of the humic acid-type melanins of Eurotium echinulatum, Aspergillus glaucus sp. and other fungi. Soil Sci. Soc. Am. J. 42: 738.
105. Linhares, L.F., and J.P. Martin. 1979a. Decomposition in soil of emodin, chrysophanic acid and a mixture of anthraquinones synthesized by an Aspergillus glaucus isolate. Soil Sci. Soc. Amer. J. 43: 940-945.
106. Linhares, L.F., and J.P. Martin. 1979b. Carbohydrate content of fungal humic acid-type polymers (melanins). Soil Sci. Soc. Amer. J. 43: 313-318.
107. Ludemann, H.D., H. Lentz, and J.P. Martin. 1982. Carbon-13 nuclear magnetic resonance spectra of some fungal melanins and humic acids. Soil Sci. Soc. Amer. J. 46: 957-962.
108. McGill, W.B., H.W. Hunt, R.G. Woodmansee, and J.O. Reuss. 1981. Phoenix, a model of the dynamics of carbon and nitrogen in grassland soils, pages 49-115, in F.E. Clark and T. Rosswall, editors. Terrestrial nitrogen cycles. Ecol. Bull./NFR. (Stockholm) 33.
109. Manners, D.J. 1979. The enzymic degradation of starches, pages 75-91, in J.M.V. Blanchard and J.R. Mitchell, editors. Polysaccharides in foods. Butterworths, London.

110. Marshall, K.C. 1980. Reactions of microorganisms, ions and macromolecules at interfaces, pages 93-106, in D.C. Ellwood, J.N. Hedger, M.J. Latham, J.M. Lynch and J.H. Slater, editors. Contemporary Microbial Ecology. Academic Press, London.

111. Marshman, N.A., and K.C. Marshall. 1981. Bacterial growth on proteins in the presence of clay minerals. Soil Biol. Biochem. 13: 127-134.

112. Martin, J.P. 1971. Decomposition and binding action of polysaccharides in soil. Soil Biol. Biochem. 3: 33-44.

113. Martin, J.P., and D.D. Focht. 1977. Biological properties of soils, pages 115-169, in Soils for management of organic wastes and waste waters. ASA, CSSA and SSSA publication Madison, Wisconsin, USA.

114. Martin, J.P., and K. Haider. 1968. Phenolic polymers of Stachybotrys atra, Stachybotrys chartarum and Epicoccum nigrum in relation to humic acid formation. Soil Sci. 107: 260-270.

115. Martin, J.P., and K. Haider. 1971. Microbial activity in relation to soil humus formation. Soil Sci. 111: 54-63.

116. Martin, J.P., and K. Haider. 1976. Decomposition of specifically carbon-14-labelled ferulic acid; Free and linked into model humic acid-type polymers. Soil Sci. Soc. Amer. J. 40: 377-380.

117. Martin, J.P., and K. Haider. 1977. Decomposition in soil of specifically ^{14}C-labelled DHP and cornstalk lignins, model humic acid type polymers and coniferyl alcohols, pages 23-32, in Soil organic matter studies, vol. 2. IAEA-FAO, Vienna.

118. Martin, J.P., and K. Haider. 1979a. Effect of concentration on decomposition of some ^{14}C-labelled phenolic compounds, benzoic acid, glucose, cellulose, wheat straw and Chlorella protein in soil. Soil Sci. Soc. Amer. J. 43: 917-920.

119. Martin, J.P., and K. Haider. 1979b. Biodegradation of ^{14}C-labelled model and cornstalk lignins, phenols, model phenolase humic polymers and fungal melanins as influenced by a readily available C source and soil. App. Environ. Microb. 38: 283-289.

120. Martin, J.P., and K. Haider. 1980a. A comparison of the use of phenolase and peroxidase for the synthesis of model humic acid-type polymers. Soil Sci. Soc. Amer. J. 44: 983-988.

121. Martin, J.P., and K. Haider. 1980b. Microbial degradation and stabilization of ^{14}C-labelled lignins, phenols and phenolic polymers in relation to soil humus formation, pages 77-100, in K. Kirk, T. Higuchi and H-M. Chang, editors. Lignin biodegradation: microbiology, chemistry and potential applications. vol. 1. CRC Press, Boca Raton, Florida.

122. Martin, J.P., and D. Stott. 1981. Microbial transformations of herbicides in soil. Proc. Western Soc. Weed Sci. 34: 39-54.

123. Martin, J.P.,J.O Ervin, and R.A. Shepherd. Decomposition of iron, aluminium, zinc and copper salts or complexes of some microbial and plant polysaccharides in soil. Soil Sci. Soc. Amer. Proc. 30: 196-200.

124. Martin, J.P., K. Haider, and E. Bondietti. 1972. Properties of model humic acids synthesized by phenoloxidase and autoxidation of phenols and other compounds formed by soil fungi, pages 171-186, in D. Povoledo and H.L. Gottermann, editors. Humic substances. Proc. Int. Meeting, Nieuwersluis, The Netherlands, Pudoc, Wageningen.

125. Martin, J.P.,K. Haider, and L. Linhares. 1979. Decomposition and stabilization of ring-^{14}C-labelled catechol in soil. Soil Sci. Soc. Amer. J. 43: 100-104.

126. Martin, J.P.,K. Haider, and G. Kassim. 1980. Biodegradation and stabilization after 2 years of specific crop, lignin and polysaccharide carbons in soils. Soil Sci. Soc. Amer. J. 44: 1250-1255.

127. Martin, J.P.,K. Haider,and C. Saiz-Jimenez. 1974. Sodium amalgam reductive degradation of fungal and model phenolic polymers, soil humic acids and simple phenolic compounds. Soil Sci. Soc. Amer. Proc. 38: 760-765.

128. Martin, J.P.,K. Haider, and D. Wolf. 1972. Synthesis of phenols and phenolic polymers by Hendersonula toruloidea in relation to soil humus formation. Soil Sci. Soc. Amer. Proc. 36: 311-315.

129. Martin, J.P., A.A. Parsa, and K. Haider. 1978. influence of intimate association with humic polymers on biodegradation of {[14]C} labelled organic substrates in soil. Soil Biol. Biochem. 10: 483-486.

130. Martin, F., C. Saiz-Jimenez, and A. Cert. 1977. Pyrolysis-gas chromatography of soil humic fractions. 1. The low boiling point compounds. Soil Sci. Soc. Amer. J. 44: 1114-1118.

131. Martin, J.P., K. Haider, W.J. Farmer, and E. Fustec-Mathon. 1974. Decomposition and distribution of residual activity of some ^{14}C-microbial polysaccharides and cells, glucose, cellulose and wheat straw. Soil Biol. Biochem. 6; 221-230.

132. Martin, J.P.,W.P. Martin, J.B. Page, W.A. Raney, and J.P. DeMent. 1955. Soil aggregation. Adv. Agron. 7: 1-37.

133. Martin, J.P., H. Zunino, P. Peirano, M. Caiozzi, and K. Haider. 1982. Decomposition of ^{14}C-labelled lignins, model humic acid polymers, and fungal melanins in allophanic soils. Soil Biol. Biochem. 14: 289-293.

134. Mason, H.S. 1953. The structure of melanins, pages 277-303, in M. Gordon, editor. Pigment cell growth. Academic Press, New York.

135. Mason, H.S. 1955. Comparative biochemistry of the phenolase complex. Advances in Enzymology 16: 105-173.

136. Mathur, S.P. 1971. Characterization of soil humus through enzymatic degradation. Soil Sci. 111: 147-157.

137. Mathur, S.P., and M. Schnitzer. 1978. A chemical and spectroscopic characterization of some synthetic analogues of humic acids. Soil Sci. Soc. Amer. J. 42: 519-596.

138. Mayaudon, J., and P. Simonart. Etude de la decomposition de la matiere organique dans le sol au moyden de carbone radioactiv: 5. Plant Soil 11: 181-192.

139. Maximov, O.B., T.V. Shvets, and Yu. N. Elkin. 1977. On permanganate oxidation of humic acids. Geoderma 19: 63-78.

140. Meuzelaar, H.L.C., K. Haider, B.R. Nagar, and J.P. Martin. 1977. Comparative studies of pyrolysis-mass spectra of melanins, model phenolic polymers and humic acids. Geoderma 17: 239-252.

141. Musso, H. 1967. Phenol coupling, pages 1-94, in W.I. Taylor and A.R. Battersby, editors. Oxidative coupling of phenols. Marcel Dekker, New York.

142. Muzzarelli, R.A.A. 1977. Chitin. Pergamon, Oxford.

143. Nelson, D.W., J.P. Martin, and J.O. Ervin. 1979. Decomposition of microbial cells and components in soil and their stabilization through complexing with model acid-type phenolic polymers. Soil Sci. Soc. Amer. J. 43: 84-88.

144. Neuhauser, F.F., R. Hartenstein, and W.J. Connors. 1978. Soil invertebrates and the degradation of vanillin, cinnamic acid, and lignins. Soil Biol. Biochem. 10: 431-475.

145. Nicolaus, R.A. 1968. Melanins, Hermann. Paris.

146. Oades, J.M., and G.H. Wagner. 1971. Biosynthesis of sugars in soils incubated with ^{14}C glucose and ^{14}C dextran. Soil Sci. Soc. Amer. J. 35: 914-922.

147. Oglesby, R.T. R.F. Christman, and C.H. Driver. 1967. The biotransformation of lignin to humus. Adv. App. Microbiol. 9: 171-184.

148. Paul, E.A., and J.A. Van Veen. 1978. The use of tracers to determine the dynamic nature of organic matter, pages 61-102, in 11th Congress International Society of Soil Sci. Vol. 3.

149. Paulson, K.N., and L.T. Kurtz. 1969. Locus of urease activity of soil. Soil Sci. Soc. Amer. Proc. 33: 897-901.

150. Payne, J.W. 1980. Microorganisms and nitrogen sources. John Wiley and Sons, Chichester.

151. Payne, J.W., and C. Gilvarg. 1978. Transport of peptides in bacteria, pages 325-383, in B.P. Rosen editor. Bacterial transport. Marcel Dekker, New York.

152. Pinck, L.A., and F.E. Allison. 1951. Maintenance of soil organic matter: III. Influence of green manures on the release of native soil carbon. Soil Sci. 71: 67-75.

153. Piper, T.J., and A.M. Posner. 1972. Sodium amalgam reduction of humic acid. II. Application of the method. Soil Biol. Biochem. 4: 525-531.

154. Powell, D.A. 1979. Structure, solution properties and biological interactions of some microbial extracellular polysaccharides, pages 117-160, in R.C.W. Berkeley, G.W. Gooday and D.C. Ellwood, editors. Microbial polysaccharides and polysaccharases. Academic Press, London.

155. Reisinger, O., and G. Kilbertus. 1974. Biodegradation et humification. IV. Microorganismen intevenant dams la decomposition des cellules d' Aureobasidium pullulans. (De Bary) Arnaud. Can. J. Microbiol. 20: 299-306.

156. Robinson, T. 1963. The Organic constituents of higher plants. Burgess Publishing Co., Minneapolis.

157. Rogers, H.J., H.R. Perkins and J.B. Ward. 1980. Microbial cell walls and membranes. Chapman and Hall, London.

158. Rombouts, F.M., and W. Pilnik. 1980. Pectic enzymes, pages 227-282, in A.H. Rose, editor. Economic microbiology. vol. 5. Academic Press, London.

159. Rosenberg, S.L. 1978. Cellulose and lignocellulose degradation by thermophilic and thermotolerant fungi. Mycologia 70: 1-13.

160. Rosenberg, S.L. 1979. Physiological studies of lignocellulose degradation by the thermotolerant mould Chrysosporium pruinosum. Dev. Ind. Microbiol. 20: 133-142.

161. Saiz-Jimenez, C., K. Haider, and J.P. Martin. 1975. Anthraquinones and phenols as intermediates in the formation of dark-colored, humic acid-like pigments by Eurotium echinulatum. Soil Sci. Soc. Amer. Proc. 39: 649-653.

162. Saiz-Jimenez, C., K. Haider, and H.L.C. Meuzelaar. 1979. Comparison of soil organic matter and its fractions by pyrolysis-mass spectrometry. Geoderma 22: 25-37.

163. Sauerbeck, D. 1966. A critical evaluation of incubation experiments on the priming effect of green manure, pages 199-207, in Report of FAO/IAEA Tech. Meet. on The Use of Isotopes in Soil Organic Matter Studies, Braunschweig, Germany, 1963.

164. Sauerbeck, D. 1968. Die Umsetzung markierter organischer Substanzen in Boden in Abhangigkeit von Art, Menge, und Rottegrad. Landwirtsch. Forsch. 21: 91-102.

165. Sauerbeck, D. and F. Fuhr. 1968. Alkali extraction and fractionation of labelled plant material before and after decomposition, a contribution to the technical problems in humification studies, pages 3-11, in Isotopes and Radiation in Soil Organic Matter Studies. IAEA, Vienna.

166. Sauerbeck, D., and M.A. Gonzalez. 1977. Field decomposition of carbon-14-labelled plant residues in various soils of the Federal Republic of Germany and Costa Rica, pages, 117-132, in Soil Organic Matter Studies, IAEA-FAO, Vienna. Vol.1.

167. Sauerbeck, D.R., and B.G. Johnen. 1974. Radiometrische Untersuchungen zur Humusbilanz. Landv. Forsh. 27: 137-145.

168. Schaller, F. 1968. Soil animals. Univ. of Michigan Press, Ann Arbor.

169. Schlesinger, W.H. 1977. Carbon balance in terrestrial detritus. Ann. Rev. Ecol Systems 8: 51-81.

170. Schnitzer, M. 1977. Recent findings on the characterization of humic substances extracted from soils from widely differing climatic zones, pages 117-130, in Proc. of Symposium Soil Organic Matter Studies, Braunschweig, Germany.

171. Schnitzer, M., and S.U. Khan. 1972. Humic substances in the environment. Marcel Dekker, New York.

172. Schnitzer, M., and J.A. Neyroud. 1975. Further investigations on the chemistry of fungal "humic acids". Soil Biol. Biochem. 7: 365-371.

173. Shields, J.A., and E.A. Paul. 1973. Decomposition of C-14-labelled plant materials under field conditions. Can. J. Soil Sci. 53: 297-306.

174. Shimada, M. 1980. Stereobiochemical approach to lignin biodegradation: possible significance of nonstereospecific oxidation catalysed by laccase for lignin decomposition by white-rot fungi, pages 195-213, in T.K. Kirk, T. Higuchi and H-M. Chang, editors. Lignin biodegradation : microbiology, chemistry and potential applications. vol. 1. CRC Press, Boca Raton, Florida.

175. Sims, J.J., A.F. Rose and R.R. Izac. 1978. Applications of ^{13}C NMR to marine natural products, pages 297-378, in P.J. Scheuer, editor. Marine natural products. Academic Press, New York.

176. Sjoblad, R.D., and J-M. Bollag. 1981. Oxidative coupling of aromatic compounds by enzymes from soil microorganisms, pages 113-152, in E.A. Paul and J.N. Ladd, editors. Soil biochemistry vol.5. Marcel Dekker New York.

177. Skujins, J. 1976. Extracellular enzymes in soil. CRC Crit. Rev. Microbiol 4: 383-421.

178. Sørensen, L.H. 1975. The influence of clay on the rate of decay of amino acid metabolites synthesized in soils during decomposition of cellulose. Soil Biol. Biochem. 7: 171-177.

179. Sørensen, L.H. 1981. Carbon-nitrogen relationships during the humification of cellulose in soils containing different amounts of clay. Soil Biol. Biochem. 13: 313-321.

180. Sowden, F.J., S.M. Griffith, and M. Schnitzer. 1976. The distribution of nitrogen in some highly organic tropical volcanic soils. Soil Biol. Biochem. 8: 55-60.

181. Staaf, H., and B. Berg. 1981. Plant litter input to soil, pages 147-162, in F.E. Clark and T. Rosswall, editors. Terrestrial nitrogen cycles. Ecol. Bull. NFR (Stockholm) 33.

182. Stevenson, F.J. 1957. Investigations of amino polysaccharides in soil. Soil Sci. 83: 113-122.

183. Stevenson, F.J. 1982. Humus chemistry. John Wiley, New York.

184. Stewart, J.W.B., and R.B. McKercher. 1982. Phosphorous cycle, pages 221-238, in R.G. Burns and J.H. Slater, editors. Experimental microbial ecology. Blackwell Scientific Publications, Oxford.

185. Stott, D.E., J.P. Martin, D. Focht, and K. Haider. 1983. Biodegradation, stabilization in humus and incorporation in biomass of 2,4-D and Chlorocatechol carbons. Soil Sci. Soc. Amer. J. 47: 66-70.

186. Stotzky, G. 1967. Clay minerals and microbial ecology. N.Y. Acad. Sci., Trans. (Ser. 11) 30: 11-21.

187. Stotzky, G. 1974. Activity, ecology, and population dynamics of microorganisms in soil, pages 57-135, in A. Laskin and H. Lechevalier, editors. Microbial ecology. CRC Press, Cleveland, Ohio.

188. Stotzky, G. 1980. Surface interactions between clay minerals and microbes, viruses and soluble organics and the probable importance of these interactions to the ecology of microbes in soil, pages 231-247, in R.C.W. Berkeley, J.M. Lynch, J. Melling, P. R. Rutter and B. Vincent, editors. Microbial adhesion to surfaces. Ellis Horwood, Chichester.

189. Stotzky, G., and R.G. Burns. 1982. The soil environment: clay-humus-microbe interactions, pages 105-133, in R.G. Burns and J.H. Slater, editors. Experimental microbial ecology. Blackwell Scientific Publications, Oxford.

190. Suflita, J.M., and J-M. Bollag. 1981. Polymerization of phenolic compounds by a soil-enzyme complex. Soil Sci. Soc. Amer. J. 45: 297-302.

191. Swift, M.J., O.W. Heal, and J.M. Anderson. 1979. Decomposition in terrestrial ecosystem. Blackwell Scientific Publications, Oxford.

192. Swincer, G.D. J.M. Oades, and D.J. Greenland. 1968. Studies on soil polysaccharides. 1. The isolation of polysaccharides from soil. II. The composition and properties of polysaccharides in soils under pasture and under fallow wheat rotation. Aust. J. Soil Res. 6, 211-239.

193. Swincer, G.D., J.M. Oades, and D.J. Greenland. 1969. The extraction, characterization, and significance of soil polysaccharides. Adv. Agron. 21: 195-235.
194. Thomson, R.H. 1971. Naturally occurring quinones. Academic Press, New York.
195. Thorpe, J., and G.D. Smith. 1949. Higher categories of soil classification: order, suborder, and great soil groups. Soil Sci. 67: 117-126.
196. Tsao, G.T. 1978. Cellulose material as a renewable resource. Process Biochem. 13: 12-14.
197. Turchenek, L.W., and J.M. Oades. 1979. Fractionation of organo-mineral complexes by sedimentation and density techniques. II. Organic and mineral components. Geoderma 21: 311-343.
198. Varadi, J. 1972. The effect of aromatic compounds on cellulase and xylanase production of fungi Shizophyllum commune and Chaetomium globosum, pages 129-135, in A.H. Walters and E.H. Hueck-van der Plas, editors. Biodeterioration of materials. vol. 2. Applied Science Publishers, London.
199. Verma, L., and J.P. Martin. 1976. Decomposition of algal cells and components and their stabilization through complexing with model humic acid-type phenolic polymers. Soil Biol. Biochem. 8: 85-90.
200. Verma, L., J.P. Martin, and K. Haider. 1975. Decomposition of carbon-14-labelled proteins, peptides, and amino acids, free and complexed with humic polymers. Soil Sci. Soc. Amer. Proc. 39: 279-284.
201. Vohra, R.M., C.K. Shirkot, S. Dhawan, and K.G. Gupta. 1980. Effect of lignin and some ot its components on the production and activity of cellulase(s) by Trichoderma reesei. Biotech. Bioeng. 22: 1497-1500.
202. Wada, K., and T. Higashi. 1976. The categories of aluminum- and iron- humus complexes in ando soils determined by selective dissolution. J. Soil Sci. 27: 357-368.
203. Wagner, G.H. 1975. Microbial growth and carbon turnover, pages 269-305, in E.A. Paul and A.D. McLaren, editors. Soil biochemistry vol. 3. Marcel Dekker, New York.
204. Wagner, G.H., and V.K. Mutatkar. 1968. Amino components of soil organic matter formed during humification of C^{14} glucose. Soil Sci. Soc. Am. Proc. 32: 683-686.
205. Westermark, V., and K-E. Eriksson. 1974. Carbohydrate-dependent enzymic quinone reduction during lignin degradation. Acta Chem. Scand. 28: 204-208.
206. Westermark, V., and K-E. Eriksson. 1975. Purification and properties of cellobiose : quinone oxidoreductase from Sporotrichum pulverulentum. Acta Chem. Scand. 29: 419-424.

207. Wolf, D.C., and J.P. Martin. 1976. Decomposition of fungal mycelia and humic-type polymers containing carbon-14 from ring and side-chain labelled 2,4-D and chlorpropham. Soil Sci. Soc. Amer. J. 40: 700-704.

208. Wolfinbarger, L. 1980. Transport and utilization of peptides by fungi, pages 281-300, in J.W. Payne, editor. Microorganisms and nitrogen sources. John Wiley and Sons, Chichester.

209. Zeikus, J.G. 1980. Fate of lignin and related aromatic substrates in anaerobic environments, pages 101-109, in T.K. Kirk, T. Higuchi and H-M. Chang, editors. Lignin biodegradation : microbiology, chemistry, and potential applications. vol. 1. CRC Press, Boca Raton, Florida.

210. Zunino, H., F. Borie, S. Aguilera, J.P. Martin, and K. Haider. 1982. Decomposition of ^{14}C-labelled glucose, plant and microbial products and phenols in volcanic ash-derived soils of Chile. Soil Biol. Biochem. 14: 37-43.

ROOT AND SOIL MICROBIAL INTERACTIONS WHICH INFLUENCE THE AVAILABILITY OF PHOTOASSIMILATE CARBON TO THE RHIZOSPHERE

A.J.M. SMUCKER and G.R. SAFIR[1]

INTRODUCTION

Extensive plant root systems, reported to achieve lengths of up to 71,000 m per wheat (<u>Triticum</u> <u>aestivum</u> L.) plant (Pavlychenko 1937) provide a vast surface area which supplies a significant amount of metabolic energy for the microbial populations of the soil rhizosphere. Although a complete description of the complex rhizosphere is beyond the scope of this chapter, it is clear that the enhanced microfloral populations of this dynamic root and soil interface both enhance root function by increasing the absorption of water and nutrients and promote root dysfunction by infection and disease. Discriminating factors which control the delicate mutualistic and/or pathogenic interactions between the rhizosphere microflora and the plant root appear to be a function of genotype and environmental conditions.

Recently, the root systems of plants have been investigated by researchers interested in their genetic modification, their physiology when stressed, and their membrane and hormone responses to the environment. This interest combined with the application of recently developed techniques provide an excellent opportunity for quantifying those mechanisms which regulate the ecology of

[1]Professor, Associate Professor, Departments of Crop & Soil Sciences and Botany and Plant Pathology, respectively, Michigan State University, East Lansing, MI 48824, MSU Agric. Expt. Sta. Journal Article No. 11385.

the rhizosphere. This chapter will emphasize the partitioning and utilization of photoassimilates by plant root systems. The adverse and beneficial effects of the abiotic and biotic environments of the soil will be emphasized. Recent methods for measuring plant root system responses and associated rhizosphere modifications will be reported. Since general root functions will not be addressed, readers are referred to the reviews published by Arkin and Taylor (1981), Brouwer et al. (1981), Carson (1974), Harley and Russell (1974), Russell (1979) and Torrey and Clarkson (1975).

RHIZOSPHERE

A finite soil zone associated with plant roots which has a greater rate of microbial activity may be defined as the rhizoplane (the root surface) or the rhizosphere (soil volume adjacent to plant roots). These and other soil areas adjacent to germinating seeds (spermosphere or sporosphere), at the base of the plant (laimosphere) and roots growing in epiphyte mats of rain forests (Jordan 1982 and Nadkarni 1981) or the soil-humus mats of turf thatch (epiphyteosphere) compose the complex and dynamic interface between the plant and soil.

Since the quality and quantity of root exudates and total root length or surface area primarily determine the extent of the rhizosphere, those factors contributing to the losses of C and N compounds have the greatest effect upon rhizosphere activities. Compounds lost from the roots are primarily controlled by plant species, plant nutrition, age of plant roots, environmental conditions (e.g., light, temperature of roots and shoots, soil moisture, soil aeration, soil matrix, etc.), presence or absence of soil microorganisms, and foliar sprays (Hale and Moore 1979). Root exudates include amino acids, carbohydrates, enzymes and some small molecular weight proteins, organic acids, phytohormones, and other compounds which attract, stimulate or inhibit microflora and micro- and macrofauna. Soil organisms populating the rhizosphere include actinomycetes, bacteria, fungi, microarthropods, nematodes, and protozoa (Curl 1982). The antagonistic or allelopathic interactions among roots of different contiguous plant types constitute a complex of chemical and microbial interactions which are essentially unknown components of the rhizosphere effect upon plant growth.

ALLOCATION OF PLANT CARBON

Although the translocation of C to specific plant organs is essential for providing metabolic energy, the process of photoassimilate distribution and utilization by whole plant systems is not well defined. Extensive reviews emphasizing the C budgets for various above ground plant fractions of soybean (Glycine max L. Merr.) (Chatterton and Silvers 1979) and Lupinus, sp. (Pate et al. 1979) indicated that the distribution of photoassimilates is controlled by the assimilate demands of distant metabolic sinks which change with ontogenetic development or in response to environmental signals originating from biochemical mechanisms in leaves (Geiger and Giaquinta 1982). Therefore, the crucial process of C allocation appears to be a function of the competitive interactions among multiple plant sinks which are attenuated by the environment. Since there is a strong metabolic demand for C at the root tips of rapidly growing plants, it is suggested that the irreversible loss of photoassimilates to the expansive and leaky root systems of contemporary cultivars may account for the absence of a high correlation between CO_2 exchange per unit area of leaves and dry weight accumulation or crop yield.

The quantity of organic compounds released from roots has been underestimated primarily because results were based upon the measurements of plant roots grown in nutrient solutions and/or axenic conditions. Barber and Gunn (1974) showed that mechanical stresses caused roots of cereals to release from 5-9% of the plant sugars and amino acids. Subsequent studies by Barber and Martin (1976), Martin (1977), Sauerbeck and Johnen (1977), and Shadan (1980) reported that gaseous and soluble C released by root systems growing in soils or solution culture accounted for 20-39% of the C translocated to the roots. These high rates of C lost to the rhizosphere agree with other $^{14}CO_2$ studies where 34-45% of the labeled compounds exported to legume roots were lost by respiration and exudation (Carr and Pate 1967).

Carbon transfer from the root to the rhizosphere results in a net loss of photosynthates from the plant. Short-term and long-term labeling of plants with C ratio isotopes have made it possible to separate root and microbial activity in the rhizosphere. Warembourg and Billes (1979) showed a diphasic succession of $^{14}CO_2$ evolution from non-axenic roots of labeled wheat plants. Liberation of ^{14}C-substances in the solution was greatest at 3-6 h after the labeling of plants. Roots of 14 day-old dry bean (Phaseolus vulgaris L.) plants lost up to 21% of the soluble ^{14}C-label which was translocated from the source leaf during a two hour pulse and one hour chase period (Schumacher and Smucker 1984a). From 26-33% of the label transported to the root was lost to the surrounding media primarily as CO_2. Even greater quantities of C may be lost by the respiring root as $^{14}CO_2$ accounted for less than 5% of the total CO_2 respired by the root system during the three hour period of measurement.

Additional measurements of ^{14}C-label within the roots of dry beans indicated that 29% remained in the mature roots and 11% was utilized for growth by the actively growing components of roots (Schumacher and Smucker 1984a). These photoassimilates transported to the roots appeared to be used mainly for maintenance respiration, modification of cell walls, cell division and extension, and the synthesis of primary metabolites which produce large amounts of metabolic energy and CO_2.

The loss of root exudates appears to be proportional to the concentration of these organic compounds at the root surface since removal of the compounds increases exudation. Prikryl and Vancura (1980) reported that frequent replacements of the nutrient solution of wheat roots increased the exudation of axenic cultures from 20 to 34% during an eight day period. Additions of bacteria (Pseudomonas putida) to the formerly axenic culture increased the loss of carbon from 26 to 52% during the same treatment period. Organic C lost by the wheat seedlings of these experiments was directly correlated to root length or dry mass. Their data indicated that leakage by young roots may have resulted from the simple diffusion of substances away from the apoplastic free space of roots. This mechanism for C loss is supported by others who reported increased losses from roots when the permeability of cell membranes was disturbed by environmental stresses (Smucker and Adler 1980, Smucker and Erickson 1976). Johnen and Sauerbeck (1977) reported that from emergence to tillering, approximately 40% of the C fixed by wheat plants became available to soil microorganisms. At later stages, the amount of photosynthetically fixed C accessible to soil microorganisms exceeded the harvested root mass by 2.5 fold. These data indicate that much of the C lost by wheat plants after tillering resulted in the senescence and decomposition of large segments of the root system and may account for the large quantities of C lost by agronomic crops. Minchin and Pate (1973) also reported that when shoots of 21-day-old peas (Pisum sativa L.) were labeled with $^{14}CO_2$, 26% of the ^{14}C was incorporated directly into shoot dry matter and 74% was translocated to the root. Nitrogen fixation and nodule growth consumed 20% of the ^{14}C-photoassimilates. Root growth required 7%. The remaining 47% of the fixed C was lost to the soil.

CARBON LOSSES BY ROOT RESPIRATION

Photoassimilate requirements for the development of roots are essentially unknown for most plants, yet it appears that they use less C than that invested per unit of leaf as roots have fewer specialized tissues which form the complex structure of leaves. Maintenance respiration costs, however, appear to be higher for roots due to the greater distance of photoassimilate transport (Moldau and Karolin 1977), vulnerability of relatively unspecialized cells which interface with the environment resulting in large turnover rates, and the presence of an inefficient active cyanide resistant respiration pathway (Lambers et al. 1983).

Although the metabolic intermediates produced by respiration are essential to basic functions of the plant root system, it is generally accepted that root biomass would be increased if respiration were minimized (DeWit et al. 1978). The rate-limiting components of root respiration appear to be a function of carbohydrate catabolism from reserves (dry mass) and the current rate of photosynthesis (Heaketh et al. 1982). Carbon losses by root respiration have been reported to be 25% of the current photosynthetic rate and 1.5% of the total dry mass for clover (Trifolium sp.) (DeWit et al. 1978) and up to 13.5% of the root mass of ryegrass (Lolium multiflorum L.) (Hansen and Jensen 1977). Carbon losses by the respiration of nodulated roots appear to approach 50% (Minchin and Pate 1973) for peas and have been reported to require as much as 11.1 g of C per g of N_2 fixed or over 4 mol of glucose per mol N_2 (Dixon et al. 1981). If one assumes the N content of the plant to be approximately 3.5%, then a plant with a relative growth efficiency of 0.75 could respire 70 g of carbohydrate for each growth increment of 100 g of root dry matter. As the relative growth efficiency is reduced (e.g., during periods of environmental stress) to 0.5, the carbohydrate requirement could exceed 100% of each succeeding growth increment until the stress was removed. The savings to be gained by improving the carbon efficiency of nodulated legume roots responding to stress environments appear to be substantial. Reductions in the allocation of carbon to nodulated roots appear to be detrimental to the integrated plant system. For example, nodulated roots of plants, Phaseolus cv. which developed in a compensatory manner, by the accelerated root growth in those areas of the soil profile which exhibited less compacted conditions, generally had smaller shoots (Smucker, unpublished data). Additional information regarding the nitrogen dynamics of the soil may be found in Chapter 6.

Compensatory root growth reflects a greater allocation of C to a portion of the root system. Although this response mechanism is thought to be an advantage for root systems, it causes a greater expenditure of C to accomplish the same task, had there been no need for a compensatory response. Recent results from our laboratories indicate that ion absorption, respiration and water uptake are also greater in the aerated portion of compensatory roots than in the aerated controls (Schumacher and Smucker 1984a). These combined increases, however, resulted in the less efficient absorption of ions. Oxygen requirements were an average of 23% greater for the absorption of cations by the compensatory roots of three dry bean cultivars. Oxygen requirements for anion absorption were 7% greater than the control roots. Therefore, it would appear that since compensatory morphological responses provide the geneticist with a characteristic for selecting plants with superior root systems, specific morphological differences among genotypes should not be made which favor only a portion of the plant to the ultimate detriment of other mechanisms of the multiple plant functions.

Kuiper (1983) reported that the genetic variation and phenotypic plasticity of individual root responses to nutritional levels appear to be an ecological strategy for plant survival. In addition, his data showed a direct relationship between root respiration activities and growth characteristics (e.g., number of seeds) of the plant. Yet, the plasticity of root respiration and these growth characteristics differed significantly among four inbred lines of grass (Plantago major, L.) and with plant age. He concluded that genotypes producing a low number of seeds represent species with little plasticity of their physiological characteristics (e.g., growth functions, ion stimulated ATPase activities, root respiration, etc.) and that genotypes producing a high number of seeds were associated with greater plasticity of these physiological characteristics. However, he did not measure the efficiencies of photoassimilate translocation responses to different ionic strengths of the nutrient solution. Therefore, we are unable to conclude whether the plasticity of root respiratory responses to an environmental change was due to alternative metabolic pathways of stressed roots or the limitations of available photoassimilate energy to the root system.

The direct relationship between crop yield and the efficient utilization of the limited amount of photoassimilates allocated to the harvested portion of the plant, indicates that the rate of photoassimilate flow in plants appears to be controlled by the combined influences of source, path and sink while the partitioning of these compounds may be a function of assimilate potentials (Lang and Thorpe 1984). These conceptual models suggest that the rate of C flow is a function of the number and activity of metabolic sinks which are attenuated by changes in the environment. Results from our laboratories indicate that root systems with fewer actively respiring apical meristems appear to utilize their assimilates more efficiently, especially during periods without environmental stress. However, plant tolerance of soil stress appears to be related to the rapid allocation and transport of soluble carbon to the meristems of roots with a concomitant increase in the branching and expansion of the root system (Schumacher and Smucker 1984a and b).

Recent reports indicate that the net respiration of plant roots is influenced by the cyanide (CN)-resistant alternative pathway which appears to predominate when the cytochrome oxidase pathway is either substrate saturated or adenylate restricted (Lambers et al. 1983). Since the CN-resistant alternative path represents a seemingly unregulated by-pass of two-thirds of the conventional and regulated energy-conserving cytochrome-mediated electron path (Laties 1982), the mechanisms which engage this energetically inefficient alternate pathway in roots are of substantial physiological significance. The relative efficiency of carbon utilization by plant root systems which are subjected to controlled biological, chemical or physical environments is an area of study which merits greater emphasis as we improve our technological capabilities for investigating an in situ rhizosphere.

Root turnover or the death and decomposition of fine roots, may also account for a large loss of plant photoassimilates. Coleman (1976) demonstrated that the production and maintenance of plant root systems results in the largest C input to the ecosystem. He reported that 54% of grass root tips appear to survive for less than one month. Roots may require greater investments of photoassimilates than shoots even though the remaining quantity of roots at any given time are less than or equal to those of the shoots which have accumulated over the life of the plant.

A schematic of C flow through root systems is summarized in Figure 1. Some C is obviously necessary for the growth, maintenance and absorption processes of extensive root systems. However, a substantially larger portion appears to be lost by the large turnover rates of roots (St. Jon and Coleman 1983), uncontrolled CO_2 respiration and perhaps leakage of HCO_3^- ions (Smucker unpublished data). The longer C compounds are retained by the plant, either in the form of metabolic recycling, as temporary storage which may be remobilized, or as structural tissue, the greater the efficiency of carbon use.

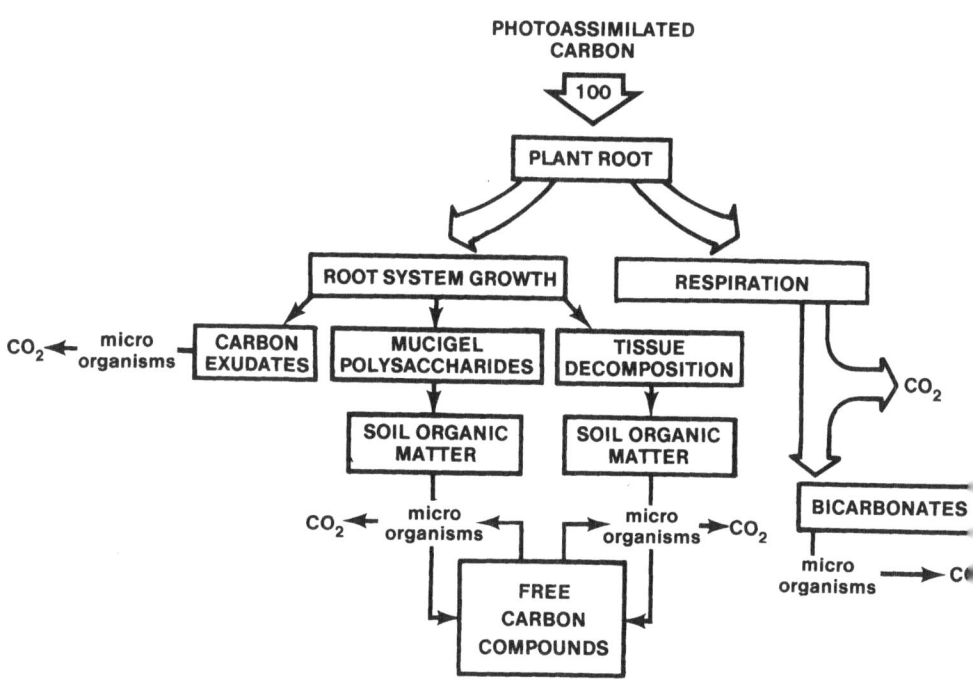

Figure 1. Flow diagram of the utilization of photoassimilates allocated to the root system. Nearly all (70-90%) of the carbon allocated to the root system is retained by the young seedling. Respiratory demands for carbon increase during subsequent weeks until nearly all (80-90%) of the carbon allocated to the maturing root system is respired and lost to the rhizosphere in the forms of carbon dioxide and bicarbonate.

These reports of excessively high losses of carbon by the roots of contemporary cultivars indicate that one or more of the following characters must be incorporated into the root system of plants before superior cultivars can be developed: 1) minimal losses of soluble exudates, 2) a lower proportion of CN-resistant respiration, 3) a minimum number of actively respiring apical meristems, 4) lower turnover rates through increased resistance to environmental decomposition and 5) readsorption and utilization of CO_2 or bicarbonate ions.

PHOTOASSIMILATE UTILIZATION DURING ANAEROBIC STRESS

An analysis of eight major crops produced in the United States indicated that there is a large unrealized genetic potential for plant yield, especially if these plant types were better adapted to the physiochemical environments in which they are grown (Boyer 1982). Although genetic selection for adaptation to adverse environments has contributed significantly to increased agricultural production, the fundamental mechanisms of root tolerance to soil stresses have been rarely understood. It may be an understatement to suggest that plant roots which consume so much of the plant C have received too little attention by the scientific community.

Historically, secondary assessments have been used to evaluate plant tolerance to adverse environments. Although, shoot to root ratios, yield, and other morphological parameters reflect the relative sensitivity of a plant to environmental stress, little new information can be gained by these studies unless there are concurrent assessments of the physiological and biochemical mechanisms. This approach to our understanding of plant adaptation to stress environments would expedite the development of tolerant cultivars and conserve our resources by enhancing nutrient and water acquisition, avoidance of toxic ions and disease, conservation of limited supplies of plant C, growth regulation and many other components of plant tolerance.

Soil flooding or the large number of ubiquitous anaerobic microsites, estimated to approach 30% of the total pore space of many well-drained soils (Smith 1976), denotes an environmental limitation to the maximum genetic expression of most major crops. Crawford (1982) reported that despite extensive research on the flooding injury of agricultural crops, it is difficult to conclude what are the initial causes of flooding damage to unadapted plants. Since there are many physiological responses to flooded conditions which result in the collapse of associated tissue, those species which apparently have either the metabolic capacity to shunt the flow of electrons through the anaerobic tissue without accumulating large quantities of toxic anaerobic metabolites (e.g., acetaldehyde, ethanol, lactate, etc.) or the morphological adaptations which improve the oxygen supply to anaerobic tissue (e.g., disrupted tissue and the formation of aerenchyma pores) appear to tolerate long and short-term soil anaerobic stress. Since this chapter will be limited to the adverse effects of anaerobiosis on carbon use by crop plants, readers interested in the specifics of tolerance to anaerobiosis are referred to the excellent reviews by Jackson et al. (1982), Crawford (1982), and Hook and Crawford (1978).

Anaerobic metabolism of C compounds in the roots of stressed plants always decreases the efficient use of photoassimilates. A comparison of aerobic and anaerobic respiration indicates that oxidatively phosphorylated metabolic energy, which produces 300 kcal for each glucose molecule, is reduced to 16 kcal when oxygen is absent (Russell 1977). This incomplete catabolism of substrates results in the production of many reduced organic compounds (Crawford 1982). Some of these anaerobic metabolites are reported to be toxic. The accumulation of ethanol and perhaps other reduced endogenous compounds appears to be toxic to plant tissues as they modify cellular membranes at high concentrations (Chin and Goldstein 1977; Ruben and Rottenberg 1982). Accumulations of ethanol resulted in the plasmolysis of root tip cells of dry edible beans

(Phaseolus vulgaris L.) subjected to anaerobiosis for 24-48 h (Smucker, et al. 1978). Ethanol appears to relax plasma membranes by interfering with the sequestration and active transport of Ca (Garrett and Ross 1983). This relaxation of the membrane increases their fluidity similar to that outlined in Figure 2. Since most cell membranes become more rigid when subjected to an anaerobic environment, the metabolic administration of endogenous ethanol or perhaps other products of anaerobic metabolism, appears to temporarily relax the membrane causing the gel-liquid matrix to return to a relatively more functional state. The net accumulation rate (i.e., production minus excretion) of ethanol appears to influence the relative tolerance of adapted or unadapted plants to soil flooding (Crawford 1978). Once stressed membranes are aerated via soil drainage or the formation of gas pores in plant tissue (Jackson, et al. 1982), the matrix of plant membranes may remain fluid (susceptible cultivars), return to the former state (resistant cultivars), or achieve a more rigid matrix (characteristic of conditioned and more tolerant cultivars) (Figure 2).

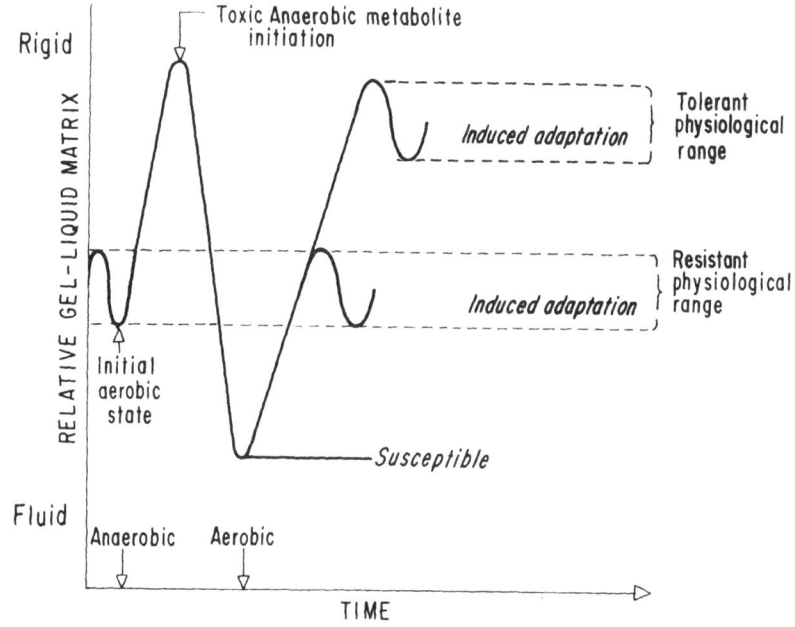

Figure 2. Diagram of root plasma membrane responses to short-term anaerobic soil conditions. Suggested membrane fluidity responses to anaerobic metabolite accumulations are summarized before, during and following anaerobic soil conditions.

It has also been reported that the maintenance of gel-liquid structures of human erythrocyte membranes depends upon glycolytic metabolism (Haest and Deuticke 1975) indicating that membrane asymmetry and fluidity were directly related to the rate of glycolysis. Applying this mechanism to root membranes suggests that most stress-induced metabolic changes altering glycolytic metabolism could result in membrane dysfunction (e.g., leakiness, cytoplasmic pH changes, excessive respiratory consumption of soluble carbohydrates, etc.). Therefore it appears that both the toxic anaerobic metabolite accumulation hypothesis of Crawford (1978) and the tissue generation of ethylene which results in the organ modification hypothesis of Jackson et al. (1982) are just two of the many mechanisms participating in the complex process of root adaptation to anaerobic soil conditions.

Roots subjected to anaerobic stresses lose greater quantities of photoassimilates (Hale et al. 1971; Rovira 1969). Although ethanol and acetaldehyde production is usually associated with anaerobic stress, these products also accumulate in plant tissue subjected to other types of stress (Kimmerer and Kozlowski 1982). Crawford (1982) reported that more than 17 hydrocarbons are accumulated and may be exuded by plants subjected to anaerobic conditions. The loss of these metabolites represents a substantial C drain from already stressed plants.

Two dry bean cultivars cultured in aseptic mist chambers (Smucker and Erickson 1976) for 17 days and labeled with $^{14}CO_2$, lost from 19-34% of their labeled compounds through the roots (Table 1). These data indicate that stressed bean plants have very leaky and actively respiring root systems which lose up to one third of their soluble photoassimilates. Although there appears to be some genetic variability in the retention capacities of nonstressed roots, adverse environments minimized the genetic influence on root leakage as anaerobic stress increased the carbon losses by the roots of these two genotypes to greater than 33%.

The most damaging effects of ethanol and other C losses to stressed root systems may be the combined attractant and stimulative effects these C sources have on motile zoospores of pathogenic microorganisms (Young et al. 1977) and fungal growth (Cannell and Jackson 1981; Smucker and Erickson 1976). The cause and effect relationships between the organic compounds of root exudates and plant disease have been summarized by Schroth and Hildebrand (1964). It is generally accepted that plant roots exude sufficient quantities of organic compounds to support large populations of microorganisms in the rhizosphere and that growth of certain microorganisms are influenced by the quality of these root exudates. Cochrane et al. (1963) reported that conidial

Table 1. Distribution of ^{14}C-label in 17 day-old Phaseolus vulgaris L. cultivars cultured in aseptic mist chambers and subjected to two 72 hour periods of anaerobic conditions. Tops were continuously labeled with $^{14}CO_2$ from day 6 to day 17 and roots were subjected to N_2 gas from days 6-9 and 13-15. Values are expressed as percent of total label in the plant and mist chamber system. Replicated experiments were repeated once (Shadan, 1980).

Treatment	Plant		Soluble Root Exudates	Respired Root Exudates
	$Bq \times 10^{-6}$ gdw^{-1}	%	%	%
Aerobic Control				
Seafarer	4.77	81.3	0.14	18.6
San Fernando	3.01	75.1	0.09	24.8
Anaerobic Control				
Seafarer	3.70	66.7	1.5	31.8
San Fernando	3.12	65.7	1.3	33.0

germination of the root pathogen Fusarium solani F. sp. phaseoli required exogenous C and N as well as a growth factor from yeast which could be replaced by ethanol, acetaldehyde, or one of several amino acids. The synergistic effects of environmental stress and root disease hypothesized by Smucker and Erickson (1976) have been confirmed for the Fusarium root rot complex of dry edible beans grown on compacted soils (Burk et al. 1980). These reports suggest that those response mechanisms controlling the disease tolerance of host plants and the pathogenicity of causative organisms appear to be modified by the degree of environmental stress. For example, a study by Duniway (1977) demonstrated that infection and disease of safflower roots by Phytophthora cryptogea was dramatically increased by a water stress of the safflower plants before inoculation with the fungus. Schneider and Pendery (1983) also showed that a preconditioning water stress of corn (Zea maize L.) early in the season combined with a second stress of the roots after infection by Fusarium moniliforme greatly increased the severity of corn stalk rot. A recent review of the subject of predisposition was presented by Colhoun (1979). Additional interdisciplinary research of abiotic and biotic plant stresses and microbiol ecology is necessary before we will be able to elucidate with confidence the complex interactions which control the dynamic root and soil interface.

VESICULAR-ARBUSCULAR MYCORRHIZAS

Mycorrhizal associations between fungi and the roots of higher plants appear to be ubiquitous. Nearly all mature plant root systems have such associations in their natural ecosystems. Vesicular-arbuscular mycorrhizas (VAM) are the most common mycorrhizal associations and are usually mutualistic as plant growth is generally, but not always, stimulated by the fungi. In addition, the fungus will neither sporulate nor grow extensively in the absence of roots.

Accordingly, the VAM fungi, which are in the family Endogonaceae (Trappe 1982), have extremely wide host ranges. Additional taxonomy of this family is reviewed by Hall and Fish (1979). Their associations appear to be present on the roots of essentially all agricultural crops grown throughout the world. VAM associations increase the growth of plants largely by enabling the infected plants to take up nutrients more efficiently. Increased uptake of K, Zn, Fe, Cu, N, Ca and P have been reported. There are several recent published discussions of the nutrition of

VAM plants (Hayman (1982), Hayman (1983), Rhodes and Gerdemann (1980), and Safir and Nelsen (1981).

It is generally believed that the extensive network of mycorrhizal hyphae is able to absorb large quantities of highly immobile soil nutrients, such as P, and then translocate these nutrients to plant roots. This phenomenon was postulated for P by Sanders and Tinker (1973), based on calculations from diffusion theory and first demonstrated experimentally by Hattingh et al. (1973). Other investigators have shown that Zn, N and S are also taken up by the hyphae and translocated to roots. The quantities of P translocated by hyphae are adequate to account for most of the increased P uptake by mycorrhizal plants. It is important to point out that the improved uptake of nutrients by mycorrhizal plants nearly always occurs when levels of the nutrients are low. Highly immobile nutrients in the soil also have low diffusion capabilities. Therefore, a network of hyphae reduces the distance nutrients must diffuse before being absorbed. In some cases, mycorrhizal roots are more efficient absorbers of nutrients than nonmycorrhizal roots. However, this increased absorbing power, is likely to be of major importance only for ions with low soil mobility.

It has been suggested that mycorrhizal roots are capable of absorbing P from different ion sources than nonmycorrhizal plants. One reason for this suggestion is that roots of mycorrhizal plants generally absorb more rock phosphate when the P content of the soil is low. Detailed studies by several workers investigating the phosphorus-specific activities in roots, indicate that both mycorrhizal and nonmycorrhizal roots are absorbing P from the soil solution. It is interesting to note that mycorrhizal hyphae have been shown by Ames et al. (1983) to be able to transport N to celery (Apium graveolens L.) roots as either nitrate or ammonium. These data suggested that the high nitrate-N, in fertilized agricultural soils, would readily diffuse to roots whereas in natural ecosystems, where ammonium is the major source of N, mycorrhizal hyphae may be necessary for increasing the uptake of the less mobile ammonium-N carrier.

The exact mechanisms controlling the rates of mycorrhizal root colonization are unknown. However, high soil P levels (about 30 µg per g) will generally result in reduced fungal colonization. It is at these high soil P levels that root P levels are also high. Generally, high root P levels are inversely correlated with root colonization by VAM.

Currently we are unable to obtain significant growth of the VAM fungi in aseptic agar culture, although several workers have been able to obtain limited growth of VAM fungi on agar with resulting infection of the roots. These infections, however, have never resulted in the production of viable fungal spores. Our inability to culture VAM fungi in vitro is probably one of the major limitations to our understanding of the biochemistry, metabolism, and genetics of both the fungi involved and the factors controlling host infection.

Well Watered Conditions Safir et al. (1971) found that mycorrhizal soybean had hydraulic conductivities 40% higher than nonmycorrhizal controls which were grown in well watered soils low in nutrition. These differences in hydraulic conductivity were eliminated by the addition of Hoagland's nutrient solution to the soil at the time of seeding. These results are also supported by the work of Nelsen and Safir (1982a) using onion (Allium cepa L.) plants, in that mycorrhizal plants had higher hydraulic conductivities than nonmycorrhizal controls when low levels of P were present in the soil at seeding time. The differences in conductivity were much larger, approximately 4-fold for onions than soybeans. These differences were eliminated by the addition of P to the soil of nonmycorrhizal plants at levels which eliminated growth differences between mycorrhizal and nonmycorrhizal plants. Levy and Krikun (1983) also found that mycorrhizal and nonmycorrhizal rough lemon seedlings of similar size had similar conductivities, transpiration rates, and water potentials, when both were supplied with a commercial nutrient solution.

It was suggested by Safir et al. (1972) that roots were the sites of the increased hydraulic conductivities of mycorrhizal plants based on comparisons of recovery rates from moderate water stress of whole plants versus plants with severed roots. Hardie and Leyton (1981) calculated root hydraulic conductivities from rates of water forced through detopped root systems using a pressure bomb. Mycorrhizal roots were much more conductive than were nonmycorrhizal roots, which indirectly supports the conclusions of Safir et al. (1972). The site of the increased conductivities of mycorrhizal plants, if present, is still uncertain, however, since other studies by Duniway (personal communication), showed that mycorrhizal safflower roots had lower hydraulic conductivities than nonmycorrhizal roots and that the conductivities decreased as mycorrhizal infection increased.

It is possible that the hyphae of mycorrhizal roots promote the uptake of soil water. Sanders and Tinker (1973) calculated potential inflow rates of water through VA mycorrhizal hyphae. These calculations indicated that if the increased transpirational water flux of mycorrhizal plants was conducted to the roots only through the hyphae, the hyphal flux rates per hyphal strand would be unrealistically high. Their calculations did not argue against a small amount of water moving through the hyphae to the roots. Indirect evidence of some hyphal water movement to the roots has been provided by Cooper and Tinker (1981) who demonstrated that a higher transpiration rate increased P translocation by two to three times the rate measured at lower transpiration rates. Nelsen and Safir (unpublished) used a split plate system which separated the hyphae from the roots. After allowing the leaves to dehydrate slightly by withholding water from the soil, small portions of the soil containing only hyphae were rehydrated. Leaf water status was monitored in a thermocouple psychrometer designed for intact leaves. Soil rehydration did not affect leaf water potentials for periods up to ten hours, after which the experiments were terminated. Experiments with onions, using tritiated water, also indicated there was no translocation by hyphae. These experiments suggest that hyphal transport of water to roots, if occurring, is not capable of altering the water balance of leaves.

Another possibility for the increased conductivity of water by mycorrhizal roots is that the mycorrhizal hyphae within the root could provide a lower resistance to water movement than the root epidermal and cortical tissues. Safir et al. (1972) argued against this mechanism by reporting that the addition of the fungal inhibitor, parachloronitrobenzene (PCNB) to the soil had no effect on the differences in the hydraulic conductivities of mycorrhizal and nonmycorrhizal soybean plants. PCNB has been shown to eliminate the increased growth and the uptake of P by mycorrhizal plants when compared to nonmycorhizal plants (Gray and Gerdemann 1969). If the hyphae within the root were providing low resistance channels, the hydraulic conductivities of mycorrhizal roots should have decreased after the application of PCNB.

Available evidence strongly suggests that hydraulic conductivities of mycorrhizal plants will be higher than those of nonmycorrhizal plants at the time of growth stimulation, especially when infection substantially increases growth over that of nonmycorrhizal plants. The higher conductivities of larger mycorrhizal plants should cause leaf water potentials to be higher than those of nonmycorrhizal controls under conditions of moderate to high evaporative demand. Stomatal conductances for vapor flux and transpiration rates should also be higher for mycorrhizal plants under the same conditions if leaf water status is at least partially controlling stomatal aperature. The work of Nelsen and Safir (1982a) supports the above hypothesis in that mycorrhizal onions had higher hydraulic conductivites, lower leaf water potentials and higher transpiration rates than nonmycorrhizal controls. Sufficient application of P to the soil of nonmycorrhizal plants eliminated size differences of the nonmycorrhizal and mycorrhizal plants as well as differences in transpiration, water potential, and hydraulic conductivity. Similar findings have been reported for rough lemon (Citrus jambhiri Lush) seedlings (Levy 1983).

Allen (1982) has shown that plant species may exhibit different water balance responses after mycorrhizal colonization. Mycorrhizal blue grama (Bouteloua gracilis) had 50% lower stomatal conductances than nonmycorrhizal controls under (well watered) conditions. These differences were present in the absence of growth differences between mycorrhizal and nonmycorrhizal plants. Allen (1982) postulated that increased cytokinin levels of mycorrhizal plants may have contributed to an increase in their stomatal conductances. Allen and Boosalis (1983) have also shown that when wheat was inoculated with two different mycorrhizal fungi, mycorrhizal plants had higher stomatal conductances than nonmycorrhizal plants when grown in either wet or dry soil.

Limited Water Conditions Mycorrhizal colonization of roots appears to modify the recovery mechanisms of plants subjected to short and long term or multiple drought cycles. Safir et al. (1971 and 1972) demonstrated that leaves of mycorrhizal soybean recovered from moderate water stress faster than those of nonmycorrhizal controls after soil rehydration. These differences were eliminated when high nutrient levels were applied to the soil at seeding time. Levy and Krikun (1980) found that mycorrhizal rough lemon plants had higher transpiration rates, stomatal conductances, and photosynthetic rates than controls of similar size during recovery from water stress. They did not find differences in water potential between mycorrhizal and nonmycorrhizal plants and suggested that hormonal differences were responsible for the larger stomatal conductances of mycorrhizal plants during recovery. Allen et al. (1981) showed that stomatal conductances of mycorrhizal blue grama plants decreased at a slower rate, and transpiration rates remained higher than smaller nonmycorrhizal plants as soil water potentials decreased. This indicates that stomatal conductances and hydraulic conductivities may be higher for mycorrhizal than nonmycorrhizal B. gracilis plants over a wide range of soil water potentials. Nelsen and Safir (1982b), exposed onion plants having mycorrhizal and nonmycorrhizal roots to seven cycles of water stress. The regime began four weeks after seeding and lasted eight weeks in order to determine the long term effects of water stress and to simulate, certain field conditions. Mycorrhizal plants at 8 and 12 weeks, which were given low (12 kg/ha) levels of P, were more drought tolerant than were nonmycorrhizal onions given high levels of P (114 kg/ha). The levels of P given to the nonmycorrhizal plants were sufficient to enable these plants to be the same size as mycorrhizal plants fertilized, minus P, under well watered conditions Nelsen and Safir (1982a). Mycorrhizal plants appeared to be more drought tolerant in that they had greater fresh and dry mass as well as higher tissue concentrations of P. The P concentrations of nonmycorrhizal plants were at very low levels (0.1 to 1.3% dry mass) which is critically low for the growth of onions (Bolgiano et al. 1983). Although phosphorus was not limiting to the mycorrhizal plants, water limited the growth of both mycorrhizal and nonmycorrhizal plants. Since leaf water potentials and transpiration rates for mycorrhizal and nonmycorrhizal plants were similar at 8 and 12 weeks, it was concluded that P deficiency was responsible for the decreased growth of nonmycorrhizal plants when compared to mycorrhizal plants.

It is not surprising that mycorrhizal plants are able to absorb P during drought, considering their extensive fungal network in the soil. It has also been shown that nonmycorrhizal onions have reduction in capacity to absorb P when water stressed (Dunham and Nye 1976). This inability to absorb P is probably magnified by the decreased diffusion rate of P in dry soil.

Two species of mycorrhizal fungi Glomus fasciculatus and Glomus mosseae differ in their effects on the water relations of wheat under drought condition (Allen and Boosalis 1983). Stomatal conductances of plants infected by both fungi were greater than those of nonmycorrhizal plants under wet or dry conditions but some leaves of G. mosseae-infected plants continued to transpire at water potentials as low as -4.1 MPa. Therefore the predominant species of mycorrhizal fungus in the soil may have considerable influence on the drought tolerance of plants.

It has been demonstrated that water availability will not only alter the effects of mycorrhizal fungi on infected plants but will influence the fungi themselves. Reid and Bowen (1979) showed the number of infections per root length of barrel medic (Medicoto truncatula G.) doubled when the soil matric potential was decreased from -0.10 to -0.19 MPa and was nearly 40 times greater at -0.19 MPa than at -1.4 MPa. Sieverding (1981) showed that infection of sorghum (Sorghum bicolor L.) and crucita (Eupatorium odoruim L.) by Glomus macrocarpus was enhanced by dry soils. Although enhancement varied with the soil type used, plant growth, which was promoted by the fungus, was greater in dry than for wet soils. Sieverding (1981) postulated that lowered nutrient availability in dry soils could be responsible for the increased infection and growth promotion. Bolgiano et al. (1983) confirmed this hypothesis by showing that mycorrhizal infection of onion in the field occurred at higher soil P levels when soil water availability decreased. Similar findings have been reported for mycorrhizal citrus plants in that decreased irrigation increased mycorrhizal infection in the field (Levy et al. 1983). It should be pointed out that water-nutrient interactions are likely to influence not only infection and growth promotion by VAM fungi but also fungal spore germination. It is also suggested that the multiple mycorrhizal fungal species and host cultivar relationships respond differently to dynamic soil nutrient and moisture conditions.

CARBON BUDGETS OF VAM ROOTS

VAM associations do improve plant growth when soil nutrients are in short supply. There have been reports, however, of significantly decreased plant growth as a result of VAM root colonization. It is, therefore, of great importance to determine the additional photoassimilate energy costs required for the maintenance of this colonization. The fact that high light intensities favor the establishment of VAM suggests that plant photosynthates are factors controlling infection (Daft and El-Giahmi 1978, Furlan and Fortin 1977). Ho and Trappe (1973) also demonstrated that some of the ^{14}C applied to the leaves of infected plants was translocated into the associated mycorrhizal fungus.

Pang and Paul (1980) reported that mycorrhizal faba bean (Vicia faba L.) transferred 47% of the fixed ^{14}C and the nonmycorrhizal plants transferred 37% of the fixed ^{14}C to the soil. Most of the difference was accounted for by increased rhizosphere respiration in that the mycorrhizal root system of faba bean produced twice as much $^{14}CO_2$ as did the nonmycorrhizal root system. Both mycorrhizal and nonmycorrhizal root systems contained active Rhizobium and there were no differences in whole plant dry weights. The authors suggested that since the mycorrhizal plants fixed more CO_2 than did the nonmycorrhizal plants they were able to compensate for the increased root respiration. Increased photosynthesis (up to 68%) of mycorrhizal versus nonmycorrhizal B. gracilis plants has also been demonstrated by Allen et. al. (1981) in the absence of mycorrhizal stimulated plant growth. Approximately two thirds of the photosynthate partitioned to the split-roots of citrus seedlings was allocated to mycorrhizal roots (Koch and Johnson 1984).

Snellgrove and Splittstoesser (1982) found with mycorrhizal and nonmycorrhizal leek (Allium porrum) of similar size that 7% more of the total fixed C was transferred from shoot to root in the mycorrhizal versus nonmycorrhizal plants. The mycorrhizal plants had a higher photoassimilation rate on a leaf dry matter basis than did nonmycorrhizal plants; however, both had similar photoassimilation rates on a leaf area basis. The authors suggested that increased hydration and leaf area of mycorrhizal plants may help to offset the C drain imposed by the fungus. It should be pointed out the nonmycorrhizal plants were given additional P in order to increase their size.

Silsbury et al. (1983) found that when CO_2 fluxes were measured on a whole plant basis for 24 mycorrhizal and nonmycorrhizal subterranean clover swards (Trifolium subterraneum L.) had similar C economies and responded to changes in light intensity in a similar manner. Infection of the swards was already well established and there were no differences in size between mycorrhizal and nonmycorrhizal swards. They also suggested that decreases in dark respiration with time are considerable and that exposure to prolonged darkness before measurement will further depress CO_2 evolution. Therefore, respiration measurements should be conducted under both standardized conditions and over time periods of 24 h or more if accurate C balances are to be made.

Additional complications arise when infection by the mycorrhizal fungus results in plant growth depressions. Bowen (1978) and Smith (1980) suggested that when soil P is limiting, mycorrhizal infection enhances plant growth and when P is abundant, mycorrhizal infection is depressed. If P availability is adequate for maximum plant growth but is not high enough to depress mycorrhizal infection completely, plant growth depressions are likely to occur. It is suggested that these reductions in growth are caused in part by the competition for carbohydrates between the fungus and the plant. It is uncertain whether the fungal colonization increases the demand of the root system, reducing the amount of photoassimilates to the shoot, or whether the allocation of C to the root system remains constant and the carbon utilized by the fungus is at the expense of the root system. The conversion of photosynthetic products into fungal metabolites such as trehalose or mannitol has not been demonstrated for VAM systems. It has been demonstrated that soybean plants, which are not P limited at early growth stages, will be decreased in size by mycorrhizal infection initially but as soil P levels decrease mycorrhizal plants will become similar or even exceed the size of nonmycorrhizal plants Bethenfalvay et al. (1982). It is clear that water and P availability, along with infection stage, photosynthetic rates, and plant growth stage, interact in determining the C balance of mycorrhizal plants. It is essential, then, that these factors be accounted for when physiological comparisons are made between mycorrhizal and nonmycorrhizal plants.

The effects of mycorrhizal infection on root exudation are also important in analyzing C budgets; however, only a few studies relating to this subject have been conducted. Initial work by Graham et al. (1981) and Ratnayke et al. (1978) showed that increased tissue P concentrations reduced the rate of metabolite leakage of plant roots. Since reduced VAM infection is correlated with increased tissue P levels, it was postulated that the organic nutrients which leaked from cortical cells were critical in determining the degree of VAM infection. As VAM infection increases tissue P in the root leakage of organic compounds from infected roots might be expected to decrease. This hypothesis, however, needs to be investigated further before the combined and separate influences of tissue P levels and mycorrhizal infection upon root exudation are known.

CURRENT DEVELOPMENTS IN RHIZOSPHERE ANALYSES

Future developments in our collective knowledge of plant root systems will be directly related to the creative combinations of current techniques and the development of new and innovative methods of measurement which may be economically incorporated into multidisciplinary research programs. There exists a myriad of direct and indirect methods for studying root systems which are summarized in the timely publication by Bohm (1979). Renewed interest in root research during the past two decades has resulted in the completion of three international root symposia. The First and Second International Root Symposia were held at Potsdam, Germany in 1971 and Brataslava, Czechoslovakia in 1981 (Brouwer et al. 1981), and the First International Root and Soil Interface Symposia was convened at Oxford, England in 1978 (Harley and Russell 1978). Many specific methods for measuring root ecology, root morphology or root physiology have been elegantly reported elsewhere. However, there are only a few direct methods for the field evaluation of root morphology which can be economically applied to genetic and environmental stress experiments involving large plant root communities. Since there are very few plant breeding programs which select superior genotypes under maximum yield management (Lambert 1983), it is suggested that the direct measurement of plant root systems and their associated microflora must be incorporated into current plant breeding and soil management research programs before superior genetic cultivars and maximum-yield management systems can be developed.

Recent developments in the economical and quantitative evaluation of spatial and temporal root development have the potential for partially resolving the dilema of expensive root research. Combinations of the field installation of many minirhizotron tubes with current state-of-the-art video microcammeras developed by Saunders and Brown (1978) and improved by Upchurch and Ritchie (1983) have tremendous potential for evaluating in situ growth at depths greater than 10-15 cm for many plant and soil communities. Although the initial investment for equipment ranges from $15,000 - $20,000, the manual and technical costs of operation are minimal (Ritchie, personnal communication). An additional problem with this approach is the unknown reason for the absence of roots in the surface few centimeters of soil above minirhizotrons, installed at angles which are less than perpendicular to the soil surface.

The direct method of extracting root and soil samples from field experiments throughout the growing season is laborious. Yet, if the roots are quantitatively separated trom the soil, this approach provides excellent information. Destructive sampling procedures have been accepted in the past and, we believe, will continue to be the best approach for evaluating interactions between plant roots and their soil environments. This rationale prompted the development of a root and soil profile sampler (Srivastava et al. 1982) and a rapid and quantitative method for separating roots and other organic debris from mineral soils (Smucker et al. 1982). The hydropneumatic elutriation manifold system provides a relatively inexpensive apparatus which improves the efficiency for recovering roots from mineral soils, reported to be 60-80% (Martin and Puckridge 1982) to greater than 99%. This system also removes operator errors, is equally efficient tor all types of mineral soils and preserves the original morphology of most root samples. It has also been used to quantify mycorrhizal colonization of root systems growing

in different soil types receiving various tillage
treatments (Mulligan et al. 1984). Root and soil profiles
of large field experiments can be sampled and the roots
quantitatively separated from the soil in two to four days.

Currently, with the application of these methods, the
greatest impediment to our root research programs is the
qualitative analyses of randomly branched root morphology.
Although computer aided measurements of root length have
been reported (Costigan et al. 1982, Voorhees et al.
1980), there is a need to establish image processor
rhizometers which will discriminate between plant roots,
their associated microflora and the accompanying organic
debris. This method would also assist in the measurement
of viable and nonviable morphological components of roots
which have been quantitatively extracted from field
experiments. The combination of direct methods of
extraction and quantitative recovery with innovative
methods of analyses will certainly assist plant and soil
scientists in delineating the combined and separate
influences of genotype and the environment on the
development and maintenance of plant root systems.

IMPERATIVES FOR RHIZOSPHERE RESEARCH

The establishment of a quantitatively based root
analysis component to many of our plant and soil research
programs has great economic and production potential. It
is essential that the most efficient C balance between
roots and shoots be determined for specific environmental
conditions. The dynamics of C utilization in the
rhizosphere, root respiration pathways, and the
interactions of inorganic ions and hormones which influence
root branching are essentially unknown. It is important to
measure C losses by plant roots resulting from complex
associations of root and soil microorganisms which
contribute both positively and negatively to the efficiency

of the whole plant. It appears reasonable to assume that plant root systems including associated microorganisms, could be developed to retain greater proportions of C compounds.

Since root development can be altered by fertilization and nutrient availability, efforts to improve the soil physical conditions and enhance the symbiotic microbial relationships should be designed to investigate the C efficiency and yield of agricultural crops. Genetic control of root branching, membrane elasticity of root cells, enzyme induction, hormone ratios and the ideal C and N balance of whole plant systems and their associated symbiotic relationships with soil microorganisms are essentially unknown for most economically important crop species.

Discovery of the relative efficiencies of the different orders of branching and other components of plant root systems in relation to the absorption energies of soil ions, water and gases as outlined by Clarkson (1974) will provide valuable information for designing the ideal root system. The use of systems models for predicting the relative efficiencies of these root components with respect to their diurnal and seasonal requirements for C, ions, water, air, etc. would greatly contribute to plant breeding and soil management programs. Knowledge of the ontogenetic and phenotypic developments of root systems grown under static and dynamic environmental conditions would provide valuable information which could be used to determine the mechanisms involved in root tolerance. The dynamics of root turnover rates during stress and nonstress conditions should be investigated to determine the relative efficiency of water absorption by viable and physiologically dead roots. Since a substantial number of reports indicate that dead roots absorb water, more research should be directed toward the relationship between plant or leaf water potential and the absorption surfaces of viable and nonviable root systems. The specific photoassimilate requirements, cytokinin production, ion absorption, water absorption, C leakage and respiration rates for each order of root branching during development, maintenance, death and disappearance should also be investigated. Until these and many more biochemical, morphological and physiological details of the plant root and their associated microbial ecology are understood, little progress will be made in genetically designing a carbon-efficient root system for specific soil types and specific levels of soil management.

SUMMARY AND CONCLUSIONS

Past morphological investigations of plant root systems and their associated microorganisms have contributed to our knowledge of root responses to soil environments. Reports of C utilization by plant roots and the influence of stress environments on the allocation of C have been summarized to emphasize another component of the complex root-soil interface. A greater understanding of the contributions by mycorrhizal fungi and other associated microorganisms to plant C efficiency and root system functions is essential. Our knowledge of the biochemical and physiological responses of roots and their microbial ecology to specific environmental conditions has been limited by the paucity of standardized measurements. Many semiquantitative yet expedient techniques used to measure the morphological root responses of genotypes to specific soil environments are incomplete and in many cases highly inaccurate. Therefore, it is necessary to develop new and quantitative methods for measuring ontogenetic developments of root systems and the microbial ecology of the rhizosphere so that the mechanisms controlling root tolerance to adverse soil environments may be more completely defined. The development of superior cultivars and high yielding management systems will be greatly curtailed until these mechanisms are better understood.

LITERATURE CITED

1. Allen, M. F. 1982. Influence of vesicular-arbuscular mycorrhizae on water movement through Bouteloua gracilis (H.B.K) Lag Ex Steud. New Phytol. 91:191-196.

2. Allen, M. F. and M. G. Boosalis. 1983. Effects of two species of V. A. mycorrhizal fungi on drought tolerance of winter wheat. New Phytol. 93:67-76.

3. Allen, M. F., W. K. Smith, T. S. Moore, Jr., and Christensen. 1981. Comparitive water relations and photosynthesis of mycorrhizal and nonmycorrhizal Bouteloua gracilis H. B. K., Lag Ex. Steud. New Phytol. 88:683-693.

4. Ames, R. N., C. P. P. Reid, L. K. Porter, and C. Cambardella. 1983. Hyphal uptake and transport of nitrogen from two ^{15}N-Labelled sources by Glomus mosseae, a vesicular-arbuscular mycorrhizal fungus. New Phytol. 95:381-396.

5. Arkin, G. F. and H. M. Taylor. 1981. Modifying the root environment to reduce crop stress. ASAE Monograph No. 4. Amer. Soc. Agri. Eng. St. Joseph, MI. 407 pp.

6. Barber, D. A. and J. K. Martin. 1976. The release of organic substances by cereal roots into soil. New Phytol. 76:69-80.

7. Barber, D. A. and K. B. Gunn. 1974. The effect of mechanical forces on the exudation of organic substances by the roots of cereal plants grown under sterile conditions. New Phytol. 73:39-45.

8. Bethenfalvay, G. J., M. S. Brown and R. S. Pacovsky. 1982. Parasitic and mutualistic associations between a mycorrhizal fungus and soybean: development of the host plant. Phytopathology. 72:889-893.

9. Bohm, W. 1979. Methods of studying root systems, Springer-Verlag, Berlin. 188 pp.

10. Bolgiano, N. C., G. R. Safir, and D. D. Warncke. 1983. Mycorrhizal infection and growth of onion in the field in relation to phosphorus and water availability. Amer. Soc. Hort. Sci. 108(5):819-825.

11. Bowen, G. D. 1978. Dysfunction and shortfalls in symbiotic responses. In Plant Disease Vol. III. J. C. Horsfall, and E. B. Cowling (eds.). Academic Press, NY. pp. 231-256.

12. Boyer, J. S. 1982. Plant productivity and environment. Science 218:443-448.

13. Brouwer, R., O. Gasparikova, J. Kolek and B. C. Loughman. 1981. Structure and function of plant roots. Proceedings of the 2nd Int. Symposium, Bratislava, Czech. Martinus Nijhoff. 415 pp.

14. Burke, D. W., D. E. Miller and A. W. Barker. 1980. Effects of soil temperature on growth of

beans in relation to soil compaction and Fusarium root rot. Phytopathology. 70:1047-1049.

15. Cannell, R. Q. and M. B. Jackson. 1981. Alleviating aeration stresses. In H. M. Taylor and G. F. Arkin, (eds). Modifying the root environment to reduce crop stresses. Chapt. 5. Amer. Soc. Agr. Eng., St. Joseph, MI. pp. 141-192.

16. Carr, D. J. and J. S. Pate. 1967. Aspects of the biology of aging. Symp. Soc. Exp. Biol. 21:559-562.

17. Carson, E. W. 1974. The plant root and its environment. Univ. Press of Virginia. 683 pp.

18. Chatterton, N. J. and J. E. Silvius. 1979. Photosynthetic partitioning into starch in soybean leaves. Plant Physiol. 64:749-753.

19. Chin, J. H. and D. B. Goldstein. 1977. Effects of low concentrations of ethanol on the fluidity of spin-labeled erythrocyte and brain membranes. Mol. Pharmacol. 13:435-441.

20. Clarkson, D. T. 1974. Ion transport and cell structure in plants. Wiley and Sons, New York. 334 pp.

21. Cochrane, J. C., et al. 1963. Spore germination and carbon metabolism in Fusarium solani. I. Requirements for spore germination. Phytopathology. 53:1155-1160.

22. Coleman, D. C. 1976. A review of root production processes and their influence on soil biota in terrestrial ecosystems. In J. M. Anderson and A. Macfodyen (eds.). The role of terrestrial and acquatic organisms in decomposition processes. Blackwell Scientific Publications. Oxford. pp. 417-434.

23. Colhoun, J. 1979. Predisposition by the environment. In Plant Disease, Volume IV. Ed., J. G. Horsfall and E. B. Cowling. Academic Press, N.Y. pp. 75-96.

24. Cooper, K. M. and P. B. Tenker. 1981. Translocation and transfer of nutrients in vesicular-arbuscular mycorrhizas. IV. Effects of environmental variables on movement of chosphorus. New Phytol. 88:327-339.

25. Costigan, P. A., J. A. Rose and T. McBurney. 1982. A micro-computer based method for the rapid and detailed measurement of seedling root systems. Plant and Soil 69:305-309.

26. Crawford, R. M. M. 1978. Metabolic adaptations to anoxia. In D. D. Hook and R. M. M. Crawford (eds.).Plant life in anaerobic environments. Ann Arbor Science, Ann Arbor, MI. pp. 119-136.

27. Crawford, R. M. M. 1982. Physiological responses to flooding. In Encyclopedia of plant physiology, New series, Physiological plant ecology II. Vol. 12B. pp. 453-477.

28. Curl, E. A. 1982. The rhizosphere: Relation to pathogen behavior and root diesease. Plant Dis. 66:624-630.

29. Daft, M. J. and A. A. El-Giahmi. 1978. Effect of arbuscular mycorrhiza on plant growth VIII. Effects of defoliation and light on selected hosts. New Phytol. 80:365-372.

30. DeWit, C. T., et al. 1978. Simulation of assimulation, respiration and transpiration of crops. Wiley and Sons, N.Y.

31. Dixon, R. O. D., Y. M. Berlier and P. A. Lespinat. 1981. Respiration and nitrogen fixation in nodulated roots of soybean and pea. Plant Soil. 61:135-143.

32. Dunham, A. J. and P. H. Nye. 1976. The influence of soil water content on the uptake of ions by roots. III Phosphate, potassium, calcium and magnesium uptake and concentration gradients in soil. J. Appl. Ecol. 13:967-984.

33. Duniway, J. M. 1977. Predisposing effect of water stress on the severity of Phytophthora root rot in safflower. Phytopathology. 67:884-889.

34. Furlan, V. and J. A. Fortin. 1977. Effects of light intensity on the formation of vesicular-arbuscular endomycorrhizas on Allium cepa by Gigaspora calospora. New Phytol. 79:335-340.

35. Garrett, K. M. and D. H. Ross. 1983. Effects of in vivo ethanol administration on Ca^{2+}/Mg^{2+} ATPase and ATP-dependent Ca^{2+} uptake activity in synaptosomal membranes. Neurochem. Res. 8:1013-1028.

36. Geiger, D. R. and R. T. Giaquinta. 1982. Translocation of photosynthate. In Govindjee (ed.) Photosynthesis Vol. II. Development, carbon metabolism, and plant productivity. Academic Press. pp. 345-386.

37. Graham, J. H., R. T. Leonard and J. A. Menge. 1981. Membrane-mediated decrease in root exudation responsible for phosphorus inhibition of vesicular-arbuscular mycorrhiza formations. Plant Physiol. 68:548-552.

38. Gray, L. E. and J. W. Gerdemann. 1969. Uptake of phosphorus-32 by vesicular-arbuscular mycorrhizae. Plant Soil 30:415-422.

39. Haest, C. W. M., and B. Deuticke. 1975. Experimental alteration of phospholipid-protein interactions within the human erythrocyte membrane. Biochim. Biophys. Acta. 401:468-480.

40. Hale, M. G. et al. 1971. Factors affecting root exudation. Adv. Agron. 23:89-109.

41. Hale, M. G. and L. D. Moore. 1979. Factors affecting root exudation II: 1970-1978. Adv. Agron. 31:93-124.

42. Hall, I. R. and B. J. Fish. 1979. A key to the Endogonaceae. Trans. Br. Mycol. Soc. 73:261-270.

43. Hansen, G. K. and C. R. Jensen. 1977. Growth and maintenance respiration in whole plants, tops and roots of Lolium multiflorum. Physiol. Plant 39:155-164.

44. Hardie, K. and L. Leyton. 1981. The influence of vesicular-arbuscular mycorrhiza on growth and water relations of red clover. I. In phosphate deficient soil. New Phytol. 89:599-608.

45. Harley, J. L. and R. S. Russell. 1978. The root soil interface. Proc. of the First Inter. Root-Soil Interface Symposium, Oxford. Academic Press, London. pp. 442

46. Hayman, D. S. 1982. Influence of soils and fertility on activity and survival of vesicular-arbuscular mycorrhizal fungi. Phytopathology. 72:1119-1125.

47. Hayman, D. S. 1983. The physiology of vesicular-arbuscular endomycorrhizal symbiosis. Can. J. Bot. 61:944-963.

48. Hattingh, M. S., L. E. Gray and J. W. Gerdemann. 1973. Uptake and translocation of ^{32}P-labeled phosphate to onion roots by endomycorrhizal fungi. Soil Sci. 116:383-387.

49. Hesketh, J. D., J. T. Woolley and D. B. Peters. 1982. Predicting photosynthesis. In Govindjee (ed.) Photosynthesis Vol. II. Development, carbon metabolism, and plant productivity. Academic Press. pp. 387-418.

50. Ho, I. and J. M. Trappe. 1973. Translocation of ^{14}C from Festica plants to their endomycorrhizal fungi. Nat. New Biol. 244:30-31.

51. Hook, D. D. and R. M. M. Crawford. 1978. Plant life in anaerobic environments. Ann Arbor Science, Ann Arbor. 548 pp.

52. Jackson, M. B., B. Herman and A. Goodenough. 1982. An examination of the importance of ethanol in causing injury to flooded plants. Plant Cell Environ. 5:163-172.

53. Johnen, B. G. and D. R. Sauerbeck. 1977. A tracer technique for measuring growth, mass and microbial breakdown of plant roots during vegetation. Soil organisms as components of ecosystems. Ecol. Bull. Stockholm 25:366-373.

54. Jordan, C. F. 1982. Amazon rain forests. Am. Sci. 70:394-401.

55. Kimmerer, T. W. and T. T. Kozlowski. 1982. Ethylene, ethane, acetaldehyde, and ethanol production by plants under stress. Plant Physiol. 69:840-847.

56. Koch, K. E. and C. R. Johnson. 1984. Photoassimilate partitioning in split-root citrus

seedlings with mycorrhizal and nonmycorrhizal root systems. Plant Physiol. 75:26-30.

57. Kuiper, D. 1983. Genetic differation in Plantago major: Growth and root respiration and their role in phenotypic adaptation. Physiol. Plant. 57:222-230.

58. Lambers, H., D. A. Day and J. Azcon-Bieto. 1983. Cyanide-resistant respiration in roots and leaves. Measurements with intact tissues and isolated mitochondria. Physiol. Plant. 58:148-154.

59. Lambert, R. J. 1983. Hybrid selection under maximum yield management. Better Crops With Plant Food. Fall p. 20.

60. Lang, A. and M. R. Thorpe. 1984. Analyzing partitioning in plants. Plant, Cell and Environ. (preprint).

61. Laties, G. G. 1982. The cyanide-resistant alternative path in higher plant respiration. Ann. Rev. Plant Physiol. 33:519-55.

62. Levy, Y., J. Dodd and J. Krikun. 1983. Effect of irrigation water salinity and rootstock on the verticle distribution of vesicular arbuscular mycorrhiza in citrus roots. New Phytol. 95:397-403.

63. Levy, Y. and J. Krikun. 1980. Effect of vesicular-arbuscular mycorrhiza on citrus jambhiri water relations. New Phytol, 85:25-31.

64. Martin, J. K. 1977. Factors influencing the loss of carbon from wheat roots. Soil Biol. Biochem. 9:1-7.

65. Martin, J. K. and D. W. Puckridge. 1982. Carbon flow through the rhizosphere of wheat crops in south Australia. In J. R. Freney and I. E. Galbally (eds).Cycling of carbon, nitrogen, sulfur and phosphorus in terrestrial and acquatic ecosystems. Springer-Verlag, Berlin. pp. 77-82.

66. Minchin, F. R. and J. S. Pate. 1973. The carbon balance of a legume and the functional economy of its root nodules. J. Exp. Bot. 24:259-271.

67. Moldau, H. and A. Karolin. 1977. Effect of reserve pool on the relationship between respiration and photosynthesis. Photosynthetica 11:38-47.

68. Mulligan, M. R., A. J. M. Smucker, G. F. Safir. 1984. Tillage modifications of root colonization by VAM fungi of dry edible bean root systems. Crop Science (In Press).

69. Nadkarni, N. M. 1981. Canopy roots: Convergent evolution in rainforest nutrient cycles. Sci. 214:1023-1024.

70. Nelsen, C. E. and G. R. Safir. 1982a. The water relations of well-watered mycorrhizal and non-mycorrhizal onion plants. Amer. Soc. Hort. Sci. 107(2):231-234.

71. Nelsen, C. E. and G. R. Safir. 1982b. Increased drought tolerance of mycorrhizal onion plants caused by improved phosphorus nutrition. Planta. 154:407-413.

72. Pang, P. C. and E. A. Paul. 1980. Effects of vesicular-arbuscular mycorrhiza on ^{14}C and ^{15}N distribution in nodulated faba beans. Can. J. Soil Sci. 60:241-250.

73. Pate, J. S., D. B. Layzell and D. L. McNiel. 1979. Modeling the transport and utilization of carbon and nitrogen in a nodulated legume. Plant Physiol. 63:730-737.

74. Pavlychenko, T. K. 1937. Quantitative study of the entire root system of weed and crop plants under field conditions. Ecology 18:62-79.

75. Prikryl, Z. and V. Vancura. 1980. Root exudates of plant:VI wheat root exudation as dependent on growth, concentration gradient of exudates and the presence of bacteria. Plant Soil 57:69-83.

76. Ratnayke, M., R. T. Leonard and J. A. Menge. 1978. Root exudation in relation to supply of phosphorus and its possible relevance to mycorrhizal formation. New Phytol. 81:543-552.

77. Reid, C. P. P. and G. D. Bowen. 1979. Effects of soil moisture on V/A mycorrhizae formation and root development in medicago. In The soil-root interface. J. L. Harley and R. Scott Russell. Academic Press, New York. pp. 211-219.

78. Rhodes, L. H. and J. W. Gerdemann. 1980. Nutrient translocation in vesicular arbuscular mycorrhizae. In C. B. Cook, P. W. Pappas, and E. D. Rudolph (eds.). Cellular interactions in symbiosis and parasitism. Ohio State University Press. pp. 173-195.

79. Rovira, A. D. 1969. Diffusion of carbon compounds away from wheat roots. Aust. J. Biol. Sci. 22:1287-1290.

80. Ruben, E. and H. Rottenberg. 1982. Ethanol-induced injury and adaptation in biological membranes. Fed. Proc. 41:2465-2471.

81. Russell, R. S. 1977. Plant root systems, their function and interaction with the soil. McGraw-Hill, London. 298 pp.

82. Safir, G. R., J. S. Boyer, and Gerdemann. 1971. Mycorrhizal enhancement of water transport in soybeans. Science 172:581-583.

83. Safir, G. R., J. S. Boyer and J. W. Gerdemann. 1972. Nutrient status and mycorrhizal enhancement of water transport in soybean. Plant Physiol. 49:300-303.

84. Safir, G. R. and C. E. Nelsen. 1981. Water and nutrient uptake by vesicular-arbuscular mycorrhizal plants. In R. F. Myers, R. F. Bartha, and W. Busscher (eds.) Mycorrhizal associations and crop production. . New Jersey

Agricultural Experiment Station Report No. RO4400-01-81. pp. 25-31.

85. Sanders, J. L. and D. A. Brown. 1978. A new fiber-optic technique for measuring root growth of soybeans under field conditions. Agron. J. 70:1073-1076.

86. Sauerbeck, D. R. and B. G. Johnen. 1977. Root formation and decomposition during plant growth. In Soil organic matter studies - Int. symp. on soil organic matter studies. Brunswick F.D.A. 1976.

87. Saunders, F. E. and P. B. Tinker. 1973. Mechanism of absorption of phosphate from soil by Endogone mycorrhizas. Nature. 233:278-279.

88. Schneider, R. W. and W. E. Pendry. 1983. Stalk rot of corn: Mechanism of predisposition by an early season water stress. Phytopathology. 73:863-871.

89. Schroth, M. N. and D. C. Hildebrand. 1964. Influence of root exudates on root infecting fungi. Annu. Rev. Phytopathol. 2:101-132.

90. Schumacher, T. E. and A. J. M. Smucker. 1984a. Carbon transport and root respiration responses of Phaseolus vulgaris genotypes to localized anoxia (Submitted to Plant Physiol).

91. Schumacher, T. E. and A. J. M. Smucker. 1984b. Ion uptake and respiration of dry bean roots subjected to localized anoxia. (Submitted to Plant Soil).

92. Shadan, M. M. 1980. Fixation, translocation and root exudation of ^{14}C-labeled assimilates by two genotypes of Phaseolus vulgaris, L. subjected to root anaerobiosis. M. S. Thesis. Mich. State Univ. 54 pp.

93. Sieverding, E. 1981. Influence of soil water regimes on VA mycorrhiza I. Effect on plant growth, water utilization and development of mycorrhiza. J. Agron. & Crop Sci. 150:400-421.

94. Silsbury, J. H., S. E. Smith and A. J. Oliver. 1983. A comparison of growth efficiency and specific rate of dark respiration of uninfected and vesicular-arbuscular mycorrhizal plants of Trifolium subterraneum L. New Phytol. 93:555-566.

95. Smith, A. M. 1976. Ethylene production by bacteria in reduced microsites and some implications to agriculture. Soil Biol. Biochem. 8:293-298.

96. Smith, Sally S. E. 1980. Mycorrhizas of autotrophic higher plants. Biol. Rev. 55:475-510.

97. Smucker, A. J. M. and F. Adler. 1980. Accumulation and loss of toxic assimilates by plant roots. Agron. Abstr. 72:93.

98. Smucker, A. J. M. and A. E. Erickson. 1976. An aseptic mist chamber system: A method for measuring root processes of peas. Agron. J. 68:59-62.

99. Smucker, A. J. M., B. D. Knezek and G. R. Hooper. 1978. Influence of short-term oxygen stress upon the translocation of zinc in navy beans. (Phaseolus vulgaris L.) In J. L. Harley and R. S. Russell. (eds).The root-soil interface. Academic Press, London. p. 434.

100. Smucker, A. J. M., S. L. McBurney and A. K. Srivastava. 1982. Quantitative separation of roots from compacted soil profiles by the hydropneumatic elutriation system. Agron. J. 74:500-503.

101. Snellgrove, R. C., W. E. Splittstoesser. 1982. The distribution of carbon and the demand of the fungal symbiont in Leek plants with vesicular arbuscular mycorrhizas. New Phytol. 92:75-87.

102. Srivastava, A. K., A. J. M. Smucker and S. L. McBurney. 1982. A mechanical sampling method of multiple soil-plant root studies. ASAE transactions 25:868-871.

103. St. Jon, T. V. and D. C. Coleman. 1983. The role of mycorrhizae in plant ecology. Can. J. Bot. 61:1005-1014.

104. Stribley, D. P., Tinker, P. B., and Rayner J. H. 1980a. Relation of internal phosphorus concentration and plant weight in plants infected by vesicular-arbuscular mycorrhizas. New Phytol. 86:261-266.

105. Stribley, D. P., Tinker, P.B., and Snellgrove, R. C. 1980b. Effects of vesicular-arbuscular mycorrhizal fungi on the relation of plant growth, internal phosphorus concentration and soil phosphate analyses. J. Soil Sci. 31:655-672.

106. Torrey, J. G. and D. T. Clarkson. 1975. The development and function of roots. Academic Press, London.

107. Trappe, J. M. 1982. Synoptic keys to the genera and species of Xygomycetous mycorrhizal fungi. Phytopathology. 72:1102-1108.

108. Upchurch, D. R. and J. T. Ritchie. 1983. Root observations using a video recording system in minirhizotrons. Agron. J. 75:1009-1015.

109. Voorhees, W. B., V. A. Carlson and E. A. Hallauer. 1980. Root length measurement with a computer-controlled digital scanning microdensitometer. Agron. J. 72:847-851.

110. Warembourg, F. R. and G. Billes. 1979. Estimating carbon transfers in the plant rhizosphere. In J. L. Harley and R. S. Russell. (eds). The soil-root interface. Academic Press. p. 183.

111. Young, B. R., F. J. Newhook and R. N. Allen.
1977. Ethanol in the rhizosphere of seedlings of
<u>Lupinus</u> <u>angustifolius</u>. L. New Zeland J. of Bot.
<u>15</u>:189-191.

THE ROLE OF MICROORGANISMS IN THE SOIL NITROGEN CYCLE

M. SCOTT SMITH AND CHARLES W. RICE*

In attempting to review microbial transformations of soil N in one brief chapter, ambition is perhaps as important a qualification as good sense and knowledge. This subject currently occupies the attention of a large percentage of all those active in soil microbiology and microbial ecology. Monographs, symposia, and stacks of research articles related to N transformations appear at a rate considerably beyond any one individual's capacity to absorb them. At least some of what is reported here will probably be made obsolete by current developments in the more active research areas: certain aspects of N_2-fixation, denitrification, and whole system analysis. Multiple volumes have been devoted to individual segments of the cycle. We begin by abandoning all pretense of comprehensive coverage. We will describe some of the significant microbial transformations and fluxes of soil N, which relate to the main themes of this book. Our review will concentrate on specific processes, for which mechanisms and organisms will be considered. Although some consideration of overall ecosystem N dynamics will be presented here, the reader interested in an integrative approach to N cycling is referred to several excellent reviews in a recent book (Clark and Rosswall 1979). We could not hope to provide a more eloquent or entertaining overview of N cycle studies than that by F. E. Clark in the volume just cited.

One might wonder what motivates all of this excitement about N cycling. We can offer economic, environmental and scientific justifications. Economically, we should note that one of the largest monetary costs of crop production is

*Department of Agronomy, University of Kentucky, Lexington, Kentucky 40546

N fertilizer. For example, approximately 60 dollars per
hectare is the cost of N fertilizer for corn production in
the U.S. In less developed countries, where production of
staple foods and protein is often a more immediate concern
than profits, the cost of fertilizer N may be prohibitive.
Environmentally, soil N transformations have been related to
several problems in the past decade. These include contamination of water by NO_3^- run-off and leaching from soils,
formation of carcinogenic nitrosamines from NO_2^-, and depletion of the atmosphere's protective ozone layer by soil-
evolved N_2O (reviewed below). Scientifically, the soil N
cycle has captured the interest of both basic and applied
microbiologists, soil scientists, and ecologists, in part
because of the incredible diversity of the participating
organisms, the fascinating complexity of their interrelationships and their role in N ecosystem dynamics.

The Nitrogen Cycle

The importance of N in biological systems and the
complex diversity of N transformations can be attributed to
the chemical properties and the multiple biochemical functions of this element. First, because the N atom can form
multiple covalent bonds (like carbon) it serves well in a
structural role. Molecules containing N including the
structural protein, keratin, and the polymerized amino
sugar, chitin, help to determine and maintain the gross
structure of organisms. Also on a finer scale, covalent and
ionic bonds of N are, in part, responsible for maintaining
secondary and tertiary structure of proteins and nucleic
acids. Second, N is suited to catalytic functions. This
role of N is partly indirect, since it maintains enzyme
structure as just described, but more immediate involvement
of N in catalysis also occurs in Schiff's base reactions and
in ligand formation with metal cofactors. Third, and perhaps
most significant for our purposes, the nature of the N atom
permits it to exist in a range of stable oxidation states,
from reduced NH_4^+ to oxidized NO_3^-. The oxidation state of N

not only influences its biotic and abiotic relationships, but also because many of the oxidation states are kinetically (but not necessarily thermodynamically) stable, numerous oxidation-reduction reactions are available for biological exploitation. Redox reactions, of course, are a source of biological energy. Some of the reduced forms of N, specifically NH_4^+ and NO_2^-, can be oxidized with the generation of useful forms of energy. A rather unique and limited group of bacteria, the nitrifiers, have evolved to exploit this energy source. Oxidized forms of N, on the other hand, are available as electron acceptors and have a function analogous to O_2 in aerobic organisms. A limited number of bacterial genera make use of oxidized N as an alternate electron acceptor to permit efficient energy conversion in the absence of O_2. In this way the N cycle may seem to be filled with biochemical and microbial peculiarities. Yet these are not esoteric oddities, they are required for the cycling of one of the most important plant and animal nutrients.

Our version of the N cycle is offered in Figure 1. The large number of reactions involved is apparent, but the diagram actually has been simplified slightly to avoid chaos. Omitted processes which should be mentioned include: some of the chemical reactions of inorganic N forms (particularly NO_2^-), additions of N gases to the atmosphere from vulcanism, N_2O production by nitrifiers during NH_4^+ oxidation, and N export and import by fauna. An important deficiency in the present context is that the critical role of soil fauna is not indicated. This role could be best diagrammed as internal cycles and regulators within the pool labeled "soil organic N and biomass." Fauna are important in the degradation of plant residues and microbial biomass, and therefore influence N mineralization and immobilization (Gould et al. 1981, Sinclair et al. 1981, Syers et al. 1974, Ineson et al. 1982, Woods et al. 1982). Some of these faunal roles are considered in other chapters of this book, particularly chapters 8 and 9.

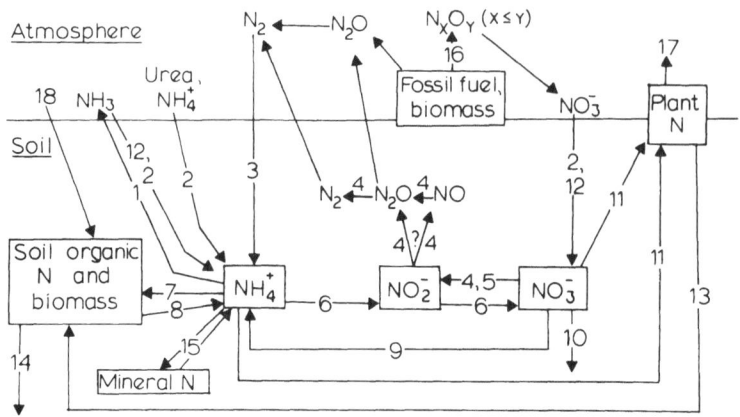

FIGURE 1. The nitrogen cycle: 1. NH₃ volatilization,
2. fertilization, 3. N₂-fixation, 4. denitrification, 5. NO₃⁻
respiration, 6. nitrification, 7. immobilization, 8. mineralization, 9. assimilatory and dissimilatory NO₃ reduction
to NH₄⁺, 10. leaching, 11. plant uptake, 12. atmospheric
deposition, 13. residue decomposition, exudation, 14. soil
erosion, 15. NH₄⁺ fixation and release by minerals, 16. combustion, 17. crop harvest and grazing, 18. addition of
organics.

Transformations involving oxidation or reduction of N,
with the important exception of plant assimilation of NO_3^-,
are generally the exclusive province of soil microorganisms.
Several key reactions are carried out only by procaryotes.
These include N_2-fixation, denitrification, nitrate respiration, chemoautotrophic nitrification, and probably dissimilatory NO_3^- reduction to NH_4^+. (See however, Finlay et al.
(1983) for a report of eucaryotic nitrate respiration.)
Eucaryotic microbes, particularly fungi, participate with the
procaryotes in immobilization, mineralization, assimilatory
NO_3^- reduction to NH_4^+, and residue degradation. Although
these processes may be directly catalyzed primarily or
entirely by microbes, it should not be assumed that larger
organisms of the soil ecosystem have no effect. As this
review suggests, soil microbes represent only a potential;
the expression of this potential is controlled by the abiotic
and biotic environment. Soil fauna have important effects on

this environment and therefore indirect effects on microbial N transformations. An example is the increase in water infiltration and soil aeration caused by earthworm tunneling which may promote aerobic processes such as nitrification but inhibit anaerobic processes such as denitrification.

In this chapter we will consider three reactions in which the activity of soil microbes is required: N_2-fixation, the major biological mechanism of soil N input; denitrification, the predominant biological pathway of soil N loss; and nitrification, which is in a sense at the center of the soil N cycle. We will attempt to describe when and why these reactions occur, and will emphasize the relationship between the microbes involved and their chemical, physical and biotic environment.

Denitrification and Reduction of Nitrogen Oxides

Because N is the mineral nutrient most commonly limiting plant production, the N-supplying fixation reaction is generally perceived as being desirable, and N-removing denitrification detrimental. It is apparent, however, that in any system in a steady state gains and losses must balance. The biosphere as a whole seems, in fact, to balance N_2-fixation (biological plus chemical) with an approximately equivalent annual rate of N_2 production by denitrification (Knowles 1982). Thus, from a global perspective it is appropriate to consider microbial denitrification as being unavoidable, if not necessary. Yet in managed agricultural ecosystems the additional input of fixed N as fertilizers has a considerable cost both energetically and economically. The portion of this fertilizer lost from the soil via denitrification, and so unavailable to crops, cannot presently be determined with precision, but it is certainly a significant portion. Economy and productivity of agricultural ecosystems would be enhanced by reducing denitrification.

Types of N oxide reduction and organisms involved

The ability to reduce NO_3^- and NO_2^- is widely distributed among organisms, but pathways, mechanisms and purposes vary. Higher plants and some eucaryotic and procaryotic microbes reduce NO_3^- to NH_4^+, then incorporate NH_4^+ into amino acids and the N-containing constituents of their cells. This assimilatory reduction of NO_3^- to NH_4^+ may occur under aerobic or anaerobic conditions. It is generally regulated by exogenous NH_4^+; NH_4^+ inhibits assimilatory NO_3^- reduction. At least in many aerobic organisms, assimilatory reduction requires the diversion of reducing power from energy-yielding metabolism and so it is energetically expensive. For more detailed comparison of assimilatory and dissimilatory NO_3^- reduction consult Payne (1973, 1981).

Although assimilatory reduction has significant effects on the fate of soil NO_3^-, these effects are in some ways less dramatic than those of the dissimilatory types of NO_3^- reduction. In dissimilatory reduction, the nitrogenous end-products are not incorporated into cells, and this creates a potential for loss of N from the ecosystem. Also, dissimilatory reduction usually is associated with energy generation, rather than energy consumption as in assimilatory reduction. Therefore, under favorable conditions a given quantity of biomass may reduce NO_3^- more rapidly by dissimilatory than by assimilatory mechanisms. For these reasons, and because dissimilatory NO_3^- reduction is a unique attribute of certain procarotic microorganisms, we will limit further discussion to dissimilatory reactions.

Bacteria in several, but by no means all, genera are able to execute the first step, NO_3^- reduction to NO_2^-, but lack the ability to further reduce NO_2^- to N gases. For many of these organisms NO_3^- is known to serve in the absence of O_2 as a terminal electron acceptor in respiration. Accordingly, these bacteria are often referred to as NO_3^- respirers, or less frequently NO_2^- accumulators. The enteric Escherichia coli is the organism in which this type of NO_3^- reduction has been most thoroughly characterized, but this species is probably not important in soils. Payne (1973) lists 44 additional genera containing NO_3^- respirers. Notable soil inhabitants include <u>Bacillus</u>, <u>Arthrobacter</u>, <u>Rhizobium</u>, and <u>Streptomyces</u>. Most, or perhaps all, of these organisms may carry out a slow, incomplete reduction of NO_2^- to N_2O; the significance of this process is unknown (Smith 1982).

A second group of bacteria also reduce NO_3^- to NO_2^-, but apparently do so in a different way. These organisms are generally obligate anaerobic fermenters, in contrast to most NO_3^- respirers which grow most efficiently in aerobic atmospheres, and may lack the electron transport machinery necessary for respiration. The reduction of NO_3^- increases growth in these organisms also. <u>Clostridium</u> may be the most common soil organism in this category (Caskey and Tiedje 1979).

Some of the bacteria which reduce NO_3^- primarily to NO_2^- are able to further reduce NO_2^- under certain conditions. Unlike the denitrifiers, these organisms produce NH_4^+ as the end-product of NO_2^- reduction, not N gases (Yordy and Ruoff 1981, Cole and Brown 1980). Dissimilatory reduction of NO_3^- and NO_2^- to NH_4^+ is of interest because this reductive, anaerobic process results in the conservation of soil N, not the loss of gaseous N. There have been some reports that a significant portion of the NO_3^- reduced in anaerobic soils is converted to NH_4^+ rather than gas (Stanford <u>et</u> <u>al</u>. 1975, Buresh and Patrick 1978). Bacteria able to produce extracellular NH_4^+ are probably about as abundant in soils as denitrifiers (Smith and Zimmerman 1981). However, reduction to NH_4^+ does not appear to proceed as rapidly as denitrification except perhaps under highly anoxic, substrate rich conditions. One microbial habitat in which reduction to NH_4^+ does seem to be favored is the bovine rumen (Kaspar and Tiedje 1981). It is interesting to speculate that the same reaction may occur in the gut of soil animals, but the fate of NO_3^- ingested by these organisms is not known.

Some soil organisms are able to reduce NO_3^- and NO_2^- completely to N gas. These are the denitrifying bacteria. The products of denitrification are, listed in the usual

Table 1. Genera of denitrifying and chemoautotrophic nitrifying bacteria in soils[†]

Denitrification	$NH_4^+ \longrightarrow NO_2^-$	$NO_2^- \longrightarrow NO_3^-$
Acinetobacter	Nitrosomonas	Nitrobacter
Agrobacterium	Nitrosolobus	
Alcaligenes	Nitrosospira	
Arthrobacter	Nitrosococcus	
Azospirillum		
Bacillus		
Chromobacterium		
Corynebacterium		
Cytophaga		
Flavobacterium		
Hyphomicrobium		
Paracoccus		
Pseudomonas		
Rhizobium		
Thiobacillus		

[†] Only taxonomically accepted genera for which soil is a
common habitat are included. Principal sources are Payne
(1981) and Buchanan and Gibbons (1974).

sequence of their appearance and in order of increasing
abundance: NO, N_2O, and N_2. Payne (1981) lists 22 genera
containing organisms reported to denitrify but it is probable
that only a few of these are important in soils (Table 1).
One survey of many soils showed that Alcaligenes, Flavobacte-
rium, and particularly Pseudomonas are the most common soil
denitrifiers (Gamble et al. 1977).

Environmental factors determining denitrification rates

The most important determinant of denitrification rates
in soils is aeration. The denitrifying enzymes will be
synthesized and retain activity only when O_2 is at a low
concentration. With only rare exceptions (Kaneko and Ishimoto
1977), which are probably of minimal significance in soils,
denitrifying bacteria are capable of, and grow more rapidly
using, aerobic respiration. Some Bacillus species may also
grow anaerobically with no N-oxide reduction by fermentative
processes, but most are like Pseudomonas in that they grow

only by respiration using either O_2 or N-oxides as terminal electron acceptors.

In spite of the strict requirement for very low O_2 tensions observed in the physiological characterization of pure cultures, there have been several observations of denitrification and functional denitrifying enzymes in largely aerobic soils (Smith and Tiedje 1979a, Starr et al. 1974, Broadbent and Clark 1965). This apparent anomaly is explained by hypothesizing the existence of anaerobic micro-sites within an aerobic soil. Denitrification could persist in small zones where biological O_2 consumption is more rapid than O_2 replenishment by diffusion. It has been suggested that the center of soil aggregates would be likely to harbor anaerobic microbial activity (Smith 1977, Greenwood and Goodman 1967), but these models do not consider the limited energy substrate supply for microbes in a stable aggregate center. Perhaps a more likely place to seek anaerobic microsites would be within large pieces of decomposing organic material.

Although denitrification may occur in aerobic soils, rates are greatly accelerated when the soil is poorly aerated (Focht and Verstraete 1977, Bremner and Shaw 1958). Soil aeration is determined by the relative rates of O_2 diffusion and O_2 consumption. Oxygen diffusion rates are influenced by soil texture, structure and moisture content; thus a heavy clayey soil is more prone to denitrify than a sandy or well-structured soil. Water content is of primary importance since the rate of gas movement through a water-filled soil pore is orders of magnitude less than through an air-filled pore. When soils become water-saturated following heavy rainfall or irrigation, denitrification can rapidly remove a large quantity of soil NO_3^-.

In continuously anaerobic habitats, such as some aquatic sediments, denitrification will be limited most often not by aeration, but by the supply of NO_3^-. Under these conditions NO_3^- is depleted by denitrification and is not regenerated from NH_4^+ by nitrification. The NO_3^- supply from external sources, such as the aerobic overlying waters, then determines denitrification rate (Patrick and Reddy 1976). In non-flooded soils denitrification rate sometimes may be related to NO_3^- concentration (Nommik 1956, Wijler and Delwiche 1954) or more precisely, NO_3^- diffusion rates (Phillips et al. 1978). However, soil aeration and energy supply seem more likely to be primary rate-controlling factors.

The availability of energy has a dual effect on soil denitrification. First, the oxidation of substrates will accelerate O_2 consumption and so promote establishment of anaerobiosis. Second, electron donors are required for reduction of N-oxides and growth of denitrifiers. A few denitrifying chemolithotrophs have been discovered: Thiobacillus denitrificans uses the oxidation of reduced sulfur compounds to drive denitrification (Ishaque and Aleem 1973) and Rhodopseudomonas spheroides is a photosynthetic denitrifier (Satoh et al. 1974). However, most denitrifiers and all of those believed to be predominant in soils, grow by oxidation of organic matter. Most of the soluble, low molecular weight organics which can be decomposed during aerobic growth can also be used during denitrification. The exception may be compounds such as hydrocarbons for which O_2 is a required reactant for degradation (Evans 1977). The degradation of complex polymers by denitrifiers has not been investigated in detail. When soils are made anaerobic and supplied with adequate NO_3^-, the rate of denitrification is observed to be closely related to the availability of organic substrate (Bremner and Shaw 1958, Burford and Bremner 1975). Manure addition to soils can enhance denitrification (Rolston et al. 1979). Plowing down plant residues may create favorable conditions for denitrification. When flooded peat soils are drained for agricultural production, the accumulated organic material deep in the soil profile becomes susceptible to decomposition and large quantities of NO_3^- are produced by mineralization and nitrification. In these situations denitrification rates may be very high (Guthrie and Duxbury 1978).

In the rhizosphere organic compounds sloughed off and exuded by roots and the associated increase in oxygen consumption by root and microbial respiration would be expected to favor denitrification. Denitrifying bacteria can be 1 to 2 orders of magnitude more abundant in the rhizosphere than in soil without roots (Garcia 1973, Woldendorp 1962). Although cropped soils generally denitrify more than fallow soils (Rolston 1979, Stefanson 1972, Volz et al. 1976), sometimes the plant may compete with the denitrifier for NO_3^- and when this is in short supply, denitrification may actually be reduced in the rhizosphere (Smith and Tiedje 1979b). The latter effect would be expected to occur more in natural ecosystems than in N-fertilized agronomic soils.

Temperature and pH also influence denitrification, much as they do most other microbiological processes. Denitrification has been observed to occur over a broad pH range; from pH 11 (Prakasam and Loehr 1972) to 4 (Nommik 1956). However, neutral or slightly alkaline soils favor the reaction. In highly acidic soils apparent denitrification actually may proceed by chemical reactions rather than biological catalysis. Nitrite, and perhaps to a small extent NO_3^-, participate in several decomposition reactions which are favored by acidity, and yield N gas (Knowles 1982). The reported temperature optimum for soil denitrification is surprisingly high, greater than 60°C (Bremner and Shaw 1958). Thermophilic denitrifiers are known, but non-biological reactions may be a better explanation for rapid denitrification at high temperatures. However, low soil temperatures are certainly more often restrictive than high temperatures (Knowles 1982).

Magnitude and significance of denitrification

Soil scientists have devoted much research effort to obtaining accurate measurements of fertilizer lost by denitrification in agricultural soils. The resultant data may seem meager, but not if the methodological difficulties are considered (Focht 1978). Early measurements of denitrification under field conditions were made by the N balance approach, in which sources and sinks of soil N were measured and the amount of N which disappeared was equated with denitrification. Because of the large background pool of organic N present in virtually all soils, meaningful results usually are obtained only if isotopically labelled N, either ^{15}N enriched or depleted, is used. Even with labelled N, this approach is indirect, insensitive, generally not applicable to short-term measurements, requires soil amendment with N, and combines the analytical errors associated with each of the required measurements. Direct analysis of the gaseous denitrification products would be preferred, but since N_2 is the major constituent of the atmosphere it is impossible to measure production of this gas under realistic conditions. One solution to these problems would be the development of techniques for the direct measurement of labelled N_2 produced from ^{15}N-enriched fertilizer (Rolston et al. 1978, Siegel et al. 1982). A second approach to direct measurement was provided by the discovery that C_2H_2 inhibits the reduction of N_2O to N_2 but has little effect on the rate of total N gas production by denitrifiers (Federova et al. 1973). In the presence of C_2H_2 only N_2O measurement is necessary, a relatively simple task because of the low atmospheric content of N_2O and availability of sensitive analytical techniques. Numerous investigators have used this C_2H_2 inhibition technique in laboratory studies. Although some difficulties may be associated with this method in field studies, Ryden et al. (1979) were able to make direct estimates of denitrification in California soils.

Several of the available estimates of denitrification in agricultural soils are summarized in Table 2. It can be seen that the percentage of fertilizer lost by denitrification is extremely variable. Perhaps the best current estimate of average fertilizer lost in agricultural soils is Hauck's (1981) value of 30% of the amount applied. Unfortunately, almost no information of comparable reliability is available on denitrification rates in non-agricultural soils. One difficulty in these cases is that the addition of labelled

Table 2. Summary of selected in-field measurements of denitrification rate and percent of N fertilizer lost in agricultural soils

Reference	Comments	Denitrification in kg N/ha per time	Denitrification as percent of N applied
Allison 1955	51 lysimeter balances, no label, cropped soils	--	20
Hauck 1981	34 selected N balance studies	--	30
Broadbent and Carlton 1978	^{14}N (depleted) balance, irrigated corn	53-329/3 years	16-25
Rolston et al. 1978	Direct measurement of $^{15}N_2$ and N_2O	0-66/day	0-73
Olson 1980	^{15}N balance, irrigated corn	12/year	24
Kissel and Smith 1978	^{15}N balance, bermudagrass	--	10
Reddy and Patrick 1980	^{15}N balance, rice paddy	21-36/growing season	36-51
Ryden et al. 1979	C_2H_2 inhibition method	51-123/days	37
Ryden and Lund 1980	C_2H_2 inhibition method	95-22/year	14-77

inorganic N represents an unrealistic perturbation. It is unlikely that denitrification rates in most non-agricultural systems are as great as those reported in Table 2. It is generally assumed that the nutrient cycles of undisturbed, natural ecosystems exhibit lower rates of net nutrient loss and gain.

In the late 1970's well-publicized reports that one of the denitrification products could have deleterious environmental effects provided a new focus for denitrification research (McElroy et al. 1977, C.A.S.T. 1976). It was hypothesized by atmospheric chemists that N_2O produced via

denitrification is photochemically converted to NO in the
stratosphere and the resulting NO catalyzes the reduction of
ozone (O_3) to O_2. Strastospheric O_3 represents an atmospheric filter for UV radiation; thus increased fertilizer use
was linked to an increased frequency of skin cancer and to
climatic perturbation. The small quantity of soil-evolved NO
and the much larger amounts of combustion-derived NO and
higher N oxides apparently do not interact with stratospheric
ozone because they are oxidized in the lower atmosphere.
These oxides contribute to another recently recognized environmental concern, acid precipitation. It is now generally
accepted that there are mechanisms of N_2O production other
than denitrification. In soils these include nitrifying
bacteria (Bremner and Blackmer 1981), non-denitrifying NO_3^-
reducing bacteria (Smith and Zimmerman 1981), and chemical
reactions. It is now estimated that combustion of fossil
fuels and biomass also releases N_2O in significant quantities
(Crutzen et al. 1979). Although this issue has not been
completely resolved, some recent analyses by atmospheric
chemists suggest that increased N_2O production via increased
fertilizer denitrification may have less immediate and
drastic effects on atmospheric O_3 than originally predicted
(Crutzen 1981).

Nitrification

The oxidation of NH_4^+ to NO_3^-, although it is an internal
soil process, can have a significant effect on the ultimate
fate of soil N. This is because the substrate, NH_4^+, is
generally immobilized by cation exchange reactions and so is
not subject to rapid leaching from the soil. The product,
NO_3^-, can be removed more rapidly by leaching and denitrification. The nitrification reaction may also affect the
assimilation of inorganic N by plants. Although most plants
are capable of incorporating either NH_4^+ or NO_3^-, the greater
mobility of NO_3^- and its transport to the plant in the water
transpiration stream usually results in greater exploitation
of the NO_3^- form by plants, at least in fertilized agricultural

ecosystems. Microbial N assimilation (immobilization) can also involve either inorganic form but assimilation of NO_3^- is generally inhibited in the presence of NH_4^+ and not all microbes can grow with NO_3^- as a N source.

The importance of nitrification to nutrient losses from forested ecosystems was illustrated at Hubbard Brook (N.H.) (Likens et al. 1968). After forest clearing, NO_3^- concentrations in stream water increased approximately 60-fold and this increase was attributed to soil nitrification. Further studies have confirmed that the effect of perturbation on nitrification is an important factor determining the magnitude of N loss from disturbed ecosystems (Vitousek et al. 1979).

Delay of nitrification is now recognized as a potentially valuable mechanism of managing N loss from agricultural ecosystems. Specific chemical inhibitors of nitrification are being marketed for application to soil. The best known of these is 2-chloro-6(trichloromethyl)-pyridine or 'N-Serve' (Goring 1962). The temporary delay in NO_3^- production caused by these chemicals can significantly reduce the potential for early season leaching and denitrification. In soils where these losses may occur equal crop yields can often be produced with smaller applications of N fertilizer if such inhibitors are used.

Chemoautotrophic nitrifying bacteria

Since the isolation of Nitrosomonas and Nitrobacter by Winogradsky (1890) soil nitrification has been attributed to a limited number of bacteria. This early work revealed the fundamentals of chemoautotrophic nitrification. Conversion of NH_4^+ to NO_3^- is a two-step process with NO_2^- as the intermediate, requiring a consortium of bacteria. Nitrifying bacteria generate energy by oxidation of NH_4^+ and NO_2^-, rather than organic materials.

One factor which has limited the characterization of nitrifier behavior in soils, and even in cultures, is the relatively slow growth rate of these bacteria. Under optimum culture conditions nitrifier doubling times are more than an order of magnitude greater than those for many other bacteria; usually in excess of 15 hours. In soils, growth is probably considerably slower. Even when substrate is not limiting, generation times of <u>Nitrosomonas</u> and <u>Nitrobacter</u> may range from 20 to over 100 hours (Morrill and Dawson 1962). However, this does not necessarily indicate that nitrification rates are slow. We have observed conversion in agricultural soils of as much as 50 kg NH_4^+-N·ha^{-1} to NO_3^- in only 5 days. Because the potential energy yield from NH_4^+ and NO_2^- oxidation is low, large amounts of substrate must be utilized; thus slow growth may be associated with rapid nitrification.

The difficulty of isolating and enumerating the slow-growing nitrifiers contributes to some uncertainty about precisely which organisms are responsible for nitrification in soils. In older texts and even in some current soils literature nitrification is assumed to be the exclusive domain of <u>Nitrosomonas</u> and <u>Nitrobacter</u>. Without doubt the bacteria which effectively nitrify are of much more limited diversity than the microbes responsible for most other soil processes (compare with denitrification in Table 1). This was best illustrated in a study by Fliermans <u>et al</u>. (1974) using specific fluorescent antibody staining techniques. It was found that all NO_2^- oxidizing isolates from a number of varied habitats fell into only 2 <u>Nitrobacter</u> serotypes. In virtually all studies employing traditional isolation procedures and media, these are the only nitrifying genera isolated.

More recent investigations suggest that the diversity of chemoautotrophic nitrifiers may not be as restricted as previously believed. Dommergues <u>et al</u>. (1978) also suggested that there is greater serological variation among soil <u>Nitrobacter</u> isolates than observed by Fliermans <u>et al</u>. (1974). Revealing studies by Belser and Schmidt (1978a,b), using a combination of fluorescent antibody staining and modified isolation procedures, indicate that at least three genera of NH_4^+ oxidizers can co-exist in soils. <u>Nitrospira</u>

was isolated as frequently as Nitrosomonas, and Nitrosolobus
isolates were common though less numerous than the other two.
They also noted serological diversity within each genus.
More important, diversity among denitrifiers with regard to
behavior in soils is probable (Josserand and Faurie 1981).
It is clear that conventional isolation and counting procedures alone are inadequate for ascertaining the role of
chemoautotrophic nitrifiers. This point was demonstrated by
Rennie and Schmidt (1974) when they compared counts of soil
nitrifier populations determined by direct microscopic
counting (using fluorescent antibody stains) and by standard
serial dilution isolation techniques. The latter counts were
four orders of magnitude less than the former.

Heterotrophic nitrification

The assumption that only chemoautotrophic bacteria are
responsible for nitrification in all soils is made question-
able by some observations. In acid soils it has been noted
that amendment with NH_4^+ fails to increase and may actually
inhibit the rate of NO_3^- production (Weber and Gainey 1962).
It would be expected that NH_4^+ would increase chemoautotrophic
nitrifier populations and so increase the nitrification rate.
Also in acid soils, nitrification was observed even though
nitrifiers were not detected (Ishaque and Cornfield 1972).
In the same soils, liming did not increase the nitrification
rate. Based on differential responses of soils and pure
cultures of nitrifiers to chlorate inhibition Ghiorse and
Alexander (1978) suggested that unknown nitrifying populations exist in soils.

Dalton (1977) showed that methylotrophs, which grow by
oxidizing methane or other one carbon compounds, are also
capable of oxidizing NH_4^+ and suggested that the activity of
these bacteria may be ecologically significant. Oxidation of
NH_4^+ to NO_2^-, NO_3^- or organic N-oxides is known to be catalyzed
by a variety of heterotrophic fungi and bacteria (Doxtader
and Alexander 1966, Verstraete and Alexander 1972). These
organisms do not necessarily derive energy from the reaction,

and heterotrophic oxidation sometimes occurs at a much
slower rate, per cell, than chemoautotrophic nitrification.
On the other hand, the heterotrophs may be more numerous than
the chemoautotrophs in soil.

The importance of heterotrophic nitrification remains
unknown. It seems this mechanism is least likely to be
significant in a soil with neutral pH, particularly if the
soil is NH_4^+-fertilized or the NH_4^+ supply is high. This
opinion is based on the in situ effectiveness of specific
inhibitors of chemoautotrophic nitrification and on the
response to NH_4^+ additions. Also unknown is the significance
of a report that in soils with high levels of Mn oxides,
transformation of NO_2^- to NO_3^- may be an abiotic, chemical
reaction (Bartlett 1981).

Environmental factors controlling nitrification

Although the variety of microorganisms involved in soil
nitrification may be greater than previously believed, it
remains clear that the nitrifiers are of relatively limited
diversity. This lack of diversity may be a factor in the
apparent sensitivity of nitrification to acidity and to some
agricultural chemicals. Inhibition by pesticide application
may be more likely for nitrification than most other soil
reactions (Mainwright 1978). Several commercial broad-
spectrum fungicides are effective nitrification inhibitors at
recommended application rates.

The sensitivity of nitrification to soil acidity is
well-known. Nitrification rates in soil are generally
optimal at neutral to slightly alkaline pH and decline
rapidly below a pH of about 6 (Morrill and Dawson 1962,
Morrill and Dawson 1967). Acidity, or the Al made soluble at
low pH, may be the primary factor limiting nitrification
rates in soils (Brar and Giddens 1968). In most soils NO_2^-
does not accumulate because NH_4^+ conversion to NO_2^- is the
rate-limiting step. However, when base-forming NH_3 (or
equivalently, urea) is added to alkaline soils, the pH in the
zone of fertilizer application may sometimes exceed 9. The

alkalinity and the resultant predominance of NH_3, rather than NH_4^+, are more inhibitory to NO_2^- oxidizers than to NH_4^+ oxidizers. In this instance, NO_2^- may accumulate to concentrations sufficient to cause plant injury (Court et al. 1962).

Rapid nitrification rates require well-aerated soils. Chemoautotrophic nitrifiers may exhibit a lower affinity for O_2 than many aerobic, heterotrophic microbes (Focht and Verstraete 1977). This suggests that nitrification can be inhibited when heterotrophic O_2 demand is high such as when decomposable organic matter is abundant, or when soils are flooded. Limited soil water is also detrimental. Rates have been observed to be maximal near 0.01 to 0.03 MPa soil moisture tension (near field capacity) and reduced by 95% at 1.5 MPa tension (near plant wilting point) (Sabey 1969).

The activity of nitrifiers in soil is often limited by substrate (NH_4^+) supply (Focht and Verstraete 1977). Therefore, increased availability of NH_4^+ should increase the rate, as well as the quantity, of NO_3^- production.

The impossibility of isolating or maintaining chemoautotrophs on organic-based media, and perhaps competition for O_2 with heterotrophs, may have contributed to the belief, widely held in the past, that organic matter is actually toxic to nitrifying bacteria. Certainly these bacteria are inhibited by some organics, as are all organisms, but it now seems clear that there is no specific toxicity by organic compounds in general. Clark and Schmidt (1967) have shown that nitrifier growth is increased in the presence of some organic compounds. Presumably these bacteria can lighten their biosynthetic burdens by assimilating growth factors or other organic compounds from external sources.

Related to the question of organic matter toxicity is the role of allelochemical inhibition of nitrifiers in the rhizosphere. This theory was first offered by Theron (1951), to account for low soil NO_3^- concentrations and apparently low nitrification rates in some ecosystems, particularly grasslands. Rice and Pancholy (1973) noted that the predominant

form of inorganic N was NO_3^- and NH_4^+, in early and climax
successional stages, respectively. Also, the numbers of
nitrifying bacteria decreased with succession. Furthermore,
it has been shown that nitrifiers in culture can be inhibited
by extracts of plant roots (Moore and Waid 1971). Direct
inhibition of nitrifiers by plant roots is teleologically
appealing because this "would aid in the conservation of
nitrogen and energy in the climax ecosystem" (Rice and
Pancholy 1973). Unfortunately, several other studies suggest
that this simple, direct mechanism of nitrification control
is not adequate to account for the observations. (Also
consult Clark and Paul [1970] or Verstraete [1981] for discussion of both sides of this issue.) Acidity (Brar and
Giddens 1968) or limited availability of NH_4^+ substrate
(Robinson 1963) may actually be the primary factors limiting
nitrification in grasslands. Purchase (1974) did not observe
inhibition of nitrifiers by washings of grass roots and noted
that nitrifier populations were low throughout the soil and
not just in the rhizosphere. His experiments suggest that
nitrification in grasslands is limited by rapid immobilization and plant uptake of NH_4^+.

Nitrogen Fixation

Few areas in science have captured as much attention or
undergone such rapid expansion in the last decade as the
study of N_2-fixation. The mass of new information available
is indicated by a partial listing of recently published books
(Hardy 1977 [a four volume set], Gibson and Newton 1981,
Stewart 1975, Nutman 1976, Newton and Orme-Johnson 1980,
Postgate 1971, Bergerson 1981, Broughton 1981, Broughton
1982, Quispel 1974, Newton and Nyman 1976 [two volumes],
Takahashi 1975, Subba Rao 1980). In the space available here
we cannot deal with such rapidly developing areas of study
as: the chemistry of inorganic catalysts for N_2-fixation
(Chatt 1981), the biochemistry of nitrogenase (Burris et al.
1981, Kennedy et al. 1981), or the genetics of N_2-fixation
(Brill 1980). Developments in the latter area and in

recombinant DNA technology have proceeded so rapidly, that
some have optimistically predicted the creation of non-legume
N_2-fixing crop plants.

Here we will consider the environmental constraints and
limitations on N_2-fixation and give particular emphasis to
one way in which microbes deal with these constraints-symbio-
tic relationships with other organisms. Only procaryotic
microbes are capable of N_2-fixation and many do so more or
less independently. Yet co-evolution of N_2-fixing associations between procaryotes and eucaryotes seems common. The
potential advantages of these associations may become apparent in the following discussion. The diversity and ubiquity
of N_2-fixing relationships (outlined in Table 3) provides
evidence that such advantages exist.

Non-symbiotic N_2-fixation

Knowles (1977) lists 15 genera of bacteria, known to
occur in soil, which are capable of fixing N_2 independently
of other organisms. The most familiar of these, and probably
the most widely distributed, are <u>Azotobacter</u>, <u>Beijerinckia</u>,
<u>Spirillum</u> (some species reclassified to <u>Azospirillum</u>),
<u>Bacillus</u>, <u>Enterobacter</u>, <u>Klebsiella</u>, and <u>Clostridium</u>. These
free-living soil N_2-fixers are heterotrophic and so are
dependent on external sources of organic matter for energy.
Energy supply is probably the primary limitation on N_2-
fixation by these organisms. Biological N_2-fixation is
energetically expensive for a microbe, as is industrial N_2
fixation for human use. In addition to the direct cost in
ATP and reducing power, one must consider energetic requirements for enzyme synthesis and maintenance of a reduced
environment for the enzyme. In an interesting comparison of
the physiology of free-living and symbiotic <u>Rhizobium</u>
N_2-fixers, Mulder (1975) indicated that free-living organisms
fix 10 to 20 mg N/g of carbohydrate consumed, while the
efficiency of symbiotic systems measured in this way is more
than an order of magnitude higher. However, Knowles (1977)
suggests that the efficiency of N_2-fixing organisms in soil

Table 3. The diversity of biological N_2-fixing relationships

Type	Procaryote -	Eucaryote	Comments
Non-symbiotic heterotrophic	Azotobacter Clostridium Klebsiella several others	none	Depend on exogenous energy supply
Non-symbiotic photosynthetic	cyanobacteria, anoxygenic photosynthetic bacteria	none	Depend on high moisture, light
Associative, rhizosphere	Azospirillum Azotobacter others	most often studied in grasses	No specific N_2-fixing structure
Extracellular symbioses	cyanobacteria	fungi, liverworts, Sphagnum, Azolla, others	Procaryote outside eucaryote cell wall, usually simple N_2-fixing structure
Intracellular symbioses	Rhizobium Frankia Nostoc Rhizobium	legumes many angiosperms Gunnera Trema	Procaryote within eucaryote cell wall, organized complex N_2-fixing structure

may be considerably higher than those in pure culture, where these efficiency values have been determined. Nevertheless, amendment of soils with an energy source usually increases non-symbiotic N_2-fixation indicating that the process is usually energy-limited (Barrow and Jenkinson 1962, Delwiche and Wijler 1956, Rice et al. 1967).

Another environmental factor influencing N_2-fixation by non-symbiotic soil organisms is O_2. Under anaerobic or water-saturated conditions N_2-fixation rates are sometimes greater than in aerobic soils (Brouzes et al. 1969, Fehr et al. 1972). Some free-living N_2-fixers are obligate anaerobes, such as Clostridium, while others, such as Klebsiella and Bacillus, are facultative anaerobes but fix N_2 only anaerobically. Free-living organisms growing aerobically, such as Azotobacter, have an additional problem of protecting nitrogenase from O_2, since sensitivity to O_2 seems to be an inherent characteristic of these enzymes. Thus, O_2 protection can be another factor favoring symbiotic N_2-fixing structures, such as nodules.

Inorganic N, NH_4^+ or NO_3^-, limits N_2-fixation primarily by inhibiting the synthesis of nitrogenase. It is considerably less expensive for a cell to use external sources of fixed N when available than to fix N_2. Free-living organisms fix only as much N_2 as they need so the addition of fertilizer N to soils will usually drastically reduce biological fixation (Delwiche and Wijler 1956, Yoshida et al. 1973). Greater than 90% of the N fixed non-symbiotically is immobilized in microbial tissue, restricting its availability to plants while as little as 5 to 10% may be microbially immobilized in symbiotic fixation (Mulder 1975).

Considering these limitations it should not be surprising that rates of non-symbiotic fixation in soils are usually reported to be relatively low. Knowles (1977) has summarized a large number of rate measurements. For plant-free unamended soils, most values fall between 0 and approximately 2 kg $N \cdot ha^{-1} \cdot day^{-1}$. These are almost all short-term measurements and are difficult to extrapolate to annual fixation rates. In a few studies where this has been attempted estimates are less than 10 kg $N \cdot ha^{-1} \cdot yr^{-1}$ (Vlassak et al. 1973, Steyn et al. 1970). To an agronomist, accustomed to dealing with fertilizer inputs and crop demands which may be greater than this by an order of magnitude or more, these rates seem insignificant. But in undisturbed, natural ecosystems, where N flux occurs at a slower rate, non-symbiotic fixation by heterotrophs should not be ignored.

The cyanobacteria, or blue-green algae, are free-living N_2-fixers which are not subject to one of the limitations discussed above, dependence on external energy sources. Under favorable conditions; high moisture, light, and neutral to alkaline pH, these photosynthetic procaryotes may fix N_2

rapidly. It should be realized that at most times, however, soil surfaces, where light is available, are desiccated. Therefore, high rates observed in short-term measurements, such as the values of up to 6.2 kg $N \cdot ha^{-1} \cdot day^{-1}$ reported by Granhall (1975), are not necessarily indicative of annual N inputs. In most temperate region soils fixation by cyano-bacteria is probably less than 10 kg $N \cdot ha^{-1} \cdot yr^{-1}$ (Mayland 1966, Cameron and Fuller 1960). In flooded or waterlogged soils, or rice paddies, this source of N may be considerably more important (Watanabe and Brotonegoro 1981).

Associative or rhizosphere N_2-fixation

In the early 1970's excitement was aroused by reports of high rates of N_2-fixation associated with the roots of some tropical grasses. Annual fixation by Paspalum and Azotobacter was estimated to be as high as 90 kg $N \cdot ha^{-1}$ (Dobereiner et al. 1972). This relationship was found to be largely un-structured; the bacteria colonize the rhizosphere, grow in the root mucigel material, or partially invade the external layers of the root. There does appear to be some specificity involved since only certain bacteria will colonize certain plants (Dobereiner et al. 1971). This associative N_2-fixing relationship has now been detected in numerous, diverse plant species but the grasses have received particular attention (Knowles 1977, Dobereiner and Day 1975). Plants with the C-4 type of metabolism seem to be capable of supporting higher rates of associative fixation, perhaps because of greater substrate availability in the rhizosphere.

There is currently some disagreement about the signi-ficance of associative N_2-fixation. The methodology employed in many of the early rate measurements has been questioned (van Berkum and Bohlool 1980, Barber et al. 1976, Lethbridge et al. 1982). In Canadian grasslands annual fixation was estimated to be only 2 to 3 kg $N \cdot ha^{-1} \cdot yr^{-1}$ (Vlassak et al. 1973). Fixation associated with roots of wheat and sorghum was never observed to be greater than 2.5 g $N \cdot ha^{-1} \cdot day^{-1}$ (Pedersen et al. 1978). Nevertheless, there is sufficient evidence to conclude that associative N_2-fixation does occur and it is reasonably certain that its magnitude is at least greater than fixation by non-rhizosphere, non-symbiotic bacteria. Most convincing are results from the venerable Broadbalk plots at Rothamsted, England. Associative N_2-fixation under mixed grass and herbaceous vegetation was estimated to be about 30 kg $N \cdot ha^{-1} \cdot yr^{-1}$ (Day et al. 1975). This was determined by following the accretion of soil N over many years and was well supported with acetylene reduction assays and other measurements.

Structured symbioses other than Rhizobium-legumes

The lack of specific structures for N_2-fixation in associative relationships may be one factor limiting N accumulation in these systems. Since there is no specific mechanism for substrate transport from the plant to the microbe, it might be expected that much of the energy source in the rhizosphere would be consumed by non-N_2-fixers. Similarly, since there is no specific mechanism for transport from the microbe to the plant, the availability of fixed N to the plant should be limited. It has become clear that numerous, diverse symbioses exist which at least partially solve these problems. Varying levels of interaction or degrees of structural complexity are observed.

Many of the structurally simpler symbioses include cyanobacteria as the procarotic N_2-fixer (Stewart et al. 1980). Perhaps the most familiar are the lichens. Several of these, but certainly not all, fix N_2. Because lichens grow slowly, their rates of fixation are also low. Only in systems where lichens are a significant portion of the total biomass are they likely to be a significant source of N. This could include rock surfaces, extremely cold regions, or areas subject to prolonged periods of desiccation.

Several plants harbor cyanobacteria in somewhat special-ized pores, glands, or spaces in the leaf (Silvester 1977). As in the lichens the procaryotes are extracellular to the eucaryotes. Although these are not true soil microorganisms, the N_2-fixed by them may ultimately become involved in the soil N cycle. Plants involved in this type of symbiosis include some liverworts, Sphagnum, and the water fern Azolla. All of these are limited to wet or flooded soils. The most significant of these probably is the Azolla-Anabaena symbiosis. This plant grows profusely in rice paddies and may support fixation rates in excess of 100 kg $N \cdot ha^{-1} \cdot yr^{-1}$ (Talley et al. 1977). It can be employed as a green manure crop in rice production and may sometimes permit sustained high grain yields with little or no input of N fertilizer.

A higher degree of structural complexity is found in the angiosperm Gunnera (Silvester 1976). In this case the cyanobacteria, Nostoc, penetrates the cell walls in glands on the plant stems. These structures have an extensive vascular system which presumably would improve the efficiency of transport of materials between the symbionts. It is unlikely that sufficient light penetrates this structure to support photosynthesis by the cyanobacteria, so the plant probably supplies the energy source. Rates of fixation in one species have been estimated at 72 kg $N \cdot ha^{-1} \cdot yr^{-1}$ (Silvester and Smith 1969).

The angiosperm Trema, which occurs mostly in the tropics and sub-tropics, forms root nodules which are similar in structure to those on legumes. The most remarkable aspect of this symbiosis is that the endophyte of this non-legume seems to be Rhizobium (Trinick 1973).

Nitrogen-fixing actinomycetes inhabit root nodules of eight different orders of angiosperms (Becking 1977). In terms of distribution and ecological significance this symbiosis is second only to that of the legumes. In terms of efficiency of N_2-fixation it may be second to none. The plants infected are generally woody species. Alnus is the host which is probably most common in the temperate Northern Hemisphere. The actinomycetes are intracellular and are within nodules which have a well-developed vascular system and highly organized structure. Only recently has it been possible to isolate the endophyte and achieve reinfection of

plants with pure cultures (Callaham et al. 1978, Benson
1982). The microorganism is usually placed in the genus
Frankia. Reported annual rates of fixation range from 2 to
362 kg N·ha^{-1}, most typically they fall near 100 kg N·ha^{-1}
(Becking 1977, Silvester 1977). Nitrogen-fixation from this
source can be of considerable significance in forestry, land
reclamation, and succession in natural ecosystems.

It is necessary to mention, particularly in this volume,
N$_2$-fixing associations between animals and the microorganisms
contained in their digestive system. Several animals, mostly
mammals and insects, have been examined for potential N$_2$-
fixation, in many cases with positive results (reviewed by
Knowles 1977). In general the rates do not seem great enough
to make a large contribution to the animal's diet. However,
termites are an exception (Benemann 1973, Breznak et al.
1973). The N$_2$-fixing bacteria inhabit the hindgut of this
insect where they receive a constant supply of substrate and
are perhaps also protected from O$_2$. The very low N content
of the termite's diet suggests obvious advantages of this
symbiosis. Breznak et al. (1973) could detect no nitrogenase
activity in 17 other insect genera, so it should not be
assumed that this is a common attribute of the soil fauna.
However, in some tropical areas termites are very abundant
and the fixation in their guts may amount to a significant
input of soil N. The importance of fauna-associated fixation
in N cycling of various ecosystems has been reviewed recently
by Waughman et al. (1981).

The Rhizobium-legume symbiosis:

Discussion of this association has been so pervasive
recently that it is not necessary (or possible) to offer a
full review here. Most of the books referenced above include
detailed discussion of N$_2$-fixation in legumes. We wish to
simply point out the relative significance of this source of
fixed N, and to briefly consider the intricacy of the symbiotic interactions.

Precise estimates of the percentage of total N_2-fixation attributable to legumes are not available. In part, this is due to the surprisingly small amount of information available on the thousands of species of wild legumes. Agricultural legumes alone probably account for between 1/3 to greater than 1/2 of all the Earth's biological N_2-fixation and from 1/5 to 1/3 of the total annual input of fixed N (NAS 1978). Fixation rates by individual legume crops can be very impressive, usually amounting to at least 50 kg $N \cdot ha^{-1} \cdot yr^{-1}$ (Alexander 1977). Well-managed alfalfa may accumulate well over 300 kg $N \cdot ha^{-1} \cdot yr^{-1}$.

This most active N_2-fixing system is representative of the most organized, specific and intricately structured symbioses. Each step: bacterial binding to the root, infection, formation of the nodule, and the fixation process itself, seems to require multiple complex interactions between the partners. Although many rhizobia will fix N_2 independently under stringently controlled laboratory conditions (Child 1980), it seems certain that only in symbiosis is fixation by Rhizobium significant.

The control of rhizobial binding and legume infection illustrates the complexity of this symbiosis. The current model of these still imperfectly understood processes has been reviewed by Hubbell (1981). The specificity of rhizobia is well known; only certain species of Rhizobium infect certain legumes and even within species there is significant variation in ability to nodulate. This is thought to be due, in part, to a double interaction between plant and rhizobial surfaces. Specificity of binding is determined by the interaction of a lectin protein on the plant surface and an antigen, probably a carbohydrate, on the bacterial surface (Bohlool and Schmidt 1974, Hamblin and Kent 1973). This binding interaction alone is not sufficient to initiate infection; a further specific interaction between rhizobially-released hydrolytic enzymes (Martinez-Molina et al. 1979) and cross-reactive antigens, again probably polysaccharides, on the plant cell wall is postulated to initiate

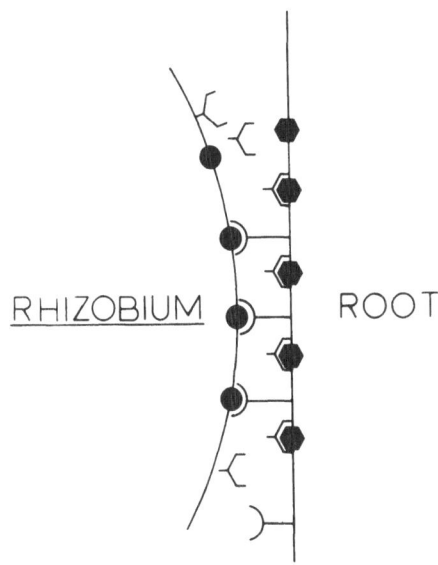

RHIZOBIUM   ROOT

FIGURE 2. Specific surface interactions between Rhizobium
and root. Circular structures indicate plant lectins and
the cross-reactive components of the bacterial cell wall.
Specificity of binding is attributed to these. Hexagonal
structures indicate rhizobial hydrolytic enzymes and cross-
reactive antigens or substrates on the plant surface. This
interaction is hypothesized to regulate the initiation of
infection. Adapted from Hubbell (1981).

bacterial penetration of the plant (Figure 2).

Further examples of the finely controlled relationship
between the partners can be observed in the functioning of
the nodule. It is remarkable that approximately one third of
the total plant photosynthate is transported to the nodule,
even though nodules constitute only a small fraction of the
total plant mass (Pate 1977). After the addition of fixed N,
perhaps one half of this photosynthate is efficiently transported back to the shoots as ureides and amino acids. The
smooth operation of this transport obviously requires a well-
developed vascular connection between the nodule and the
plant, as well as unknown biochemical regulatory mechanisms.

Leghemoglobin, a molecule in the nodule which protects
nitrogenase from O_2 inactivation but apparently supplies O_2

for rhizobial respiration, seems to be the result of a cooperative biochemical synthesis. The plant possesses the genes for synthesis of the protein portion of this molecule (Sidloi-Lumbroso et al. 1978), but these genes are expressed only in the presence of the rhizobia. However, there is evidence that the bacteria are responsible for synthesizing the heme portion of leghemoglobin (Nadler and Avissar 1977). Therefore, both symbionts seem to work together to produce a single functional molecule.

Conclusion

In this review we have taken what might be called a process, rather than an ecosystem, approach. Nitrogen cycle processes have been considered individually, rather than attempting a unification of these reactions into a description of N cycle dynamics for entire ecosystems. This approach results partly from our own background and interests, and partly from a belief that ecosystem function can be understood only with a knowledge of how individual processes are regulated. Therefore, we have emphasized the ways in which the biotic and abiotic environment limits or promotes these processes; for example the regulation of denitrification by soil water and availability of decomposable organic matter, of nitrification by NH_4^+ and O_2 supply and by pH, of N_2-fixation by energy and external sources of fixed N. The chauvinistic microbiologist's viewpoint is that bacteria and fungi are primarily responsible for operation of the N cycle. However, a possibly more useful characterization of the role of microorganisms in soil nutrient cycles is that microbes are potential catalysts. The expression of this catalytic activity is controlled by soil physical and chemical characteristics, climate, and interactions with plants and soil animals.

Literature Cited

1. Alexander, M. 1977. Introduction to soil microbiology, second edition. John Wiley and Sons, New York.
2. Allison, F. E. 1955. The enigma of soil nitrogen balance sheets. Adv. Agron. 7:213-250.
3. Barber, L. E., J. D. Tjepkema, S. A. Russell, and H. J. Evans. 1976. Acetylene reduction (nitrogen fixation) associated with corn inoculated with Spirillum. Appl. Environ. Microbiol. 32:108.
4. Barrow, N. J., and D. S. Jenkinson. 1962. The effect of water-logging on fixation of nitrogen by soil incubated with straw. Plant Soil 16:258-262.
5. Bartlett, R. J. 1981. Nonmicrobial nitrite to nitrate transformation in soils. Soil Sci. Soc. Am. J. 45:1054-1058.
6. Becking, J. H. 1977. Dinitrogen fixing associations in higher plants other than legumes. In R. W. F. Hardy and W. S. Silver (eds.), A Treatise on Dinitrogen Fixation-Section III: Biology. John Wiley and Sons, New York.
7. Belser, L. W., and E. L. Schmidt. 1978a. Diversity in the ammonia-oxidizing nitrifier population of a soil. Appl. Environ. Microbiol. 36:584-588.
8. Belser, L. W., and E. L. Schmidt. 1978b. Serological diversity within a terrestrial ammonia-oxidizing population. Appl. Environ. Microbiol. 36:589-593.
9. Benemann, J. R. 1973. Nitrogen fixation in termites. Science 181:164-165.
10. Benson, D. R. 1982. Isolation of Frankia strains from older actinorhizal root nodules. Appl. Environ. Microbiol. 44:461-465.
11. Bergersen, F. J. 1981. Methods for evaluating biological nitrogen fixation. John Wiley and Sons, New York.
12. Bohlool, B. B., and E. L. Schmidt. 1974. Lectins: A possible basis for specificity in the Rhizobium-legume root nodule symbiosis. Science 185:269-271.
13. Brar, S. S., and J. Giddens. 1968. Inhibition of nitrification in Bladen grassland soil. Soil Sci. Soc. Am. Proc. 32:821-823.
14. Bremner, J. M., and A. M. Blackmer. 1981. Terrestrial nitrification as a source of atmospheric nitrous oxide. In C. C. Delwiche (ed.), Denitrification, Nitrification and Atmospheric Nitrous Oxide. John Wiley and Sons, New York.
15. Bremner, J. M., and K. Shaw. 1958. Denitrification in soil. I. Factors affecting denitrification. J. Agr. Sci. 51:40-51.
16. Breznak, J. A., W. J. Brill, J. W. Mertins, and H. C. Coppel. 1973. Nitrogen fixation in termites. Nature 244:577-580.
17. Brill, W. J. 1980. Biochemical genetics of nitrogen fixation. Microbiol. Rev. 44:449-467.
18. Broadbent, F. E., and F. E. Clark. 1965. Denitrification. In W. V. Bartholomew and F. E. Clark (eds.), Soil Nitrogen. American Society of Agronomy, Madison, Wisconsin.

19. Broadbent, F. E., and A. B. Carlton. 1978. Field trials with isotopically labeled nitrogen fertilizer. In D. R. Nielsen and J. G. Macdonald (eds.), Nitrogen in the Environment, Vol. I. Academic Press, New York.

20. Broughton, W. J. (ed.) 1981. Nitrogen fixation, vol. 1. Ecology. Oxford University Press, New York.

21. Broughton, W. J. (ed.) 1982. Nitrogen fixation, vol. 2. Rhizobium. Oxford University Press, New York.

22. Brouzes, R. J. Lasik, and R. Knowles. 1969. The effect of organic amendment, water content and oxygen on the incorporation of $^{15}N_2$ by some agricultural and forest soils. Can. J. Microbiol. $\underline{15}$:899-905.

23. Buchanan, R. E., and N. E. Gibbons (eds.) 1974. Bergey's manual of determinative bacteriology, eighth edition. Williams and Wilkins Co., Baltimore.

24. Burford, J. R., and J. M. Bremner. 1975. Relationships between the denitrification capacities of soils and total, water-soluble and readily decomposable soil organic matter. Soil Biol. Biochem. $\underline{7}$:389-394.

25. Burris, R. H., D. J. Arp, R. V. Hageman, J. P. Houchins, W. J. Sweet, and M. Tso. 1981. Mechanism of nitrogenase action. In A. H. Gibson and W. E. Newton (eds.), Current Perspectives in Nitrogen Fixation. Elsevier, Amsterdam.

26. Buresh, R. J., and W. H. Patrick, Jr. 1978. Nitrate reduction to ammonium in anaerobic soil. Soil Sci. Soc. Am. J. $\underline{42}$:913-918.

27. Callaham, D., P. del Tredici, and J. G. Torrey. 1978. Isolation and cultivation in vitro of the actinomycete causing root nodulation in Comptonia. Science $\underline{199}$:899-902.

28. Cameron, R. E., and W. H. Fuller. 1960. Nitrogen fixation by some algae in Arizona soils. Soil Sci. Soc. Am. Proc. $\underline{24}$:353-356.

29. Caskey, W. H., and J. M. Tiedje. 1979. Evidence for clostridia as agents of dissimilatory reduction of nitrate to ammonium in soils. Soil Sci. Soc. Am. J. 43:931-936.

30. Chatt, D. 1981. Towards new catalysts for nitrogen fixation. In A. H. Gibson and W. E. Newton (eds.), Current Perspectives in Nitrogen Fixation. Elsevier, Amsterdam.

31. Child, J. J. 1980. Nitrogen fixation by free-living Rhizobium and its implications. In N. S. Subba Rao (ed.), Recent Advances in Biological Nitrogen Fixation. E Arnold, London.

32. Clark, C., and E. L. Schmidt. 1967. Growth response of Nitrosomas europaea to amino acids. J. Bacteriol. 93:1302-1308.

33. Clark, F. E., and E. A. Paul. 1970. The microflora of grassland. Advan. Agron. $\underline{22}$:375.

34. Clark, F. E., and T. Rosswall (eds.) 1979. Terrestrial nitrogen cycles: processes, ecosystem strategies and management impacts. Ecological Bulletins No. 33. United Nations Environment Program.

35. Cole, J. A., and C. M. Brown. 1980. Nitrite reduction to ammonia by fermentative bacteria: A short circuit in the biological nitrogen cycle. FEMS Microbial. Lett. 7:65-72.

36. C.A.S.T. (Council for Agricultural Science and Technology). 1976. Effect of increased nitrogen fixation on stratospheric ozone. Report No. 53.

37. Court, M. N., R. C. Stephen, and J. S. Waid. 1962. Nitrite toxicity arising from the use of urea as a fertilizer. Nature 194:1264.

38. Crutzen, P. J., L. E. Heidt, J. P. Krasnec, W. H. Pollock, and W. Seiler. 1979. Biomass burning as a source of atmospheric gases CO, H_2, N_2O, NO, CH_3, Cl and COS. Nature 282: 253-256.

39. Crutzen, P. J. 1981. Atmospheric chemical processes of the oxides of nitrogen, including nitrous oxide. In C. C. Delwiche (ed.), Denitrification, Nitrification, and Atmospheric Nitrous Oxide. John Wiley and Sons, New York.

40. Dalton, H. 1977. Ammonia oxidation by the methane oxidising bacterium, Methylococcus capsulatus strain bath. Arch. Microbiol. 114:273-279.

41. Day, J. M., D. Harriss P. J. Dart, and P. van Berkum. 1975. The broadbalk experiment. An investigation of nitrogen gains from non-symbiotic nitrogen fixation. In W. D. P. Stewart (ed.), Nitrogen Fixation by Free-living Bacteria. Cambridge University Press, London.

42. Delwiche, C. C., and J. Wijler. 1956. Non-symbiotic nitrogen fixation in soil. Plant Soil 7:113-129.

43. Dobereiner, J., and J. M. Day. 1975. Nitrogen fixation in the rhizosphere of tropical grasses. In W. D. P. Stewart (ed.), Nitrogen Fixation by Free-living Microorganisms. Cambridge University Press, London.

44. Dommergues, Y. R., L. W. Belser, and E. L. Schmidt. 1978. Limiting factors for microbial growth and activity in soil. Advances in Microbial Ecology 2:49-104.

45. Doxtader, K. G., and M. Alexander. 1966. Role of 3-Nitropropanoic acid in nitrate formation by Aspergillus flavus. J. Bacteriol. 91:1186-1191.

46. Evans, W. 1977. Biochemistry of the bacterial catabolism of aromatic compounds in anaerobic environments. Nature 270:17-22.

47. Federova, R. I., E. I. Milekhina, and N. I. Ilyukhina. 1973. Possibility of using the gas-exchange method to detect extraterrestrial life: Identification of nitrogen fixing organisms. Akad. Nauk SSR Izvestia Ser. Biol. 6:797-806.

48. Fehr, P. I., P. C. Pang, R. A. Hedlin, and C. M. Cho. 1972. Some factors affecting a symbiotic nitrogen fixation in soils as measured by ^{15}N enrichment. Agron. J. 64:251-254.

49. Finlay, B. J., A. S. W. Span, and J. M. P. Harman. 1983. Nitrate respiration in primitive eucaryotes. Nature 303:333-336.

278

50. Fliermans, C. F., B. B. Bohlool, and E. L. Schmidt. 1974. Autecological study of the chemoautotroph Nitrobacter by immunofluorescence. Appl. Microbiol. 27:124.
51. Focht, D. D. 1978. Methods for analysis of denitrification in soils. In D. R. Nielsen and J. G. McDonald (eds.), Nitrogen in the Environment, Vol. 1. Academic Press, New York.
52. Focht, D. D., and W. Verstraete. 1977. Biochemical ecology of nitrification and denitrification. Advances in Microbial Ecology 1:135-214.
53. Gamble, T. N., M. R. Betlach, and J. M. Tiedje. 1977. Numerically dominant denitrifying bacteria from world soils. Appl. Environ. Microbiol. 33:926-939.
54. Garcia, J. L. 1973. Influence de la rhizosphere du riz sur l'activite denitrifiante potentielle des sols de rizieres du Senegal. Oecol. Plant. 8:315-323.
55. Ghiorse, W. C., and M. Alexander. 1978. Nitrifying populations and the destruction of nitrogen dioxide in the soil. Microb. Ecol. 4:233-240.
56. Gibson, A. H., and W. E. Newton (eds.) 1981. Current perspectives in nitrogen fixation. Australian Academy of Science, Canberra.
57. Goring, C. A. 1962. Control of nitrification by 2-chloro-6-trichloromethyl pyridine. Soil Sci. 93:211-218.
58. Gould, W. D., R. J. Bryant, J. A. Trofymow, R. V. Anderson, E. T. Elliott, and D. C. Coleman. 1981. Chitin decomposition in a model soil system. Soil Biol. Biochem. 13:487-492.
59. Granhall, U. 1975. Nitrogen fixation by blue-green algae in temperate soils. In W. D. P. Stewart (ed.), Nitrogen Fixation by Free-living Micro-organisms. Cambridge University Press, London.
60. Greenwood, D. J., and D. Goodman. 1967. Direct measurements of the distribution of oxygen in soil aggregates and in columns of fine soil crumbs. J. Soil Sci. 18:182-196.
61. Guthrie, T. F., and J. M. Duxbury. 1978. Nitrogen mineralization and denitrification in organic soils. Soil Sci. Soc. Am. J. 42:908-912.
62. Hamblin, J., and S. P. Kent. 1974. Possible role of phytohaemagglutinin in Phaseolus vulgaris L. Nat. New. Biol. 245:28-30.
63. Hardy, R. W. F. (ed.) 1977. A treatise on dinitrogen fixation. John Wiley and Sons, New York.
64. Hauck, R. D. 1981. Nitrogen fertilizer effects on nitrogen cycle processes. In F. E. Clark and T. Rosswall (eds.), Terrestrial Nitrogen Cycles. Ecological Bulletins No. 33. United Nations Environment Program.
65. Hubbell, D. H. 1981. Legume infection by Rhizobium: A conceptual approach. BioSci. 31:832-837.
66. Ineson, P., M. A. Leonard, and J. M. Anderson. 1982. Effect of collembolan grazing upon nitrogen and cation leaching from decomposing leaf litter. Soil Biol. Biochem. 14:601-605.

67. Ishaque, M., and A. H. Cornfield. 1972. Nitrogen mineralization and nitrification during incubation of East Pakistan tea soils in relation to pH. Plant Soil 37:91-95.

68. Ishaque, M., and M. I. H. Aleem. 1973. Intermediates of denitrification in the chemoautotroph Thiobacillus denitrificans. Arch. Microbiol. 94:269-282.

69. Josserand, A. G. Gay, and G. Faurie. 1981. Ecological study of two Nitrobacter serotypes coexisting in the same soil. Microb. Ecol. 7:275-280.

70. Kaneko, M., and M. Ishimoto. 1978. A study on nitrate reductase from Propiomobacterium acidi-propionici. J. Biochem. 83:191-200.

71. Kaspar, H. F., and J. M. Tiedje. 1981. Dissimilatory reduction of nitrate and nitrite in the bovine rumen: Nitrous oxide production and effect of acetylene. Appl. Environ. Microbiol. 41:705-709.

72. Kennedy, C., F. Cannon, M. Cannon, R. Dixon, S. Hill, J. Jenson, S. Kumar, P. McLean, M. Merrick, R. Robson, and J. Postgate. 1981. Recent advances in the genetics and regulation of nitrogen fixation. In A. H. Gibson and W. E. Newton (eds.), Current Perspectives in Nitrogen Fixation. Elsevier, Amsterdam.

73. Kissel, D. E., and S. J. Smith. 1978. Fate of fertilizer nitrate applied to coastal bermudagrass on a swelling clay soil. Soil Sci. Soc. Am. J. 42:77-80.

74. Knowles, R. 1977. The significance of asymbiotic dinitrogen fixation by bacteria. In R. Hardy and A. Gibson (eds.), A Treatise on Dinitrogen Fixation, Section IV. Biology. John Wiley and Sons, New York.

75. Knowles, R. 1982. Denitrification. Microbiol. Rev. 46:43-70.

76. Lethbridge, G., M. S. Davidson, and G. P. Sparling. 1982. Critical evaluation of the acetylene reduction test for estimating the activity of nitrogen-fixing bacteria associated with the roots of wheat and barley. Soil Biol. Biochem. 14:27-35.

77. Likens, G. E., F. H. Bormann, and N. M. Johnson. 1968. Nitrification: Importance to nutrient losses from a cutover forested ecosystem. Science 163:1205-1206.

78. Mainwright, M. 1978. A review of the effects of pesticides on microbial activity in soils. J. Soil Sci. 29:287-298.

79. Martinez-Molina, E., V. M. Morales, and D. H. Hubbell. 1979. Hydrolytic enzyme production by Rhizobium. Appl. Environ. Microbiol. 38:1186-1188.

80. Mayland, H. F., T. H. McIntosh, and W. H. Fuller. 1966. Fixation of isotopic nitrogen on a semiarid soil by algae crust organisms. Soil Sci. Soc. Am. Proc. 30:56-60.

81. McElroy, M. B., S. C. Wolfsy, and Y. L. Yung. 1977. The nitrogen cycle: Perturbations due to man and their impact on atmospheric N_2O and O_3. Philos. Trans. R. Soc. London 277B:159-181.

82. Moore, D. R. E., and J. S. Waid. 1971. The influence of washings of living roots on nitrification. Soil Biol. Biochem. 3:69-83.

83. Morrill, L. G., and J. E. Dawson. 1962. Growth rates of nitrifying chemoautotrophs in soil. J. Bacteriol. 83:205-206.

84. Morrill, L. G., and J. E. Dawson. 1967. Patterns observed for the oxidation of ammonium to nitrate by soil organisms. Soil Sci. Soc. Am. Proc. 31:757-760.

85. Mulder, E. G. 1975. Physiology and ecology of free-living, nitrogen-fixing bacteria. In W. D. P. Stewart (ed.), Nitrogen Fixation by Free-living Micro-organisms. Cambridge University Press, London.

86. Nadler, K. D., and Y. J. Avissar. 1977. Heme synthesis in soybean root nodules. Plant Physiol. 60:433-436.

87. N.A.S. (National Academy of Sciences). 1978. Nitrates: An environmental assessment.

88. Newton, W. E., and C. J. Nyman (eds.) 1976. Proceedings of the First International Symposium on Nitrogen Fixation. Washington State University Press, Pullman.

89. Newton, W. E., and W. H. Orme-Johnson (eds.) 1980. Nitrogen fixation. Vol. 1: Free-living systems and chemical models. University Park Press, Baltimore.

90. Nommik, H. 1956. Investigations on denitrification in soil. Acta Agr. Scand. 6:195-228.

91. Nutman, P. S. (ed.) 1976. Symbiotic nitrogen fixation in plants. Cambridge University Press, London.

92. Olson, R. V. 1980. Fate of togged nitrogen fertilizer applied to irrigated corn. Soil Sci. Soc. Am. J. 44:514-517.

93. Pate, J. S. 1977. Functional biology of dinitrogen fixation by legumes. In R. W. F. Hardy and WS Silver (eds.), A Treatise on Dinitrogen Fixation - Section III: Biology. John Wiley and Sons, New York.

94. Patrick, W. H., and K. R. Reddy. 1976. Nitrification denitrification reactions in flooded soils and water bottoms: Dependence on oxygen supply and ammonium diffusion. J. Environ. Qual. 5:469-472.

95. Payne, W. J. 1973. Reduction of nitrogenous oxides by microorganisms. Bacteriol. Rev. 37:409-452.

96. Payne, W. J. 1981. Denitrification. John Wiley and Sons, New York.

97. Pedersen, W. L., K. Chakrabarty, R. V. Klucas, and A. K. Vidaver. 1978. Nitrogen fixation (acetylene reduction) associated with roots of winter wheat and sorghum in Nebraska. Appl. Environ. Microbiol. 35:129-135.

98. Phillips, R. E., K. R. Reddy, and W. H. Patrick, Jr. 1978. The role of nitrate diffusion in determining the order and rate of denitrification in flooded soil. II. Theoretical analysis and interpretation. Soil Sci. Soc. Am. J. 42:272.

99. Postgate, J. R. (ed.) 1971. The chemistry and bio-chemistry of nitrogen fixation. Plenum Press, London.

100. Prakasam, T. B. S., and R. C. Loehr. 1972. Microbial nitrification and denitrification in concentrated wastes. Water Res. 6:859-869.
101. Purchase, B. S. 1974. Evaluation of the claim that grass root exudates inhibit nitrification. Plant Soil 41:527-539.
102. Quispel, A. (ed.) 1974. The biology of nitrogen fixation. American Elsevier, New York.
103. Reddy, K. R., and W. H. Patrick, Jr. 1980. Losses of applied ammonium ^{15}N, urea ^{15}N, and organic ^{15}N in flooded soils. Soil Sci. 130:326-330.
104. Rennie, R. J., and E. H. Schmidt. 1974. Fluorescent antibody techniques to study the autecology of Nitrobacter in soils. Abstracts of the Annual Meeting of the American Society for Microbiology. Chicago.
105. Rice, E. H., and S. K. Pancholy. 1973. Inhibition of nitrification by climax ecosystems. II. Additional evidence and possible role of tannins. Amer. J. Bot. 60:691-702.
106. Rice, W. A., E. A. Paul, and L. R. Wetter. 1967. The role of anaerobiosis in asymbiotic nitrogen fixation. Can. J. Microbiol. 13:829-836.
107. Robinson, J. B. 1963. Nitrification in a New Zealand grassland soil. Plant Soil 19:173-183.
108. Rolston, D. E., D. L. Hoffman, and D. W. Toy. 1978. Field measurement of denitrification. I. Flux of N_2 and N_2O. Soil Sci. Soc. Am. J. 42:863-869.
109. Rolston, D. E., F. E. Broadbent, and D. A. Goldhamer. 1979. Field measurement of denitrification. II. Mass balance and sampling uncertainty. Soil Sci. Soc. Am. J. 43:703-708.
110. Ryden, J. C., L. J. Lund, K. Letey, and D. D. Focht. 1979. Direct measurement of denitrification loss from soils: II. Development and application of field methods. Soil Sci. Soc. Am. J. 43:110.
111. Ryden, J. C., and L. J. Lund. 1980. Nature and extent of directly measured denitrification losses from some irrigated vegetable crop production units. Soil Sci. Soc. Am. J. 44:505-511.
112. Sabey, B. R. 1969. Influence of soil moisture tension on nitrate accumulation in soil. Soil Sci. Soc. Am. Proc. 33:263-266.
113. Satoh, T., Y. Hoshino, and H. Kitamura. 1974. Isolation of denitrifying photosynthetic bacteria. Agric. Biol. Chem. 38:1749-1751.
114. Siegel, R. S., R. D. Hauck, and L. T. Kurtz. 1982. Determination of ^{30}N$_2$ and application to measurement of N_2 evolution during denitrification. Soil Sci. Soc. Am. J. 46:68-74.
115. Sidloi-Lumbroso, R., L. Kleiman, and H. M. Schulman. 1978. Biochemical evidence that leghaemoglobin genes are present in the soybean but not in Rhizobium genome. Nature 273:558-560.

116. Silvester, W. B. 1976. Endophyte adaptation in Gunnera-Nostoc symbiosis. In P. S. Nutman (ed.), Symbiotic Fixation in Plants. Cambridge University Press, London.

117. Silvester, W. B. 1977. Dinitrogen fixation by plant associations excluding legumes. In R. Hardy and A. Gibson (eds.), A Treatise on Dinitrogen Fixation IV. John Wiley and Sons, New York.

118. Silvester, W. B., and D. R. Smith. 1969. Nitrogen fixation by Gunnera-Nostic symbiosis. Nature 224:1321.

119. Sinclair, J. L., J. F. McClelland, and D. C. Coleman. 1981. Nitrogen mineralization by Acanthomoeba polyphaga in grazed Pseudomonas paucimobilis populations. Appl. Environ. Microbiol. 42:667-671.

120. Smith, K. A. 1977. Soil aeration. Soil Sci. 123:284-291.

121. Smith, M. S., and J. M. Tiedje. 1979a. Phases of denitrification following oxygen depletion in soil. Soil Biol. Biochem. 11:261-267.

122. Smith, M. S., and J. M. Tiedje. 1979b. The effect of roots on soil denitrification. Soil Sci. Soc. Am. J. 43:951-955.

123. Smith, M. S., and K. Zimmerman. 1981. Nitrous oxide production by non-denitrifying soil nitrate reducers. Soil Sci. Soc. Am. J. 45:865-871.

124. Smith, M. S. 1982. Dissimilatory reduction of NO_2^- to NH_4^+ and N_2O by a soil Citrobacter sp. Appl. Environ. Microbiol. 43:854-860.

125. Starr, J. L., F. E. Broadbent, and D. R. Nielsen. 1974. Nitrogen transformations during continuous leaching. Soil Sci. Soc. Am. Proc. 38:283.

126. Stanford, G., J. O. Legg, S. Dzienia, and E. C. Simpson. 1975. Denitrification and associated nitrogen transformations in soils. Soil Sci. 120:147-152.

127. Stefanson, R. C. 1972. Soil denitrification in sealed soil-plant systems. I. Effect of plants, soil water content and soil organic matter content. Plant Soil 33:113-127.

128. Stewart, W. D. P. (ed.) 1975. Nitrogen fixation by free-living microorganisms. Cambridge University Press, London.

129. Stewart, W. D. P., P. Powell, and A. N. Rai. 1980. Symbiotic nitrogen-fixing cyanobacteria. In W. D. P. Stewart and J. R. Gallon (eds.), Nitrogen Fixation. Academic Press, New York.

130. Steyn, P. L., and C. C. Delwiche. 1970. Nitrogen fixation by nonsymbiotic microorganisms in some California soils. Environ. Sci. Technol. 4:1122-1128.

131. Subba Rao, N. S. (ed.) 1980. Recent advances in biological nitrogen fixation. E. Arnold, London.

132. Syers, J. K., A. N. Sharpley, and D. R. Keeney. 1979. Cycling of nitrogen by surface-casting earthworms in a pasture ecosystem. Soil Biol. Biochem. 11:181-185.

133. Takahashi, H. (ed.) 1975. Nitrogen fixation and the nitrogen cycle. University of Tokyo Press, Tokyo.

134. Talley, S. N., E. Lim, and D. W. Rains. 1977. Application of Azolla in crop production. In J. M. Lyons, R. C. Valentine, D. A. Phillips, D. W. Rains and R. C. Huffaker (eds.), Genetic Engineering for Nitrogen Fixation. Plenum Press, New York.

135. Theron, J. J. 1951. The influence of plants on the mineral nitrogen and the maintenance of organic matter in the soil. J. Agric. Sci. Cambridge 41:289-296.

136. Trinick, M. J. 1973. Symbiosis between Rhizobium and the non-legume, Trema aspera. Nature 244:459-460.

137. van Berkum, P., and B. B. Bohlool. 1980. Evaluation of nitrogen fixation by bacteria in association with roots of tropical grasses. Microbiol. Rev. 44:491-517.

138. Verstraete, W. 1981. Nitrification. In F. E. Clark and T. Rosswall (eds.), Terrestrial Nitrogen Cycles. Ecological Bulletins No. 33. United Nations Environment Program.

139. Verstraete, W., and M. Alexander. 1972. Heterotrophic nitrification by Arthrobacter sp. J. Bacteriol. 110:955-961.

140. Vitousek, P. M., J. R. Gosz, C. C. Grier, J. M. Melillo, W. A. Reiners, and R. L. Todd. 1979. Nitrate losses from disturbed ecosystems. Science 204:469-474.

141. Vlassak, K., E. A. Paul, and R. E. Harris. 1973. Assessment of biological nitrogen fixation in grassland and associated sites. Plant Soil 38:637-649.

142. Volz, M. G., M. S. Ardakani, R. K. Schulz, L. H. Stolzy, and A. D. McLaren. 1976. Soil nitrate loss during irrigation: Enhancement by plant roots. Agron. J. 68:621-627.

143. Watanabe, I., and S. Brotonegoro. 1981. Paddy fields. In W. J. Broughton (ed.), Nitrogen Fixation, Vol. 1. Clarendon Press, Oxford.

144. Waughman, G. J., J. R. J. French, and K. Jones. 1981. Nitrogen fixation in some terrestrial environments. In W. J. Broughton (ed.), Nitrogen Fixation, Vol. 1. Clarendon Press, Oxford.

145. Weber, D. F., and P. L. Gainey. 1962. Relative sensitivity of nitrifying organisms to hydrogens ions in soils and in solutions. Soil Sci. 94:138-145.

146. Wijler, J., and C. C. Delwiche. 1954. Investigations on the denitrifying process in soil. Plant Soil 5:155-169.

147. Winogradsky, S. 1890. Sur les organismes de la nitrification. C.R. Acad. Sci. Paris 110:1013.

148. Woldendorp, J. W. 1962. The quantitative influence of the rhizosphere on denitrification. Plant Soil 17:267-270.

149. Woods, L. E., C. V. Cole, E. T. Elliott, R. V. Anderson, and D. C. Coleman. 1982. Nitrogen transformations in soil as affected by bacterial-microformal interactions. Soil Biol. Biochem. 14:93-98.

150. Yordy, D. M., and K. L. Ruoff. 1981. Dissimilatory nitrate reduction to ammonia. In C. C. Delwiche (ed.), Denitrification, Nitrification, and Atmospheric Nitrous Oxide. John Wiley and Sons, New York.
151. Yoshida, T., R. A. Roncal, and E. M. Bautista. 1973. Atmospheric nitrogen fixation by photosynthetic micro-organisms in a submerged Phillippine soil. Soil Sci. Plant Nutr. 19:117-123.

THE ROLE OF MICROFLORA IN TERRESTRIAL SULFUR CYCLING

J. P. Nakas*

SULFUR IN THE ENVIRONMENT

In most ecosystems, S is considered to be a macronutrient whose concentration can be limiting to the existing biota (Nyborg 1978). Although sulfate represents the form of S which is most readily available to plants and microorganisms, it is now generally accepted that organic forms of S represent the major S component(s) in most terrestrial ecosystems (Freney et al. 1972, Fitzgerald 1978, David et al. 1982). Soils generally contain between 10% and 25% of total S as sulfate-sulfur with the remainder as organic forms which are mostly carbon-bonded S and ester sulfates (Campbell 1975). Although the largest reservoirs of S, as well as their rates of trans-formation, are due to abiotic and microbial processes, the anthropogenic inputs (due almost exclusively to the combustion of fossil fuels and the smelting of ores) provide a substantial contribution of S to the atmosphere and thus to aquatic and terrestrial ecosystems. Kellogg et al. (1972) concluded that by A.D. 2000, the anthropogenic input of S to the atmosphere will exceed natural inputs in the Northern Hemisphere. Because of the large reservoir of organic sulfur present in most soils (Fig. 1) as well as the increased deposition of sulfate-S, it is essential to evaluate factors affecting S dynamics including S mineralization and immobilization and thus S availability to plants and soil biota. Although the full impact of anthropogenic sulfur inputs on terrestrial

    Department of Environmental and Forest Biology, S.U.N.Y.,
College of Environmental Science and Forestry, Syracuse,
New York 13210.

286

ecosystems has not yet been clarified, specific effects have been postulated (Hutchinson and Havas 1980, Mitchell and Landers 1981) and David et al. (1983) have constructed some detailed models for S transformation and movement through forest ecosystems.

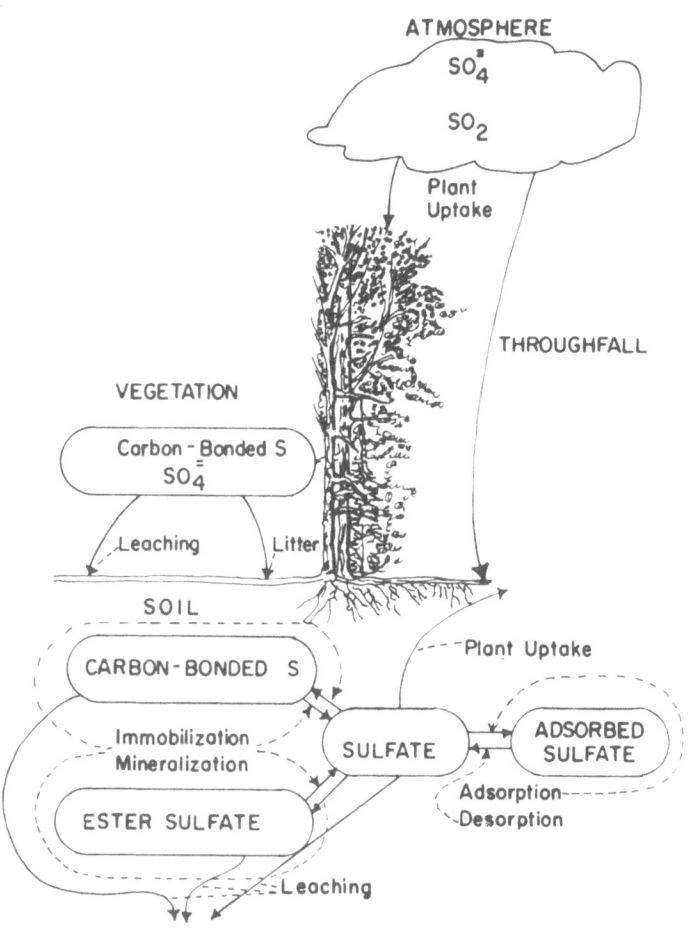

FIGURE 1. Major sulfur components present in terrestrial ecosystems.

MAJOR SOIL SULFUR COMPONENTS

Generally, the majority of sulfur in soil surface horizons is present in organic forms with lesser amounts present as sulfate (David et al. 1982). However, in subsoil horizons appreciable amounts of sulfate may accumulate and constitute a major portion of the total sulfur. This is especially true in acid soils where sulfate adsorption by soils and clays reaches a maximum in the pH range 2-4. Adsorption is known to occur on the surfaces of mineral colloids, particularly hydrated Al- or Fe-hydroxides or on clay minerals, especially kaolinite (Ulrich 1980). However, in most soils the organic sulfur component will constitute the major sulfur fraction (Freney 1967, Freney et al. 1970).

There are numerous examples of sulfur-containing organic compounds which have been isolated from soil and ultimately identified. These compounds have been categorized as amino acids, vitamins, sulfonium compounds, sulfate esters, sulfides, sulfoxides, isothiocyanates, and others (Freney 1967). However, due to the difficulty of extraction, purification, and identification procedures, as well as constant metabolic turnover, the major sulfur components of most soils remain largely unknown. Current practical methodology limits organic sulfur speciation to two categories: (i) carbon-bonded sulfur which is obtained by difference: total S-(HI-S + ZnHCl-S) (Landers et al. 1983) and (ii) ester sulfate, which is organic S reducible to H_2S with hydriodic acid (Fitzgerald 1978). The incorporation of sulfate into organic sulfur has been studied by Fitzgerald and Johnson (1983) using ^{35}S-SO_4^{-2}. The authors concluded that the metabolic fate of sulfate was similar in both agricultural and forest ecosystems and the concentration and relative availability of inorganic sulfate will regulate mineralization of the non-salt extractable S fraction (ester sulfate).

Analyses of soils taken from various locations throughout the world have revealed a pattern for organo-sulfur compounds in which the combination of compounds present as carbon-bonded S and ester sulfate S comprise approximately 90% of the total S. Inorganic S constitutes the remaining 10%, primarily as sulfate, sulfide, and thiosulfate (Fitzgerald 1978, David et al. 1982).

Several studies have demonstrated the relative importance of organo-sulfur compounds as potential sources of sulfur for plants and microorganisms (Freney et al. 1969, 1971, 1975). It is well established from pure culture studies that several microbial genera are capable of sulfate conversion to reduced organic forms such as methionine, cysteine, and cystine. Plants are also capable of the same type of assimilation (Maynard et al. 1985). Although most of the sulfur in soil organisms has been identified as cysteine and methionine residues (Freney 1976), the actual rates of sulfur reduction and incorporation have not been quantified.

It has been clearly established that organic S, in the form of ester sulfate and carbon-bonded S, represents the major S component of agricultural soils and, therefore, serves as a potentially important reservoir of S to the soil biota (Freney 1976, Freney et al. 1970). Fitzgerald (1978) summarized data from 208 different soils and found that only 5.2% of the total S was present as sulfate and only 2.8% as elemental S and sulfide. In every soil analysis where organic S components were reported, the combination of ester sulfate and carbon-bonded S exceeded 50% of the total sulfur present with several soils containing 60% to 90% of total sulfur in organic form.

Research in forest soils on S components in particular and nutrient cycling in general has lagged behind agricultural soils. David et al. (1982) have examined forest soils in the Adirondack Mountains of upstate New York and found carbon-bonded S and ester sulfate comprised an average of 93% of the total S in the upper horizons at both a hardwood and a conifer site. However, carbon-bonded S was identified as the largest individual soil S component with an average of 74% of total S. Also David et al. (1982) were the first investigators to report on the correlations among sulfur components, microbial biomass, and sulfohydrolase activity. These findings strengthen the importance of microbial activity to S transformation and availability in forest soils. These data further suggest that anthropogenic inputs of sulfate to certain terrestrial ecosystems may not exceed current estimates of internal S turnover. In Adirondack forest soils, inorganic S ranged from 5% of total S in the O1, O2, and A2 horizons of a conifer site to 10-15% of total S in the B23 and ClX horizons of a hardwood site. Mitchell et al. (1983) demonstrated that the combination of above ground and below ground S production constitutes 59% of the total S input to the soil ecosystem. The transformation of these inputs is regulated by microbial activity and thus more research is needed to elucidate which factors affect microbial S dynamics.

The forms of S which volatize from soil vary drastically with soil conditions and the particular organic S compounds being metabolized (Banwart and Bremner 1976). Although earlier investigators had hypothesized that H_2S was the dominant form of volatile S from both aquatic and terrestrial systems, recent studies have shown that other sulfur compounds may be more important (Asami and Takai 1963, Bremner 1977, Farwell et al. 1979). Through dissimilatory sulfate reduction, H_2S may be produced as the result of the utilization of sulfate as an electron acceptor. In addition, H_2S can also be formed by metabolism of organic S compounds, as can organic S gases. Organic S evolution is not restricted to reduced systems but is enhanced by oxygen limitation (Banwart and Bremner 1976). Furthermore, the evolution of S compounds is complicated by the sorption processes of soils which may result in part from the formation of metallic sulfides (Ponnamperuma 1972, Ayotade 1977). There exist numerous reports in the literature confirming the production of volatile organosulfur compounds from various substrates and under varying conditions by pure cultures of bacteria and fungi (Bremner and Steele 1978). However, several of these volatile organics have been found to be directly toxic to both plants and microorganisms as well as inhibit crucial mineralization processes such as ammonification (Ashworth et al. 1975). Banwart and Bremner (1976, 1976a, b) examined the evolution of S-containing organics from many soils exposed to aerobic and anaerobic conditions and amended with a variety of substrates. A distinct pattern of S volatilization emerged from these studies which established that the major S forms released were methyl mercaptan (CH_3SH), dimethyl disulfide (CH_3SSCH_3), dimethyl sulfide (CH_3SCH_3), carbonyl sulfide (COS), and carbon disulfide (CS_2). Also, these compounds could all be traced to organic matter decomposition. Similarly, CH_3SH, CH_3SCH_3 and CH_3SSCH_3 were the major sulfur components released (from methionine) under aerobic as well as anaerobic conditions. These compounds represent the major forms of volatile sulfur which have been identified and they are all of microbial origin.

Although the quantity and form of volatile S compounds are important for understanding the S cycle and determining the relative importance of natural and anthropogenic inputs in the global S balance (Bremner and Steele 1978, Hornor and Mitchell 1981), their contribution to overall S dynamics within a given heterotrophic system is generally less than one percent of total S.

ORGANIC SULFUR TRANSFORMATION

Most plants rely on the presence of sulfate to fulfill their S requirements. Assimilatory sulfate reduction has been found in all metaphytes studied and sulfate can serve as the sole source of S (Schiff and Hodson 1973, Maynard et al. 1985). Excluding the biogenic and anthropogenic sources of S (as SO_2 or $SO_4^=$), plants must rely upon the two major reservoirs of S present in soils: carbon-bonded S and ester sulfates. Therefore, the S requirements of plants create a dynamic equilibrium in which the microbial-mediated decomposition pathways for organic sulfur are ultimately regulated by the cellular needs of microorganisms thus resulting in S immobilization. Only during periods of excessive organic matter decomposition or cell lysis will an abundance of sulfur be available for plant use. Under conditions of sulfur stress in which the sulfur content of organic substrates is minimal, cellular immobilization of S will greatly restrict sulfate availability (Barrow 1961, Freney 1967). Freney (1975) has compiled a list of abiotic and biotic factors influencing the release of sulfate from soil organic matter. Investigations of organic S mineralization must be meticulously performed and cautiously interpreted especially if soils have been stored. Also, air drying of soils has been shown to result in increased rates of sulfate release (Barrow 1961, Freney 1967, David et al. 1982).

   C-bonded sulfur.

   The most thoroughly investigated examples of organic sulfur mineralization are the amino acids cysteine, cystine and methionine. The latter two amino acids represent the main focus of carbon-bonded mineralization since the first step in cysteine transformation is the oxidation to cystine. Freney (1967) and Ehrlich (1981) have assembled data from a number of pure culture, mixed culture, and soil habitat studies and developed the following pathway:

$$HSCH_2CH(NH_2)COOH \longrightarrow \underset{\displaystyle C-CH_2CH(NH_2)COOH}{\overset{\displaystyle S-CH_2CH(NH_2)COOH}{|}} \rightarrow \underset{\displaystyle CO_2CH_2CH(NH_2)COOH}{\overset{\displaystyle SCH_2CH(NH_2)COOH}{|}} \longrightarrow$$

cysteine   cystine   cystine disulfoxide

$$HOOSCH_2CH(NH_2)COOH \longrightarrow HOO_2SCH_2CH(NH_2)COOH \rightarrow SO_4^=$$

cystine sulfinic acid   cysteic acid   sulfate

Additionally, cysteine transformation may also yield sulfide in the form of H_2S via the following (Freney 1967):

$$HSCH_2CH(NH_2)COOH \longrightarrow \underset{\displaystyle SCH_2CH(NH_2)COOH}{\overset{\displaystyle SCH_2CH(NH_2)COOH}{|}} \longrightarrow$$

cysteine   cystine

$$HSSCH_2CH(NH_2)COOH + CH_3COCOOH + NH_3 \rightarrow \underset{\displaystyle SCH_2CH(NH_2)COOH}{\overset{\displaystyle SCH_2CH(NH_2)COOH}{|}} + H_2S$$

thiocysteine   pyruvic acid   ammonia   cystine   hydrogen sulfide

The decomposition of methionine in soil results primarily in the release of volatile organosulfur compounds, notably methyl mercaptan and dimethyl disulfide (Bremner and Steele 1978).

These data indicate the need for a greater understanding of the environmental factors affecting carbon-bonded sulfur mineralization. This class of organic sulfur is now recognized as a major sulfur component in soils (David et al. 1982) as well as some fresh water lake sediments (Mitchell et al. 1981). However, the form of carbon-bonded S in soil may be very different and much more recalcitrant than the above-mentioned amino acids. Structural identification of carbon-bonded S compounds in soil is necessary in order to establish the relative rates of formation in what can be considered the more humified fraction(s) of soil. Bettany et al. (1979) compared C, N, and S concentrations in various soil fractions selected along an organic matter gradient. Sulfur levels did not follow the general pattern of C and N concentrations indicating a unique pathway in humus formation and transformation. Consequently, S availability for plant and microbial use is to some extent regulated by the lability of a particular soil fraction. For example, HI-reducible S was associated with the aliphatic side chain components of the humic acid fraction rather than the condensed aromatic core. However, the majority of organic S for all soils was associated with the more labile components of both humic and fulvic acids. After identifying transformation inter-mediates and rates of immobilization, it will then be possible to determine meaningful rates of turnover and thus availability for plants and microorganisms. Although carbon-bonded S has been identified as one of the major S constit-uents in several types of soils (Fitzgerald 1978, Tabatabai and Bremner 1972, David et al. 1982), the impact on S cycling may be minimal due to slow rates of turnover and stability through polymerization into inaccessible forms.

Ester sulfate.

The second major component of organic sulfur

includes all organic sulfur forms which can be reduced by hydriodic acid to H_2S (Fitzgerald 1978). As originally defined by Freney (1958, 1961), a mixture of hydriodic, formic, and hypophosphorous acids will reduce all compounds containing either a C-O-S or a C-N-S linkage to H_2S. In those soils which have been analyzed for S content, the ester sulfate fraction represents approximately 25-50% of the total S (Fitzgerald 1978). Tabatabai and Bremner (1972) however, demonstrated ester sulfate concentrations in excess of 90% of the total S present in some Iowa soils.

Early work by Freney et al. (1971) demonstrated the movement of ^{35}S-$SO_4^=$ into both the carbon-bonded as well as the ester sulfate fractions. However, the incorporation of sulfate into the ester sulfate fraction of fulvic acid was predominant. Although actual rates of sulfate incorporation were not determined, a major S component was identified as a potentially important reservoir of S for both plant and microbial use.

The major mechanism for sulfate release from sulfate esters occurs via sulfohydrolase enzymes which will catalyze cleavage of the ester. The most common assay procedure relies on the release of p-nitrophenol from p-nitrophenyl sulfate (Tabatabai and Bremner 1972). Because this procedure is specific for arylsulfatases, it cannot quantify all sulfohydrolase activity. However, the relative ease of the procedure and the fact that most sulfohydrolases are active against arylsulfate substrates, has made this the method of choice. The major forms of ester sulfate which have been identified in soil and which serve as substrates for sulfohydrolases are choline sulfate, chondroitin sulfate, tyrosine sulfate, and a variety of related aryl- and alkyl sulfates (Fitzgerald 1978).

Under certain conditions, it has been observed that sulfate release from soils under non-limiting conditions was unrelated to arylsulfatase activity (Kowalenko and Lowe 1971). However, as has been pointed out by Fitzgerald (1978), sulfohydrolases are subject to end product inhibition and will therefore not be synthesized at maximal rates after an initial burst in enzyme activity and subsequent accumulation of sulfate. Therefore, it would be unrealistic to expect that sulfohydrolase activity would always be correlated with sulfate release (Kowalenko and Lowe 1975) over an extended period of time. David et al. (1982) demonstrated a high degree of correlation among soil organic matter content, carbon-bonded S, ester sulfate, microbial biomass and sulfohydrolase activity. Also, Press et al. (1985) have recently reported on the relationship between arylsulfatase activity in peat and acidic deposition. More research is needed on the temporal changes in these values under conditions of both nutritional variation and environmental stress.

The formation of ester sulfate in soil may represent a unique mechanism for storage and therefore regulation of sulfate for both plant and microbial use. Fitzgerald (1978) has shown that increased soil sulfate concentrations due to acidic deposition may result in increased ester sulfate formation and thus a mechanism for storage of sulfate in a form which will not alter the existing soil pH.

Most research on the regulation of sulfohydrolase activity has centered on enzyme synthesis in response to nutritional stress (Fitzgerald et al. 1975, Fitzgerald 1978). In general the sulfohydrolases are not constitutive in either bacteria or fungi and their synthesis is induced by low concentrations of C and S. In the microorganisms which have been studied, subtle differences exist in the induction and repression of sulfohydrolases from various sources. For example, choline sulfatase is substrate inducible in bacteria (Fitzgerald and Scott 1974), but is subject to a sulfur-mediated derepression in Aspergillus nidulans and Neurospora crassa (McGuire and Marzluf 1974). A thorough discussion of the regulation of sulfohydrolase synthesis in both bacteria and fungi can be found in reviews by Fitzgerald (1978) and Dodgson et al. (1982).

INORGANIC SULFUR TRANSFORMATIONS

Oxidation.

Several groups of microorganisms are capable of
oxidizing various forms of inorganic sulfur. The microorganisms involved can be divided into three major groups
as described by Zinder and Brock (1978): The colorless S
bacteria, heterotrophic S oxidizers, and the photosynthetic
bacteria.

The colorless S bacteria, of which members of the genus
Thiobacillus are probably best known, are capable of autotrophic growth via the oxidation of inorganic S compounds to
sulfate with a preference for H_2S rather than elemental S.
Additionally, thiosulfate and polythionates can be utilized
as a source of energy. Most members of the genus
Thiobacillus are either acid tolerant or acidophilic
microorganisms due to the generation of sulfuric acid
immediately following sulfate production (Brock 1966).

Several common genera of bacteria and fungi have been
described which are capable of oxidizing inorganic S
compounds but rely on trace levels of organics for growth
and reproduction (Ehrlich 1981). Some of these microorganisms have been identified as members of the following
genera: Arthrobacter, Bacillus, Micrococcus, Pseudomonas and
Mycobacterium. Although work on fungal contributions toward
S transformations has remained sparse, members of the genera
Saccharomyces and Debaryomyces have been found capable of
partial sulfur oxidation. Nor and Tabatabai (1976) have
developed a procedure for detecting thiosulfate and
tetrathionate in soils which has helped to identify
heterotrophic microorganisms capable of partial sulfur
transformations.

The major group of microorganisms responsible for the

anaerobic oxidation of inorganic S is the photosynthetic
bacteria. These microorganisms can always be found in
anaerobic zones of lakes and oceans where light and H_2S
concentrations are optimal (Stanier 1976). The primary end
product of sulfide oxidation is elemental S which is
deposited intracellularly by purple sulfur bacteria
(Chromatiaceae) and extracellularly by green sulfur bacteria
(Chlorobiaceae). However, these microorganisms are not
major components of the soil microflora and do not contribute
to the terrestrial cycling of S. <u>Thiobacillus</u> <u>denitrificans</u>
is the only chemoautotrophic sulfur oxidizing bacterium
capable of anaerobic oxidation of H_2S, SO_3^- and HSO_3^- using
NO_3^- as an electron acceptor (Ehrlich 1981).

The anaerobic oxidation of sulfide coupled to the reduction of fixed CO_2 constitutes the pivotal reaction in
bacterial photosynthesis. Members of the families
Chromatiaceae, Chlorobiaceae, and Rhodospirillaceae comprise
the major groups of phototrophic microorganisms capable of
anaerobic oxidation of sulfide (Ehrlich 1981). The reducing
power generated by sulfide oxidation is then coupled to
biosynthesis based on CO_2 assimilation. Castenholz (1976,
demonstrated that certain cyanobacteria are also capable of
carrying out bacterial photosynthesis utilizing H_2S and
photosystem I.

Heterotrophic oxidation of sulfide, either by true
heterotrophs or by mixotrophs, usually results in the
formation of elemental S which may subsequently be oxidized
to sulfate under conditions of H_2S limitation (Ehrlich 1981).
Although strict autotrophic growth on H_2S by normally
heterotrophic populations is often difficult to verify,
members of the genus <u>Beggiatoa</u> are capable of this type of
nutrition under conditions of H_2S limitation. Additionally,
Skerman <u>et</u> <u>al</u>. (1957) reported oxidation of H_2S by
<u>Sphaerotilus</u> <u>natans</u>, <u>Alternaria</u>, and yeast. Roy and Trudinger
(1970) have proposed a pathway for organisms capable of
oxidizing sulfide to sulfate:

$$S^{-2} \longrightarrow X \longrightarrow SO_3^{-2} \longrightarrow SO_4^{-2}$$
$$\Updownarrow$$
$$S_8$$

The intermediate (X) is thought to be a derivative of glutathione or a membrane-bound thiol.

The autotrophic oxidation of elemental sulfur to sulfuric acid has in the past been commonly associated with the genus Thiobacillus, especially T. thiooxidans and T. ferrooxidans which can both grow at a pH near 1.0. However, from 1970 until the present, many new organisms capable of autotrophic growth on sulfur at low pH and elevated temperatures have been described. Recently, some of these facultative autotrophic microorganisms have been used for the desulfurization of coal (Kargi and Robinson 1982) due to their elevated rate of reduced sulfur oxidation compared to the more common T. ferrooxidans (Hoffman et al. 1981). A fascinating and lucid description of many of these discoveries has been published in a monograph by T. D. Brock (1978) and will therefore not be further described in this chapter.

Sulfur reductions.

The reduction of oxidized forms of sulfur can proceed via assimilative reduction or dissimilative reduction. In each case, sulfur is produced in the -2 oxidation state and is available for oxidation by plants and microorganisms.

The pathway for assimilatory sulfate reduction is presented in Fig. 2. The initial reduction of sulfate to sulfite occurs after reaction with ATP to form adenylyl sulfate (Stanier et al. 1976). Although sulfate represents the major source of sulfur for most microorganisms, it necessitates reduction since most sulfur-containing compounds contain sulfur in the -2 oxidation state. After active transport into the cell, sulfate is converted to a more active form (adenosine-5'-phosphosulfate or APS) by ATP sulfurylase. APS is subsequently phosphorylated by APS phosphokinase to produce adenosine-3'-phosphate-5'-phosphosulfate (PAPS). The latter compound is then reduced by thioredoxin and results in sulfite release. The

$$SO_4{}^{2-} + ATP \xrightarrow{\text{ATP sulfurylase}} APS + PP$$

$$APS + ATP \xrightarrow{\text{APS kinase}} PAPS + ADP$$

$$NADPH + H^+ + PAPS \xrightarrow{\text{PAPS reductase}} PAP + SO_3{}^{2-} + NADP^+$$

$$3NADPH + 3H^+ + SO_3{}^{2-} + 2H^+ \xrightarrow{SO_3{}^{2-} \text{ reductase}} H_2S + 3H_2O + 3NADP^+$$

$$H_2S + serine \xrightarrow{\text{CoAS acetate}} cysteine + H_2O$$

FIGURE 2. General scheme of assimilatory sulfate reduction. Taken from Ehrlich (1981).

subsequent reduction of sulfite to sulfide requires the transfer of six electrons and is catalyzed by the enzyme sulfite reductase which differs from the dissimilatory sulfite reductase although both enzymes contain a common type of tetrahydroporphyrin-based prosthetic group (Murphy and Siegel 1973). Thus, sulfite is intracellularly reduced to sulfide. However, due to the acute toxicity of H_2S to most microorganisms, H_2S is immediately combined with O-acetylserine to form L-cysteine which is the major source for sulfur-containing organic compounds within the cell (Gottschalk 1979).

Recent studies by Cuhel et al. (1981, 1982a,b) have shown that assimilatory S metabolism can be a valuable tool for the measurement of natural population protein synthesis which is correlated to true microbial growth under in situ conditions. This method has very good application in marine systems because sulfate is always present at saturating conditions (>25mM) and the assimilation of ^{35}S-SO_4 is a genuine reflection of microbial protein synthesis and hence, growth.

Dissimilatory sulfate reduction necessitates anaerobic conditions and the use of sulfate as an electron acceptor. The rate of dissimilatory sulfate reduction will be a direct function of sulfate concentration and the particular organic moiety being oxidized (Fenchel and Blackburn 1979). Sulfate reduction in agronomic soils is limited to rice paddies although sporadic flooding of soils may result in temporary release of H_2S and volatile sulfur organics (Bremner and Steele 1978). The overall contribution of dissimilatory sulfate reduction in soils to the total sulfur budget is minimal when compared to this process in aquatic systems. The high concentration of sulfate salts in sea water ensures non-limiting diffusion of sulfate to the sediment (Fenckel and Blackburn 1979).

Dissimilatory sulfate reduction results from the metabolism by strict anaerobes which utilize sulfate as a terminal electron acceptor. The bacteria responsible for this transformation are members of the genera Desulfovibrio, Desulfotomaculum, and Desulfomonas. These microorganisms require the reduced end products of other fermentations (especially lactate, although ethanol and malate will also sustain growth) as oxidizable substrates from which reducing power can be generated for the reduction of sulfate (Stanier 1976). Lactate is oxidized to pyruvate, acetyl CoA, and finally acetate with the generation of one mole of ATP per mole of lactate (Fig. 3). The reduction of sulfate occurs in four distinct reductive steps after an initial reaction of sulfate with ATP to produce APS; this is a more active form of sulfate and is identical to the initial reaction in assimilatory sulfate reduction. APS is successively reduced to sulfite, trithionate, thiosulfate, and sulfide (Gottschalk 1979). Although ferredoxin, rubredoxin, menaquinone and various cytochromes have been shown to participate in these reactions, the exact location of each electron carrier remains unknown.

MAJOR FACTORS AFFECTING SULFUR TRANSFORMATIONS

Sorption capacity and leaching.

The ability of a soil to sorb various oxidized and reduced sulfur compounds is dependent on the soil type as well as specific chemical and physical parameters (Bremner and Banwart 1976, Ghiorse and Alexander 1976). Because of

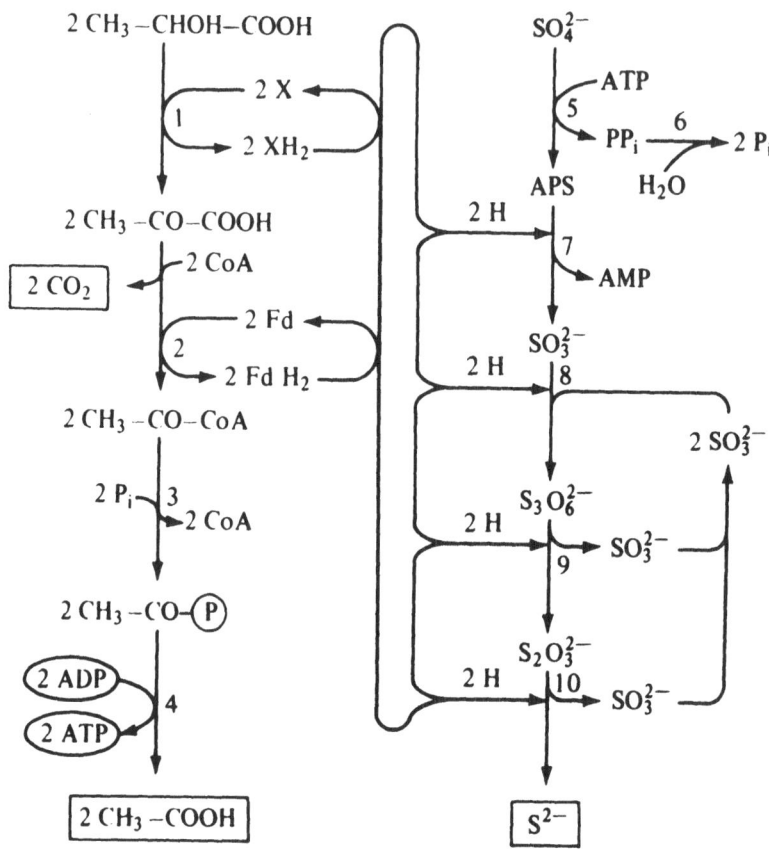

FIGURE 3. Pathway of dissimilatory sulfate reduction. 1,
lactate dehydrogenase, H-acceptor not known; 2, pyruvate-
ferredoxin oxidoreductase; 3, phosphotransacetylase; 4,
acetate kinase; 5, ATP sulfurylase; 6, pyrophosphatase; 7,
APS reductase; 8, sulfite reductase (desulfoviridin); 9,
trithionate reductase; 10, thiosulfate reductase. The
electron donors of the reductases are not known. APS,
adenosine-5'-phosphosulfate. Taken from Gottschalk (1979).

the negatively charged character of soil, the sulfate anion
may be readily leached unless adsorbed onto soil colloids
(Cronan et al. 1978, Johnson and Henderson 1979). The
resulting large input of S into ground water may be detrimental and cause a serious imbalance in available S especially
if oxygen tensions are sufficiently low to initiate sulfate
reduction.

Sulfate adsorption is affected by pH, the cations present

on exchange sites, and organic matter concentration (Fitzgerald and Johnson 1983). However, in some cases full sulfate recovery can only be achieved with phosphate or acetate due to the greater affinity of these anions for adsorption sites (Johnson and Henderson 1979, Fitzgerald and Johnson 1983).

Oxidation-reduction potential (Eh).

Eh has a major influence on S transformations within any heterotrophic system including soils. In oxic systems where Eh is greater than 350 mV, oxygen is used as the terminal electron acceptor in microbial respiration. With a depletion of oxygen, alternate electron acceptors such as NO_3^-, Mn^{+4}, Fe^{+3} and various organic compounds may be used (Patrick and Delaune 1977). Sulfate, which is in this group of anaerobic electron acceptors, is reduced to H_2S by obligate anaerobic bacteria within an Eh range of +115 to -450 mV, although an Eh of less than -95 mV is necessary to initiate this dissimilatory sulfur reduction (Zinder and Brock 1978).

Oxygen depletion causes a simultaneous decrease in decomposition rates due to the lower metabolic efficiency of anaerobiosis. In addition, some of the chemically reduced compounds, including S compounds, may be highly toxic to microflora, invertebrates and higher plants. This may result in an added depression of mineralization and C flux, although certain nutrients such as phosphate may become more available (Buckman and Brady 1969, Mah et al. 1977, Patrick and Delaune 1977, Rodriguez-Kabana et al. 1965, Skinner 1975).

Soil moisture.

Because of the low solubility of oxygen in water, it can be rapidly depleted under conditions of high oxygen demand. Within soil there is an inverse relationship between moisture content and the concentration of gases such as oxygen. In addition, moisture has a direct effect on C (Bunnel et al. 1977, Witkamp 1966) and S catabolism by its influence on the activity of soil biota.

pH.

The effects of pH on S transformations are probably
most severe on those microorganisms capable of sulfate
reduction. Because these organisms are known to be
sensitive to low pH (Alexander 1974), a decrease in their
activity would result in increasing levels of sulfate and an
accumulation of organic matter. These organisms represent a
major component of the microflora capable of organic matter
decomposition in anaerobic and sulfate-rich environments. A
gradual decline in soil pH may actually stimulate S oxidizing
organisms such as Thiobacillus which oxidize reduced inorganic
S to sulfate and thus actually generate sulfuric acid. It
is conceivable that these conditions may result in increased
solubilization and therefore availability of normally
insoluble elements. Lastly, a decrease in soil pH may
inhibit the floculation of clay particles leading to a
degeneration of soil structure (Campbell 1977).

Microflora.

Sulfur transformations are carried out by specific
microorganisms. For example, dissimilatory sulfate reduction
is accomplished by specialized, obligate anaerobic bacteria
(Desulfovibrio, Desulfuromonas, Desulfomonas and Desulfotomaculum). However, other sulfur transformations such as
assimilatory sulfate reduction can be performed by a variety
of microorganisms, and higher plants. Organic sulfur
decomposition will occur under aerobic as well as anaerobic
conditions and the process can be considered ubiquitous.
Inorganic S oxidation on the other hand, is strictly an
aerobic process carried out by autotrophic bacteria (e.g.
Thiobacillus) and a few heterotrophic bacteria (e.g.
Beggiatoa) (Stanier et al. 1976).

The microflora as a group show a wide response to
oxygen requirements (Eh) ranging from obligate anaerobes
and aerobes to facultative anaerobes (May et al. 1977,
Skinner 1975, Stotsky and Schenck 1976). In soils there is
a shift from aerobic to anaerobic metabolism when the pO_2
is less than 0.45 mm Hg and there is a depression in

respiration when pO_2 is less than 0.005 MPa (Greenwood 1961, Parr and Reuszar 1959). This shift will result in partially oxidized organic compounds such as acetic acid and $(CH_3)_2S_2$ and reduced inorganic compounds such as ferrous sulfide.

INTERACTIONS AMONG SULFUR, CARBON AND NITROGEN TRANSFORMATIONS

Organic S and C transformations may be both directly and indirectly linked. Direct linkage is especially impor- tant in those substrates in which organic S dominates since C and S mineralization rates will be related. However, the degree of S mineralization will also be dependent on the biotic demand of the system (May et al. 1968, 1972, McGill and Cole 1981). David et al. (1982) showed a high degree of correlation between organic matter content and organic sulfur constituents. Subsequently, David (1983) derived an equation to predict total S (μmoles S g^{-1}) based on organic matter content for a Spodosol in the Adirondack region of New York: Total S (μmoles S g^{-1} dry mass) = 2.17 + 0.71 (% organic matter). When total N was added to the correla- tion matrix, a high degree of correlation ($p < 0.01$) was found among N, organic matter, C and S. In addition C:N:S ratios were calculated in an attempt to predict N and S mineralization or immobilization during decomposition. Although elevated ratios would indicate immobilization, David et al. (1983) found high levels of S mineralization in a hardwood O2 horizon with a C:N:S ratio of 176:8.3:1. This would suggest that other factors need to be considered in predicting S mineralization/immobilization in forest soils.

A direct link between the N and S cycle is exemplified by the microorganism Thiobacillus denitrificans which under aerobic conditions, has the ability to oxidize partially reduced forms of elemental sulfur to sulfate. Under condi- tions of low oxygen tension, this bacterium can utilize nitrate as an electron acceptor and produce elemental nitrogen and nitrous oxide through denitrification. The use of this organism may serve as a valuable tool in determining the levels of pH and Eh which may dictate whether the release of nitrogen or the formation of sulfate will occur.

Much of the original information on S cycling was obtained from pure culture studies (Kadota and Ishida 1972) although there has been some work on S components and S transformations in soil (Freney et al. 1970, Freney et al. 1975, Bollen 1977, Wainswright and Killhan 1980, Tabatabai and Al-Khafaji 1980, David et al. 1983). Investigations in both aquatic and terrestrial ecosystems has led to the identification of major organic S components and the role of microflora in S transformation (Fenchel and Riedl 1970, Nedwell and Abram 1979, Nriagu 1968, Skopintsev et al. 1959, Zinder and Brock 1978, Bremner and Steele 1978). Although transformations or specific organic forms is lacking, Strickland and Fitzgerald (1983) have traced the decomposition and incorporation of ^{35}S-labeled sulfoquinovose in forest soils. More recently, McLaren et al. (1985) have used similar techniques to monitor sulfur transformations in organic horizons. Additional research is needed in this area to establish the major forms of S, turnover or residence times in various reservoirs, and the ultimate importance of various S forms for plant and animal use. Without such data, it is impossible to accurately assess or predict potential adverse effects (e.g. acidic deposition) on nutrient cycling.

Although the major factors such as moisture, temperature and substrate quality which affect C flux have been delineated (Bunnell et al. 1977, Dickinson and Pugh 1974, Minderman 1968, Mitchell 1979), the relationship of these parameters to sulfur transformations in forest and agricultural soils has not been determined. However, investigations of flooded soils (Connell and Patrick 1969, Patrick and Delaune 1977, Brown 1985) and aquatic sediments (Cappenberg 1974, Jorgensen and Fenchel 1974, Nriagu 1968, Windrey and Zeikus 1974) have provided some information on the importance of electron acceptors, Eh, moisture and organic matter availability on S dynamics in heterotrophic

systems. Coughenour et al. (1980) have produced a simulation model of sulfur dynamics in a grassland soil in which
some of the environmental factors which affect S dynamics
have been outlined. Similarly, although the major biotic
and abiotic factors which affect N transformation have been
examined (Alexander 1977a, Wallace and Nicholas 1969), there
has been little attempt to link these with S transformations.
Because both sulfate and nitrate constitute the major anions
present in acid precipitation, the relative abundance of
these ions may have significant effects on both S and N
transformations in systems subject to acidic inputs.
Information is also needed on the effects of C and N fluxes
and transformations on both the inorganic and organic S
constituents in all terrestrial ecosystems.

EFFECTS OF S DEPOSITION ON SOIL NUTRIENT AVAILABILITY
     The leaching of essential plant nutrients, especially
cations, from soil ecosystems has been implicated as one of
the more serious indirect effects of sulfur deposition
(Abrahamsen et al. 1976, Cronan 1978). The concentration of
excess acid in precipitation, the amount of acid water
penetrating the soil profile, and the susceptibility to
leaching of various nutrients present in the soil are
critically important factors regarding soil acidification.
Consequently, acidic soils are most often deficient in
essential nutrients and usually contain disproportionate
concentrations of all other nutrients. However, understanding the internal sulfur cycling of a particular soil ecosystem may provide insight into the true impact of acid
precipitation. For example, David et al. (1983) and Brown
(1985a) compared S dynamics in soils with different levels
of S inputs which allowed a more accurate assessment of the
effects of anthropogenic S additions on nutrient availability.
Also while working in a Douglas Fir ecosystem, Cole et al.
(1975) stated that a mobile anion and a displacing cation
were required for soil cation leaching to occur. Under
natural conditions (unstressed) the leaching process can be

controlled by carbonic acid in which cation exchange occurs
and a cation-bicarbonate solution leaches from the soil.
All acids can function in basically the same manner if the
anion involved were mobile (i.e. neither adsorbed nor
precipitated). The effectiveness of sulfuric acid as a
soil cation leaching agent would be reduced if the particular
soil were a sulfate adsorber.

A Memorandum of Intent on Transboundary Air Pollution
(Bangay and Riordan 1983) summarized the known effects of
acidic deposition on nutrient availability and nutrient
cycling via effects on the biotic component. In all cases
it is imperative to determine the rate of natural soil
acidification, especially in humid regions, and assess the
impact of anthropogenic acid deposition as a component of
this ongoing process. The real impact of atmospheric
deposition on soil nutrient availability is related to the
relative mobility of the accompanying anion (SO_4^{-2}, NO_3^-, or
an organic anion). Therefore, soils rich in Fe- and Al-
sesquioxides will demonstrate considerable resistance to
base leaching (Bangay and Riordan 1983).

A number of researchers have reported on significant
reductions in exchangeable cations and base saturation
induced by simulated acid leaching. Sulfate mobility
through the soil profile will regulate cation loss. However,
sulfate mobility is directly related to the sulfate adsorption isotherm of a given soil which is in turn positively
correlated with concentrations of aluminum and iron oxides.
Singh et al. (1980) demonstrated substantial losses of Ca^{++},
Mg^{++}, Mn^{++}, and Al^{+3} at pH 4.3 vs. pH 5.6. In each case,
the nutrient budgets for these cations registered a net
loss during the experimental period.

Abrahamsen et al.(1976) observed that the addition of
simulated acid rain to a coniferous soil significantly
increased the leaching of Ca^{++} and Mg^{++} with little effect
on other cations, nitrate, and organic nitrogen. However,
as the pH was gradually lowered below 3.0, additional cation
loss was observed. The authors concluded that severe effects

on forest soil and timber production due to acid deposition
should not be expected if the pH exceeds 4.0.

Cronan et al. (1978) demonstrated that sulfate anions
supplied approximately 76% of the charge balance in the
leaching solution of a subalpine forest in New Hampshire.
This represents the major source of both H^+ for cation
replacement and mobile anions for cation transport. In
relatively unpolluted regions, carbonic acid and organic
acids produced by microbial activity are major causes for
the replacement and transport of exchangeable cations. The
authors hypothesized that biologically controlled production
of carbonic acid and organic acids would not be as critical
to soil leaching as the input of anthropogenic mineral acids
introduced through precipitation. The results of a two year
study by Cronan et al. (1978) indicate that atmospheric
deposition of sulfate provides most of the electrical charge
balance which includes H^+ for cation replacement and mobile
anions for cation transport. Additionally, these data
indicate the pH-dependent solubility of metals such as
aluminum, may present toxicity problems for plants.

Whenever considering the effects of acid precipitation
on terrestrial ecosystems, it is essential to differentiate
between effects on biological processes responsible for
nutrient transformations, and the leaching of soil components
and plant nutrients by chemical processes. Although many
investigators have concentrated on the microbial component
when considering biological effects, the effects of
terrestrial fauna should not be overlooked. For example,
Mitchell et al. (1983) have clarified the role of the
terrestrial isopod, Oniscus asellus, in S transformations in
a forest soil. In general, faunal activity may accelerate
the conversion of inorganic sulfate to carbon-bonded and
ester sulfate forms (if sulfate is not in high concentration)
or enhance S mineralization if organic S predominates.

MICROBIAL SULFUR IN SOIL

The measurement of microbial sulfur, i.e. the amount of

S present in soil as microbial biomass, has recently been
published by Saggar et al. (1981) and is based on the
chloroform fumigation technique of Jenkinson and Powlson
(1976a, b) for the determination of biomass carbon. Bacteria
and fungi which are indigenous to the prairie soils used,
were grown in pure culture with varying levels of S and then
analyzed for S content. The S concentrations ranged between
928-1355 μg S g^{-1} dry weight of bacteria and 852-2772 μg S g^{-1}
dry wt of fungi. S recovery values were obtained by adding
known quantities of S-analyzed bacteria and fungi to soils,
applying chloroform as described by Jenkinson and Powlson
(1976a, b), and extracting with $CaCl_2$ and $NaHCO_3$. This
procedure yielded values of approximately 7.0 μg biomass
S.g^{-1} soil. Using a similar approach in the author's
laboratory yielded values between 5 and 15 μg biomass S.g^{-1}
for soils in a conifer and hardwood site, respectively
(David et al. 1982, Strick et al. 1983, Strick and Nakas
1984). Saggar et al. (1981) demonstrated in short term
(64 days) incubation experiments that under conditions of
carbon supplementation, the incorporation of sulfate was
primarily into the fulvic acid fraction. These data indicate
that most of the S-containing organic compounds synthesized
during short term incubations are relatively labile. In
contrast, a net mineralization of carbon and nitrogen
occurred during the same incubation period. The humic acid,
fulvic acid, and humin (<2 μm) fractions showed significant
increases in C and N contents while total S and HI-reducible
S increased in the fulvic acid fraction only and represented
<50% of the added ^{35}S-sulfate (Saggar et al. 1981).

Bettany et al. (1977) provided initial evidence that
assimilation, transformation, and immobilization of S in
soil does not follow the patterns which have been established
for carbon and nitrogen. These investigators also demonstrated
that HI-reducible S is primarily associated with the aliphatic
portion of the humic acid fraction rather than the condensed
aromatic core. Following C/N/S ratios and fluctuations in
percent HI-reducible S these investigators demonstrated the

importance and availability of organic S for plant use and the contribution of organic S to overall S status and general soil fertility. Recently, Maynard et al. (1983) demonstrated that HI-reducible S, expressed as a percent of total S, provided the most reliable index to evaluate the S status of soil used for the cultivation of rapeseed and wheat. Other studies, Maynard et al. (1985), have helped elucidate the role of plants in S transformation.

More information is needed on the size of the microbial S component as well as the activity of the soil microflora in the transformation of organic S into sulfate for plant use. The significance in quantifying the S constituents of soil ecosystems as well as establishing rates of transformation, lies in the ability to predict S availability as well as S deficiency due to changing environmental conditions, anthropogenic S additions, and cultivation practices.

ACKNOWLEDGMENTS

Special thanks to Penny Weiman for typing the manuscript.

REFERENCES

1. Abrahamsen, G., K. Bjor, R. Horntvedt, and B. Tveite. 1976. Effects of acid precipitation on coniferous forest. In Impact of Acid Precipitation on Forest and Freshwater Ecosystems in Norway. F. H. Braekke (ed.). SNSF-project FR/76, Oslo-As, Norway, 33-63.

2. Alexander, M. 1974. Microbial formation of environmental pollutants. Adv. Appl. Microbiol. 18:1-73.

3. Alexander, M. 1977. Introduction to Soil Microbiology, 2nd ed., John Wiley and Sons, New York, 462 p.

4. Asami, T., and Y. Takai. 1963. Formation of methyl mercaptan in paddy soils. II. Soil Sci. Plant Nutr. 9:23-28.

5. Ayotade, D. A. 1977. Kinetics and reactions of hydrogen sulfide in solution of flooded rice soils. Plant Soil 46:381-389.

6. Bangay, G. E., and C. Riordan. 1983. Memorandum of intent on transboundary air pollution. Work Group I, Impact Assessment, Final Report.

7. Banwart, W. L., and J. M. Bremner. 1982. Formation of volatile sulfur compounds by microbial decomposition of sulfur-containing amino acids in soils. Soil Biol. Biochem. 7:359-364.

8. Banwart, W. L., and J. M. Bremner. 1976a. Volatilization of sulfur from unamended and sulfate-treated soils. Soil Biol. Biochem. 8:19-22.

9. Banwart, W. L., and J. M. Bremner. 1976b. Evolution of volatile sulfur compounds from soils treated with sulfur-containing organic materials. Soil Biol. Biochem. 8:439-443.

10. Barrow, N. J. 1961. Studies on mineralization of sulfur from soil organic matter. Aust. J. Agr. Res. 12:306-319.

11. Bettany, J. R., J. W. B. Stewart, and S. Saggar. 1979. The nature and forms of sulfur in organic matter fractions of soils selected along an environmental gradient. Soil Sci. Soc. Am. J. 43:981-985.

12. Bollen, W. B. 1977. Sulfur oxidation and respiration in 54-year-old soil samples. Soil Biol. Biochem. 9: 405-410.

13. Bremner, J. M., and W. L. Banwart. 1976. Sorption of sulfur gases by soils. Soil Biol. Biochem. 8:79-83.

14. Bremner, J. M. 1977. Role of organic matter in volatilization of sulfur and nitrogen from soils. In: Proceedings of Symposium on Soil Organic Matter Studies. Vol. II, pp. 229-240. International Atomic Energy Agency, Vienna.

15. Bremner, J. M., and C. G. Steele. 1978. Role of microorganisms in the atmospheric sulfur cycle. Adv. Microb. Ecol. 2:155-201.

16. Brock, T. D. 1966. Principles of Microbial Ecology, Prentice-Hall, Englewood Cliffs, 306 p.

17. Brock, T. D. 1978. Thermophilic Microorganisms and Life at High Temperatures. Springer-Verlag, New York, 465 p.

18. Brown, K. A. 1985a. Acid deposition: effects of sulphuric acid at pH 3 on chemical and biochemical properties of bracken litter. Soil Biol. Biochem. 17: 31-38.

19. Brown, K. A. 1985b. Sulphur distribution and metabolism in waterlogged peat. Soil Biol. Biochem. 17:39-45.

20. Buckman, H. O., and N. C. Brady. 1969. The nature and properties of soils. The Macmillan Co., London, 653 p.

21. Bunnell, F. L., D. E. N. Tait, P. W. Flanagan, and K. van Cleve. 1977. Microbial respiration and substrate weight loss - I. A general model of the influence of abiotic variables. Soil Biol. Biochem. 8:33-40.

22. Campbell, R. 1977. Microbial Ecology. John Wiley and Sons, New York, 148 p.

23. Cappenberg, T. E. 1974. Interrelations between sulfate-reducing and methane-producing bacteria in bottom deposits of a fresh-water lake. I. Field observations. Ant. von Leeuwenhock 40:285-295.

24. Castenholz, R. W. 1976. The effect of sulfide on the bluegreen algae of hot springs. I. New Zealand and Iceland. J. Phycol. 12:54-68.

25. Cole, D. W., W. J. B. Crane, and C. C. Grier. 1975. The effect of forest management practices on water quality in a second-growth Douglas fir ecosystem. In Forest Soils and Land Management, Proceedings of the American Forest Soils Conference, pp. 195-207, Laval University Press, Quebec.

26. Connell, W. E., and W. H. Patrick. 1969. Reduction of sulfate to sulfide in water logged soils. Soil Sci. Soc. Am. J. 33:711-715.

27. Coughenour, M. B., W. J. Parton, W. K. Lavenroth, J. L. Dudd, and R. G. Woodmansee. 1980. Simulation of a grassland sulfur cycle. Ecological Modelling 9:179-213.

28. Cronan, C. S., W. A. Reiners, R. C. Reynolds, Jr., and G. E. Land. 1978. Forest floor leaching: contributions from mineral, organic and carbonic acids in New Hampshire subalpine forests. Science 200:309-311.

29. Cuhel, R. L., C. D. Taylor, and H. W. Jannasch. 1981. Assimilatory sulfur metabolism in marine microorganisms: sulfur metabolism, growth, and protein synthesis of Pseudomonas halodurans and Alteromonas luteo-violaceus during sulfate limitation. Arch. Microbiol. 130:1-7.

30. Cuhel, R. L., C. D. Taylor, and H. W. Jannasch. 1982a. Assimilatory sulfur metabolism in marine microorganisms: sulfur metabolism, protein synthesis, and growth of Alteromonas luteo-violaceus and Pseudomonas halodurans during perturbed batch growth. Appl. Environ. Microbiol. 43:151-159.

31. Cuhel, R. L., C. D. Taylor, and H. W. Jannasch. 1982b. Assimilatory sulfur metabolism in marine microorganisms:

312

considerations for the application of sulfate incorporation into protein as a measurement of natural
population protein synthesis.

32. David, M. B., M. J. Mitchell, and J. P. Nakas. 1982.
Organic and inorganic sulfur constituents of a forest
soil and their relationship to microbial activity.
Soil Sci. Soc. Am. J. 46:847-852.

33. David, M. B., S. C. Schindler, M. J. Mitchell, and
J. E. Strick. 1983. Importance of organic and
inorganic sulfur to mineralization processes in a
forest soil. Soil Biol. Biochem. 15:671-677.

34. David, M. B. 1983. Organic and inorganic sulfur
cycling in forested and aquatic ecosystems in the
Adirondack region of New York State. Ph.D. Thesis,
SUNY-College of Environmental Science and Forestry,
Syracuse, New York.

35. DeJong, E., and H. J. V. Shappert. 1972. Calculation
of soil respiration and activity from CO_2 profiles
in soil. Soil Science 113:328-333.

36. Dickinson, C. H., and G. J. F. Pugh (eds.). 1974.
Biology of Plant Litter Decomposition. Vol. I and II.
Academic Press, London, p. 1-775.

37. Ehrlich, H. L. 1981. Geomicrobiology. Marcel Dekker,
New York, 393 p.

38. Farwell, S. O., A. E. Sherrard, M. R. Pack, and D. F.
Adams. 1979. Sulfur compounds volatized from soils
at different moisture contents. Soil Biol. Biochem.
11:411-415.

39. Fenchel, T. M., and R. S. Riedl. 1970. The sulfide
systems: a new biotic community underneath the
oxidized layer of marine sand bottoms. Mar. Biol. 7:
255-268.

40. Fenchel, T., and T. H. Blackburn. 1979. Bacteria and
Mineral Cycling, Academic Press, New York.

41. Fitzgerald, J. W. 1978. Naturally occurring
organosulfur compounds in soil. In: J. O. Nriagu (ed.),
Sulfur in the Environment, John Wiley and Sons, New
York, p. 391-443.

42. Fitzgerald, J. W., and D. W. Johnson. 1983. Transformations of sulfate in forested and agricultural
lands. Proc. Int. Sulfur Conf. Vol. 1. pp. 411-426.

43. Freney, J. R. 1958. Determination of water-soluble
sulfate in soils. Soil Sci. 86:241-244.

44. Freney, J. R. 1961. Some observations on the nature
of organic sulfur compounds in soil. Aust. J. Agric.
Res. 12:424-432.

45. Freney, J. R. 1967. Sulfur containing organics. In:
A. D. McLaren and G. H. Peterson (eds.), Soil Biochemistry Vol. 1. Marcel Dekker, New York, p. 229-259.

46. Freney, J. R., G. E. Melville, and C. H. Williams.
1970. The determination of carbon bonded sulfur in
soil. Soil Sci. 109:310-318.

47. Freney, J. R., G. E. Melville, and C. H. Williams.
1971. Organic sulfur fractions labeled by addition of
^{35}S-sulfate to soil. Soil Biol. Biochem. 3:133-141.

48. Freney, J. R., F. J. Stevenson, and A. H. Beavers. 1972. Sulfur-containing amino acids in soil hydrolysates. Soil Sci. 114:468-476.

49. Freney, J. R., G. E. Melville, and C. H. Williams. 1975. Soil organic matter fractions as sources of plant-available sulfur. Soil Biol. Biochem. 7:217-221.

50. Ghiorse, W. C., and M. Alexander. 1976. Effect of microorganisms on sorption and fate of sulfur dioxide and nitrogen dioxide in soil. J. Environ. Qual. 5: 227-230.

51. Gottschalk, G. 1979. Bacterial Metabolism, Springer-Verlag, New York, 281 p.

52. Greenwood, D. J. 1961. The effect of oxygen concentration on the decomposition of organic materials in soil. Plant and Soil 14:360-376.

53. Hoffman, M. R., B. C. Faust, F. A. Panda, H. H. Koo, and H. M. Tsuchiya. 1981. Kinetics of the removal of iron pyrite from coal by microbial catalysis. Appl. Environ. Microbiol. 42:259-271.

54. Hornor, S. G., and M. J. Mitchell. 1981. Effect of the earthworm, Eisenia foetida (Oligochaeta) on fluxes of volatile carbon and sulfur compounds in sewage sludge. Soil Biol. Biochem. 13:367-372.

55. Hutchinson, T. C., and M. Havas (eds.). 1980. Effects of acid precipitation on terrestrial ecosystems. Plenum Press, New York, 654 p.

56. Jenkinson, D. S., and D. S. Powlson. 1976a. The effects of biocidal treatments on metabolism in soil - I. Fumigation with chloroform. Soil Biol. Biochem. 8:167-177.

57. Jenkinson, D. S., and D. S. Powlson. 1976b. The effects of biocidal treatments on metabolism in soil - V. A method for measuring soil biomass. Soil Biol. Biochem. 8:209-213.

58. Johnson, D. W., and G. S. Henderson. 1979. Sulfate adsorption and sulfur fractions in a highly-weathered soil under a mixed deciduous forest. Soil Sci. 128: 34-41.

59. Jorgensen, B. B., and T. Fenchel. 1974. The sulfur cycle of a marine sediment system. Mar. Biol. 24: 189-201.

60. Kadota, H., and Y. Ishida. 1972. Production of volatile sulfur compounds by microorganisms. Ann. Rev. Microbiol. 26:127-138.

61. Kargi, F., and J. M. Robinson. 1982. Removal of sulfur compounds from coal by the thermophilic organism Sulfolobus acidocaldarius. Appl. Environ. Microbiol. 44:878-883.

62. Kawalenko, C. G., and L. E. Lowe. 1975. Evaluation of several extraction methods and of a closed incubation method for soil sulfur mineralization. Can. J. Soil Sci. 55:1-8.

63. Kellogg, W. W., R. D. Cadle, E. R. Allen, A. L. Lazrus, and E. A. Martell. 1972. The sulfur cycle. Science 175:587-596.

314

64. Mah, R. A., D. M. Ward, L. Baresi, and T. L. Glass. 1977. Biogenesis of methane. Ann. Rev. Microbiol. 31:109-131.

65. May, P. F., A. R. Till, and A. M. Downes. 1968. Nutrient cycling in grazed pastures. I. A preliminary investigation of the use of ^{35}S gypsum. Aust. J. Agric. Res. 19:531-543.

66. May, P. F., A. R. Till, and M. J. Cummins. 1972. Systems analysis of ^{35}S sulphur kinetics in pastures grazed by sheep. J. Appl. Ecol. 9:25-49.

67. Maynard, D. G., J. W. B. Stewart, and J. R. Bettany. 1985. The effects of plants on soil sulfur transformations. Soil Biol. Biochem. 17:127-134.

68. McGuire, W. G., and G. A. Marzluf. 1974. Sulfur storage in Neurospora: soluble sulfur pools of several developmental stages. Arch. Biochem. Biophys. 161: 570-580.

69. McLaren, R. G., J. I. Keer, and R. S. Swift. 1985. Sulphur transformations in soils using sulphur-35 labelling. Soil Biol. Biochem. 17:73-79.

70. Minderman, G. 1968. Addition, decomposition and accumulation of organic matter in forests. J. Ecol. 56:355-362.

71. Mitchell, M. J. 1979. Functional relationships of macroinvertebrates in heterotrophic systems with emphasis on sewage sludge decomposition. Ecology 60: 1270-1283.

72. Mitchell, M. J., and D. H. Landers. 1981. Inorganic sulfur and organic constituents of soils and sediments and their relationships to acid precipitation and coal utilization. In Proceedings of Conference on Expanding the Use of Coal in New York State, M. H. Tress and J. G. Dawson (eds.), pp. 55-60, Res. Foundation SUNY.

73. Mitchell, M. J., D. H. Landers, and D. F. Brodowski. 1982. Sulfur constituents of sediments and their relationship to lake acidification. Water, Air and Soil Pollution 16:177-186.

74. Mitchell, M. J., M. B. David, and C. P. Morgan. 1983. Importance of organic sulfur constituents of forest soils and the role of the soil macrofauna in affecting sulfur flux and transformations. In P. Lebrun, H. M. Andre, A. DeMedts, C. Gregoire-Wilo, and G. Wauthy (eds.) New Trends in Soil Biology. Proc. VIII Int. Coll. Soil Zool., Belgium, pp. 75-85.

75. Murphy, M. J., and L. M. Siegel. 1973. Siroheme and Sirohydrochlorin. J. Biol. Chem. 248:6911-6919.

76. Nedwell, D. B., J. W. Abram. 1979. Relative influence of temperature and electron donor and electron acceptor concentrations on bacterial sulfate reduction in saltmarsh sediment. Microbial Ecol. 5: 67-72.

77. Nor, Y. M., and M. A. Tabatabai. 1976. Extraction and colorimetric determination of thiosulfate and tetrathionate in soils. Soil Sci. 122:171-178.

78. Nriagu, J. O. (ed.). 1978. Sulfur in the Environment. Part I: The atmospheric cycle. John Wiley and Sons, New York, 464 p.

79. Nyborg, M. 1978. Sulfur pollution and soils. In: J. O. Nriagu (ed.). Sulfur in the Environment, John Wiley and Sons, New York, p. 359-390.

80. Parr, J. F., and H. W. Reuszar. 1959. Organic matter decomposition as influenced by oxygen level and method of application to soil. Soil Sci. Soc. Am. J. 23: 214-216.

81. Patrick, W. H., and R. D. Delaune. 1977. Chemical and biological redox systems affecting nutrient availability in the coastal wetlands. Geoscience and Man 18:131-137.

82. Ponnamperuma, E. A. 1972. The chemistry of submerged soils. Adv. Agron. 24:29-96.

83. Press, M. C., J. Henderson, and J. A. Lee. 1985. Arylsulphatase activity in peat in relation to acidic deposition. Soil Biol. Biochem. 17:99-103.

84. Rodriguez-Kabana, R., J. W. Jordan, and J. P. Hollis. 1965. Nematodes: Biological control in rice fields: Role of hydrogen sulfide. Science 148:524-526.

85. Roy, A. B., and P. A. Trudinger. 1970. The Bio-chemistry of Inorganic Compounds of Sulfur. Cambridge University Press, New York.

86. Saggar, S., J. R. Bettany, and J. W. B. Stewart. 1981. Measurement of microbial sulfur in soil. Soil Biol. Biochem. 13:493-498.

87. Schiff, J. A., and R. C. Hodson. 1973. The metabolism of sulfate. Ann. Rev. Plant Physiol. 24:381-414.

88. Sherman, V. B. D., G. Dementjeva, and B. J. Carey. 1957. Intracellular deposition of sulfur by Sphaerotilus natans. J. Bacteriol. 73:504-512.

89. Singh, B. R., G. Abrahamson, and A. Stuanes. 1980. Effect of simulated acid rain on sulfate movement in acid forest soils. Soil Sci. Soc. Am. J. 44:75-79.

90. Skinner, F. A. 1975. Anaerobic bacteria and their activities in soil. In: Soil Microbiology, N. Walker (ed.), Halsted Press, New York, p. 1-19.

91. Skopintsev, B. A., A. V. Karpov, and O. A. Vershinna. 1959. Study of the dynamics of some sulfur compounds on the Black Sea under experimental conditions. Tr. Morok. Gidrofiz. Inst. Akad. Nauk. Ukr. SSR 16:89-111.

92. Stanier, R. Y., E. A. Adelberg, and J. Ingraham. 1976. The Microbial World, 4th ed., Prentice-Hall, Inc., Englewood Cliffs, 871 pp.

93. Stotzky, G., and S. Schenck. 1976. Volatile organic compounds and microorganisms. C.R.C. Critical Rev. Microbiol. 4:333.

94. Strick, J. E., S. C. Schindler, M. B. David, M. J. Mitchell, and J. P. Nakas. 1982. Importance of organic sulfur constituents and microbial activity to sulfur transformations in an Adirondack forest soil. Northeast Environ. Sci. 1:161-169.

95. Strick, J. E., and J. P. Nakas. 1984. Calibration of

316

a microbial sulfur technique for use in forest soils. Soil Biol. Biochem. 16:289-291.

96. Strickland, T. C., and J. W. Fitzgerald. 1983. Mineralization of sulfoquinovose by forest soils. Soil Biol. Biochem. 15:347-349.

97. Tabatabai, M. A., and J. M. Bremner. 1972. Forms of sulfur and carbon, nitrogen and sulfur relationships in Iowa soils. Soil Sci. 114:380-386.

98. Tabatabai, M. A., and A. A. Al-Khafaji. 1980. Comparison of nitrogen and sulfur mineralization in soils. Soil Sci. Soc. Am. J. 44:1000-1006.

99. Ulrich, B. 1980. Production and consumption of hydrogen ions in the ecosphere. In: Effects of Acid Precipitation on Terrestrial Ecosystems, pp. 255-282, T. C. Hutchinson and M. Havas (eds.), Plenum Press, New York.

100. Wainwright, M., and K. Kollham. 1980. Sulfur oxidation by Fusarium solani. Soil Biol. Biochem. 12:555-558.

101. Wallace, W., and D. J. Nicholas. 1969. The biochemistry of nitrifying microorganisms. Biol. Rev. 44:359-391.

102. Winfrey, M. R., and J. G. Zeikus. 1977. Effect of sulfate and carbon flow during microbial methanogenesis in freshwater sediments. Appl. Environ. Microbiol. 33:275-281.

103. Witcamp, M. 1966. Decomposition of leaf litter in relation to environment, microflora and microbial respiration. Ecology 47:194-201.

104. Withers, P. C. 1978. Models of diffusion mediated gas exchange in animal burrows. Amer. Naturalist 112: 1101-1112.

105. Zinder, S. H., and T. D. Brock. 1978. Microbial transformations of sulfur in the environment, pp. 445-466, In: Sulfur in the Environment, II. Ecological Impacts. J. O. Nriagu (ed.), J. Wiley and Sons, New York.

THE ROLE OF MICROFLORAL AND FAUNAL INTERACTIONS IN AFFECTING SOIL PROCESSES

D.C. COLEMAN*

OVERVIEW

Soils and the organisms within them are an integral part of the detritus-decomposition and nutrient-cycling that enables ecosystems to function. The physical, chemical and biotic regimes both buffer and constrain the ongoing nutrient transformations.

Using conceptual models and laboratory experiments, a considerable amount of information has been gained about interactions between primary decomposers (bacteria and fungi) and the fauna that feed on them or are predatory on the microbivores. Both bacterial feeders (protozoa and nematodes) and fungal feeders (nematodes) in microcosm experiments showed considerable facilitation of nutrient (nitrogen) return, leading to enhanced nutrient uptake and dry matter yield of test plants.

The trophic interactions mentioned above have been (and are being) tested, using selective biocides in natural ecosystems. The studies are in their early stages and require more investigation.

Recent experiments in agroecosystems are attempting to elucidate how the basic mechanisms of trophic interactions are modified or altered in zero-tillage versus conventional tillage regimes in various crops (such as dryland wheat, soybeans, and corn).

---

*Natural Resource Ecology Laboratory and Department of Zoology/Entomology, Colorado State University, Fort Collins, Colorado 80523.

With the existence of many microbivorous forms with
high production efficiency, one might expect longer food
chains than the classical 4- or 5-membered ones. This,
combined with demonstrated feeding effects on symbiotic
(mycorrhizal) fungi, means there is much further work needed
on microbial/faunal interactions in ecosystems.

INTRODUCTION

Soils as parts of ecosystems

Since the period of the International Biological Program
(IBP) and in more recent studies (1975-present), it has proven
useful to consider the abiotic and biotic components of soils
in an ecosystem context. This holistic approach was best
characterized by Pomeroy (1970), who suggested that studying
nutrient cycling is a good strategy for studying ecosystems.

Ecosystems contain organisms as well as physical and
chemical components. They are open systems with inputs of
energy and some mineral nutrients. They may range in size
from a few cubic mm of soil or water up to an entire
landscape unit extending for several hundred km^2. Ecosystems
are characterized by inputs of energy, a portion of which is
chemically fixed by autotrophs, stored as organic C for their
use and utilized, in turn, by various heterotrophs within
food webs. These principles are reviewed in introductory
text books (Odum 1971, 1983) and will not be further
discussed here.

Ecosystems also contain various amounts of nutrients
which are required by all organisms. The ones which I will
consider are principally the macronutrients: namely, C, N,
S and P. Unlike energy, some of which is dissipated at every
transformation along a food chain, many nutrients,
particularly N, P and S are returned to the system and may
cycle through it repeatedly, with some potential losses by
leaching and gaseous efflux, as well. The magnitude of
these losses varies both temporally and spatially among
ecosystems, and is a function of the biological, physical,

and chemical characteristics of the element, and the habitat
in which it occurs.

This chapter will concentrate on the roles of certain
organisms, namely the microflora and fauna existing in soil
and how they interact to affect ecosystems including plants,
which are an integral component of the soil. It is useful
when working with soil to consider not only the organisms
but also the relative availability of substrates for their
energy and nutrient demands; namely, the extent to which
compounds are labile, readily-available versus non-labile,
or nonaccessible or insoluble (Stewart and McKercher 1982).
This concept helps us view the relative contributions of
microorganisms, fauna, and plants involved. Further details
of the conjoint action of organisms, nutrients and soil are
presented in several sections of the chapter, and are also
discussed by Coleman et al. (1983).

The detritus decomposition pathway occurs on or within
the soil after plant materials (litter, roots, sloughed cells
and soluble compounds) become available through death,
senescence or other pathways (Coleman 1976). Thus, one can
envision plants, products of which are used by microbes as
the primary decomposers with the fauna eating them, and thus
affecting flows of nutrients, particularly N, P and S. The
immobilization into plants or microbes and the subsequent
mineralization which occurs principally as microbe- or
microbe- and fauna-mediated processes are critical pathways
(Coleman et al. 1983). These processes are diagrammed in
Figure 1, in which flows out from the microbes and fauna via
mineralization or direct losses into organic pools are shown.
The labile inorganic pool is the principal one that enables
subsequent microbial and plant existence to occur. Some
nutrient scarcity often limits production. Most importantly,
it is the rates of flux into and out of these labile
inorganic pools which enable ecosystems to successfully
function.

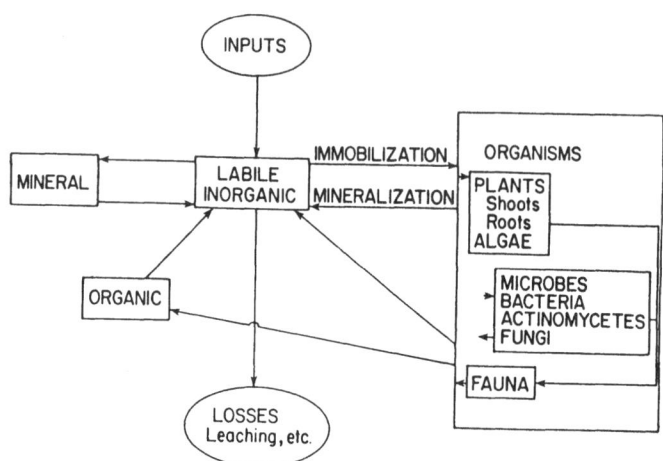

FIGURE 1. Generalized nutrient-cycling scheme in soil. The biological activities of immobilization and mineralization are of major importance in ecosystem function (from Coleman et al. 1983).

Major areas of emphasis

This review focuses on decomposition and nutrient fluxes which occur via microbial immobilization and mineralization by microbes and fauna in soil systems. The relative activities of these organisms are examined in natural ecosystems versus intensely man-managed ecosystems such as agricultural croplands and managed forest lands. The major differences occurring among managed and natural ecosystems are related to the form of litter and placement of residues. In this review I will concentrate on leaf- and root-derived materials. Very interesting developments related to wood-feeding invertebrates (xylophagy) have been reviewed recently by Breznak (1982) and are not discussed here. In addition, for an historical review on previous work in microbial and faunal interactions, Chapter 1 should be consulted.

SOIL PEDOGENESIS AND HORIZONATION

Soils are generally considered to be formed by a series of soil-forming factors which include: climate, organisms, parent material, and relief, all acting over time (Jenny 1941). These factors affect major processes (Fig. 2), which will lead to certain soil properties, such as soil organic matter status. For example, for soil P, changes in the organic and inorganic forms which result from various abiotic and biological interactions occur over many thousands of years (from 5,000 to 15,000 years), a chronosequence. There is a gradual increase and then decrease in organic P, and considerable loss of mineral P forms later in the chronosequence (Walker 1965) (Fig. 3).

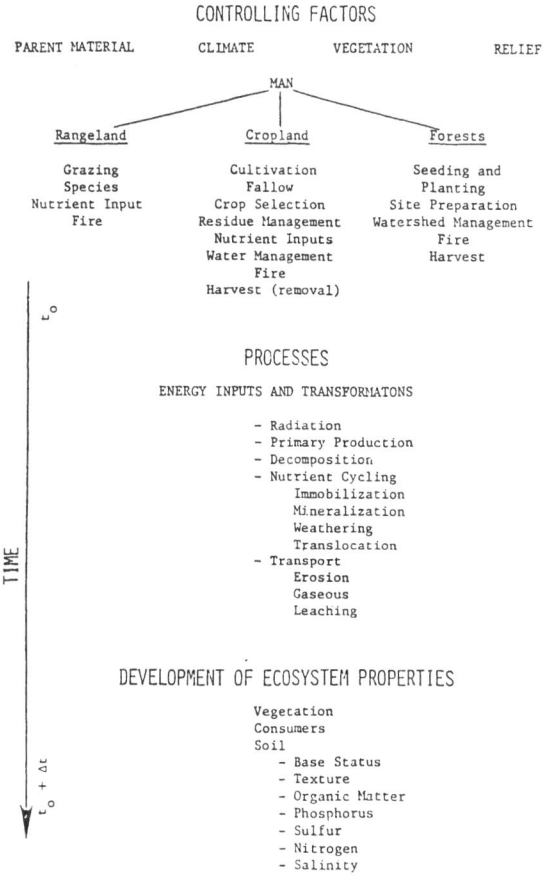

FIGURE 2. Factors influencing soil development. Controlling factors affecting processes, over time, influence ecosystem properties (from Coleman et al. 1983).

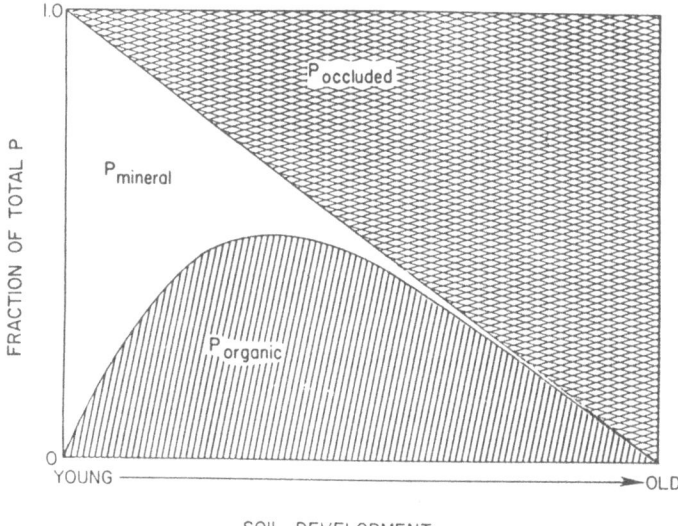

FIGURE 3. Changes of mineral, organic, and occluded forms
of phosphorus in a chronosequence. Time span from 5,000-15,000
y (modified from Walker 1965).

Over time, due to the action of biotic and abiotic factors,
a profile develops with the unconsolidated mineral material
above bedrock, the C horizon followed by a B horizon of
generally somewhat weathered material with some penetration
of roots and some deep dwelling organisms which is in turn
covered by the surface layer or the A horizon which has the
majority of plant roots, microbes and fauna residing in it.
In various forest soils, particularly those which have a
concentration of organic material in the surface layers
(litter, fermentation and humification layers, or L, A_{01},
A_{02} in North American soils terminology), there is
considerable proliferation of roots up in the organic matter.
These soils are termed "spodosols," and exist in many areas
of the temperate forests, such as temperate coniferous and
northern boreal or taiga regions. Depending on soil conditions,
spodosols also exist in parts of the Amazonian tropics. The
plant/soil interaction, as a function of geological/pedological
substrate is thus ubiquitous and compelling, as well. The
action of soil fauna, discussed further in this chapter, also

leaves species or group-specific traces of activity, adding a
faunal "character" to the organic matter, via deposition of
fecal pellets. Thus large arthropods, such as millipedes,
and small ones, such as soil mites and collembola, provide a
steady stream of fecal material to the litter and soil layers.
These are well-illustrated and described by Kubiëna (1938,
1970) and Jongerius (1964). A majority of the microbial
and faunal activity occurs in the surface few centimeters of
the soil where there is considerable organic input (Fig. 4).

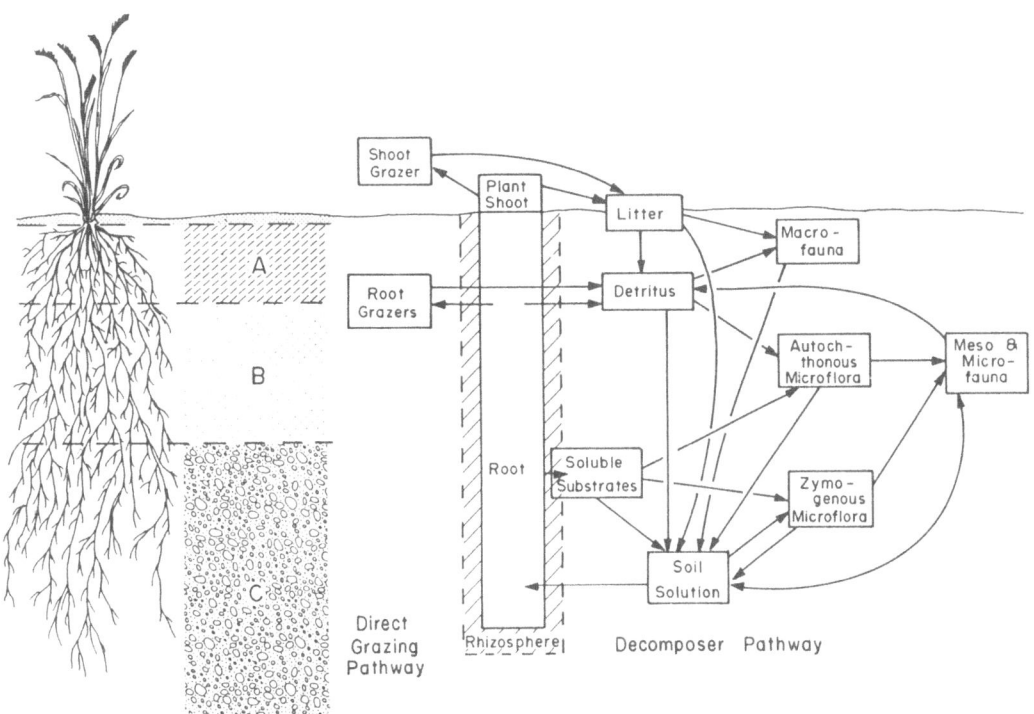

FIGURE 4. Diagram of soil profile development and root-
rhizosphere trophic pathways (from Anderson et al. 1981).

MICROBIAL-FAUNAL INTERACTIONS: THE "ACTORS" AND THE "THEATRE"

Physical environment

In considering soil and the organisms within it, we are
faced with an array of entities which extend over more than
five orders of magnitude in scale. Thus, particles range

from the very smallest size, namely clay minerals with a
diameter of 0.02 μm up to clay platelets which form
aggregates with secondary roots and fungal hyphae. These
occupy volumes of approximately 8 mm^3 and contain pores
through which small secondary roots and fungal hyphae penetrate
(Fig. 5). Depending on organic matter status and soil parent
material, aggregates may even exceed this size.

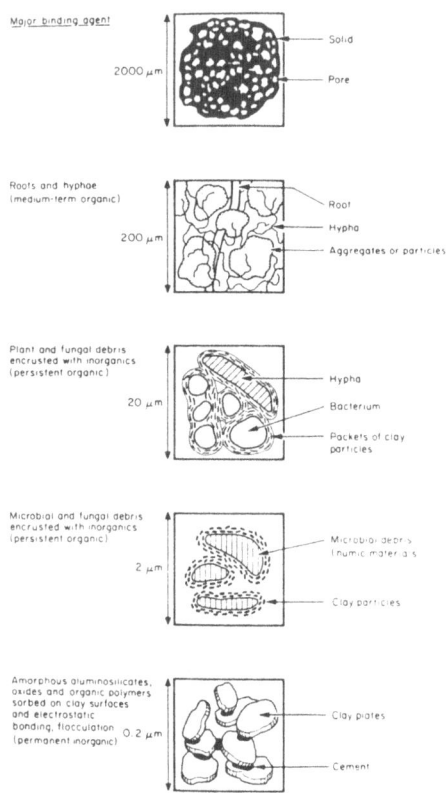

FIGURE 5. Model of aggregate organization, showing major
binding agents. Note strong affinity of organic compounds
for clay surfaces (from Tisdall and Oades 1982).

The assemblage of sand, silt and clay particles is
constantly being influenced by biological and chemical
processes, and has a strong impact on the accessibility of

food, shelter, water, oxygen, and nutrients to the soil biota.
The interaction of all components of the soil biota is as
marked here, as in any aspect of soil ecology. Thus
capability for roots to penetrate various layers of the soil,
and subsequent death of roots, leaving various channels for
access of H_2O and air is very important. In addition,
earthworms play an important role in aeration and transport
of organic matter to various parts of the soil profile.
Further aspects of their biology are discussed by Edwards
and Lofty (1978). Indeed, recent studies of the activity of
earthworms in improved pastures of ryegrass/clover in New
Zealand have shown a significant enhancement of phosphorus
mineralization in pastures containing high numbers versus
low or no populations (Sharpley et al. 1979, Syers et al. 1979).

In addition to texture, there is another property,
"structure", which is of importance to soil ecologists.
Thus a soil may have a very high clay content, but, with
considerable faunal and root penetration, and elaboration of
many kinds of carbohydrate and other organic residues, the
soil will have a good tilth, aeration, etc. More aspects of
structure are discussed in soil science textbooks, such as
that of Brady (1980).

Movement of small animals in soil is also affected by
structural and textural properties. Soil nematode movement
is restricted by fine soil granulation (Wallace 1971) and
heavier bulk densities (Jones and Thomasson 1976). Soils
with finer textures also inhibit the infiltration of fungal
spores and bacterial cells (Wilkinson et al. 1981).
Similarly, the ubiquitous soil ciliate Colpoda steinii was
restricted from using bacteria by reduction of the soil
water, which decreased the size of water filled pores below
that penetrable by the ciliate (Darbyshire 1976).

Trophic interactions among assemblages of soil
organisms (bacteria, amoebae, nematodes) isolated from the
Pawnee site in northeastern Colorado were restricted in fine
compared with coarse textured soils (Elliott et al. 1980a,
b). Bacteria growing alone developed greater numbers and

released more CO_2 in the latter soil. In the coarse soil, CO_2 evolution was greater with more biological complexity (microbes plus bacterial-feeding nematode and bacterial-feeding amoeba), whereas in the fine-textured soil there were no differences in total CO_2 output among biological treatments.

In grasslands, bacterial-feeding nematodes are more numerous in soils with greater porosity (i.e., sand content) (Ingham et al. 1982). Thus, soil texture in conjunction with organism size strongly influences trophic structure.

In addition to accessibility and specific physical effects, the proportion of coarse and fine clay particles in a given soil has a significant impact on the dynamics of particulate and dissolved organic matter. For example, Anderson (1979) showed significant amounts of very labile organic material going to fine clay particles such as those shown in Fig. 5. Adsorption/desorption processes are important in a number of processes such as anion and cation exchange, as well as humification. Burns and Martin (Chapter 4) discuss the role of these processes in humification, especially in respect to clays.

Conceptual model of microbial and faunal interactions in soil

Conceptual models, which are the precursors of various simulation models described in Chapter 11, are useful for envisioning ecosystem processes. For example, the model of immobilization and mineralization (Fig. 1) shows the major processes with the important flows occurring between the components and of which the flux of labile inorganic constituents is of key importance.

The array of microorganisms, fauna, and roots all provide organic material belowground. Principal trophic pathways are shown (Fig. 6). Pathways of organic matter utilization by primary decomposers include bacteria, actinomycetes or saprophytic fungi. In each instance, there is considerable grazing by such bacterial feeders as protozoa, nematodes, tardigrades, and enchytraeids. Little is known about primary grazers on actinomycetes; however actinomycetes

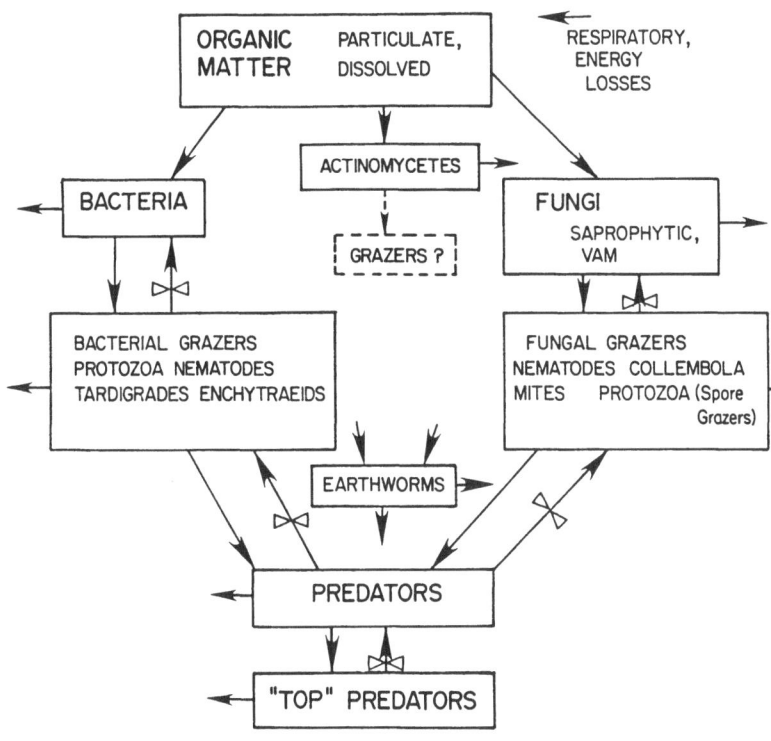

FIGURE 6. Conceptual model of breakdown of particulate and soluble organic matter in agroecosystems. Vertical arrows show material flows; horizontal arrows show respiratory energy losses; and arrows with ⋈ show return, information-feedback from trophic levels on lower ones.

can serve as food for certain nematodes (C. R. Morley, pers. commun.). The final major primary decomposer group are the fungi, a wide array of saprophytic forms and VAM (vesicular-arbuscular mycorrhizae) which take up nutrients and are mutualistic with the plant roots. The fungal grazers include stylet-bearing nematodes, collembola, mites, and certain protozoa which are spore grazers. There is a general group of predators such as large (2-3 mm) mites, nematodes, centipedes, etc., which will prey upon the bacterial and fungal feeders and presumably some top predators as well. This study of food-web behavior is discussed toward the end of the chapter.

As noted earlier, earthworms in mixing and comminuting soil play an important role, which was recognized by Darwin (1837, 1881). Earthworms ingest large amounts of soil and a single worm can ingest several times its own body mass per day in a grassland or pasture. Hence, significant impacts of earthworm feeding are expected on nutrient transformation and availability. Yeates (1981) found that there was not only direct ingestion of leaf litter and other organic debris pulled from the soil surface (Edwards and Lofty 1978), but also that earthworms will ingest and generally triturate any small bits of organic matter, including protozoa and nematodes. Thus, in a series of laboratory and field investigations, Yeates (1981) found generally only 40-50% of the nematode standing crops in soils containing earthworms versus those which had no earthworms. Similar reductions of protozoan populations due to earthworm feeding have also been observed (Piearce and Phillips 1980). The effects of such feeding on changes in secondary production and turnover of these primary grazing populations could be of marked importance, and needs further study.

For a general review of soil faunal composition, biomass, and its relationship to habitat characteristics, including amount and type of organic matter, see Petersen (1982). Petersen and Luxton recently reviewed numbers, biomass, and activity of soil fauna worldwide. They found that, while fauna in general account for less than 5% of the total detritus-decomposer respiration, their indirect, catalytic role in decomposition is considerably greater in many ecosystems. These indirect roles include feeding and its effects on microbial (prey) populations, translocation of nutrients and microorganisms to different locations in the soil profile, and even, in some cases, immobilization of N and P in feces or in nests, such as those made by termites.

TROPHIC INTERACTIONS BETWEEN MICROBES AND FAUNA

A number of studies dating from the times of Cutler et al. (1922) have demonstrated the marked effects of protozoa

on bacterial numbers and more recently on nutrient dynamics.
For instance, Cole et al. (1978) showed a very significant
enhancement (increase from 4 to 8 μg · g^{-1} soil) of
inorganic P in soil where the P initially immobilized by the
bacteria was then released by the nematodes or amoebae
feeding compared to treatments with just bacteria alone. A
detailed review of the role of protozoa in nutrient cycling
is given by Stout (1980). Stout noted that protozoa, being
among the smallest and most abundant soil fauna, often play
an important governing role in feeding upon, and promoting
turnover of, microbial populations. This has been shown in
various natural ecosystems (forest, grassland, marine) as
well as cropped-land ecosystems.

Other studies have shown that soil incubations
("microcosms") with a rhizosphere bacterium Pseudomonas
paucimobilis, growing with an omnivorous nematode,
Mesodiplogaster iheritieri, had greater respiratory activity
(CO$_2$ output) than bacteria grown alone (Coleman et al. 1977,
Anderson et al. 1978, Coleman et al. 1978). Bacterial
populations were markedly reduced when grazed upon, and
there was a significant amount of N and P mineralization
after only 10 days (Anderson et al. 1981). Other studies by
Trofymow and Coleman (1982), using a different bacterial-
feeding nematode (Pelodera), observed initial increases in
bacterial numbers; thus feeding activity by the nematodes
will have different effects depending upon some of the life
history attributes of the organisms. For example, some
species with very rapid growth rates (egg to egg in 4 days)
will produce more new body tissue and eggs than those with
lower reproductive rates (egg to egg in 10-12 days). These
experiments are corrected for temperature and other abiotic
conditions (Anderson and Coleman 1982).

Using a more enriched substrate (i.e., sewage sludge)
Abrams and Mitchell (1980) inoculated a series of incubations
with Pseudomonas fluorescens and some treatments also
received the bacterial feeding nematode Pelodera punctata.

Total respiratory activity in the presence of nematodes was greater than that of bacteria alone. Abrams and Mitchell (1980) also found some enhancement of bacterial growth rate, but changes in inorganic N and P were not determined.

Several investigators have cultured the ubiquitous nematode _Aphelenchus avenae_ on various species of fungi and monitored fungal and nematode populations. Typically, there was some inhibition of mycelial growth or respiration in cultures with nematodes versus those without nematodes (Fig. 7). This inhibition has been observed by de Soyza (1973), Wasilewska _et al._ (1975), and Trofymow and Coleman (1982) as

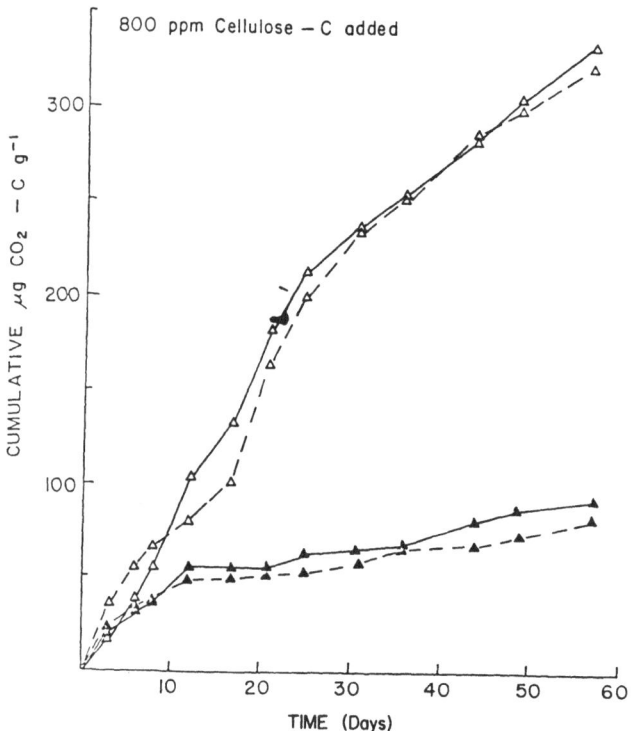

FIGURE 7. ΣCO_2 outputs in fungus alone ($\triangle$——$\triangle$) and fungus plus fungivorous nematode ($\blacktriangle$——$\blacktriangle$) with addition of 800 µg · g^{-1} soil cellulose C. ----- = CO_2 outputs with amendment of 115 µg · g^{-1} soil NH_4^+-N. NH_4^+-N amendments may have inhibited nematode activity in both experiments (from Trofymow and Coleman 1982).

reviewed by Yeates and Coleman (1982). In recent studies, microcosms with nematodes, fungi, and a pure cellulose substrate respired less CO_2 than those without the nematode. In both cases nematode grazing depressed fungal activity. More recent studies by J. A. Trofymow (pers. commun.) have shown that grazing at very low nematode population levels actually stimulate the growth of fungi slightly, which is a grazing-optimization response (Fig. 8), as reviewed by Dyer et al. (1982). The extent of occurrence of such a grazing-optimization response is still problematical in soil detritus food chains. Certainly there have been complete shifts of

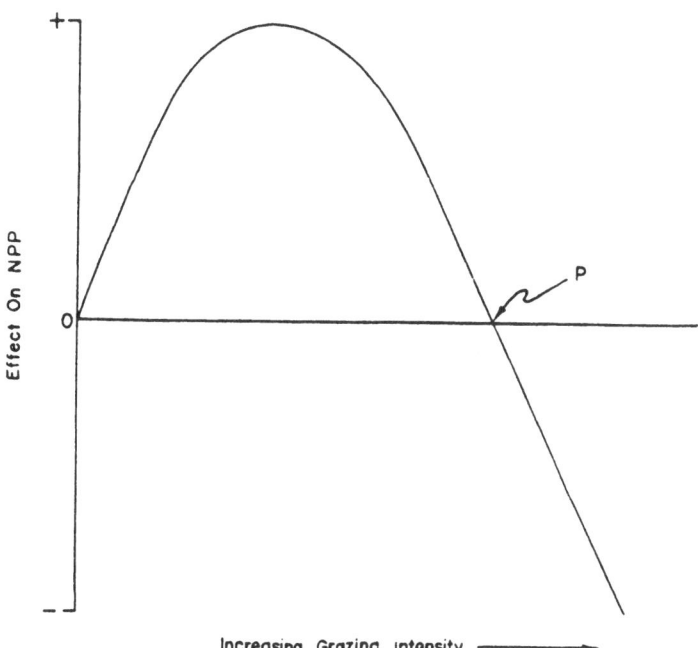

FIGURE 8. Herbivore Optimization Curve (after Dyer et al. 1982). Net production is first increased, then declines to become negative at point "P". This curve has been applied to primary production, but should be equally applicable to microbial populations, as well.

fungal species abundances due to very pronounced feeding by collembola in aspen woodland (Parkinson et al. 1979). This may move total production to the right-hand side of the optimization curve.

There have been considerable laboratory studies of feeding activities and feeding preferences of soil acari and collembola. Various investigators have found marked differences in feeding activity and in responses of microbial populations, similar to those noted above, for nematodes. Usher et al. (1982) and Coleman et al. (1983) have reported several instances where fungal populations may be stimulated or virtually exterminated, depending on nutrient condition and quality of the fungal food source. This may reflect the types of food ingested. Luxton (1982) notes that several authors have observed a greater mixture of soil particles and food in guts of deep-swelling collembola, versus, the more selective feeding of surface-dwelling forms, which evidently select for components such as pollen, spores or hyphae.

Experimental laboratory studies looking at microbial-faunal interactions and subsequent plant growth

To determine possible effects on plant growth and nutrient uptake of micro and meso-fauna feeding on microflora, Elliott et al. (1979) grew blue grama plants in soil with either bacteria or with bacteria and bacterial feeding amoebae. These investigations found an increase in total plant N at various levels of inorganic N fertilization at 20, 90, and 200 g N/g dry soil. The greatest enhancement in plant N occurred at the two highest levels of fertilization; however, there was no enhancement of dry matter yield.

Ingham et al. (1984) examined the role of both bacterial- and fungal-feeding nematodes in microcosms containing plants with chitin being an added N source. The experimental design included the following biotic components: bacteria and fungi alone; bacteria, fungi, and plants; fungal and/or bacteria-feeding nematodes; all components combined. A marked increase in nutrient uptake was detected when there was either

bacterial or fungal grazing by the nematodes. Initially,
there was an enhanced increase in total shoot N and total
shoot biomass in the treatments in which bacterial or fungal
feeders were present either singly or in combination (Fig. 9).

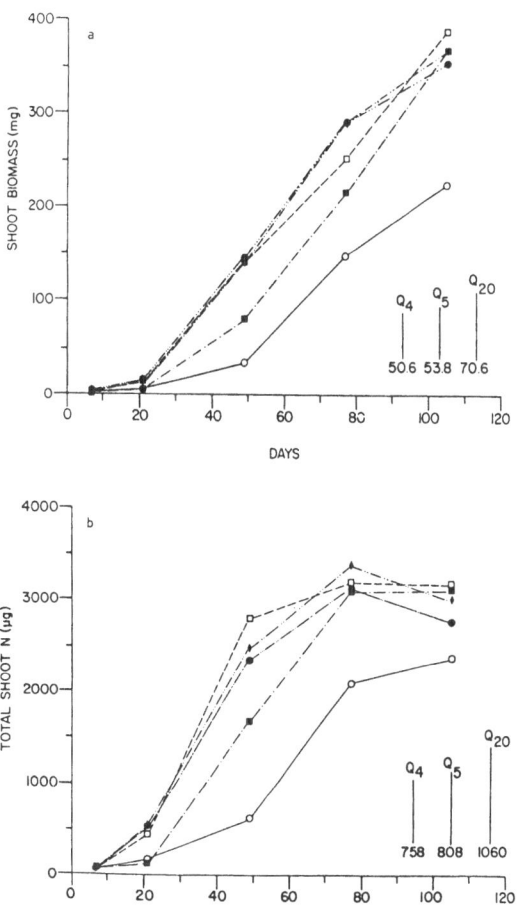

FIGURE 9. a. Shoot biomass of Bouteloua gracilis in
different biological treatments (from Ingham et al.,
submitted). b. Total shoot nitrogen in Bouteloua gracilis
in different biological treatments (from Ingham et al.,
submitted).

O———O   Plant
■—·—·—■   Plant + Bacteria
□— — —□   Plant + Bacteria + Fungus
♦—··—··—♦   Plant + Bacteria + Bacterial-feeding Nematode
●——··—●   Plant + Bacteria + Fungus + Fungal-feeding Nematode

Moreover, for treatments with the bacterial and fungal feeders, the shoot N contents did not differ at the end of the experiment (105 days), and thus the plant alone could gradually take up inorganic N but at a rate somewhat slower than when microbe feeding fauna were present.

In systems where plant growth is limited by N, a change in mineralization of microbial biomass due to grazer-mediated processes should affect plant growth. The importance of these processes should be manifested in nutrient-poor pine forests, where 60-80% per year of bacterial and fungal standing crops may be consumed (Persson et al. 1980) by protozoa (Clarholm and Rosswall 1980, Clarholm 1981) and nematodes (Persson et al. 1980).

Bååth et al. (1978 and 1981) investigated growth responses of Scots Pine seedlings (Pinus sylvestris) in an acid, peaty humus-sand mixture. The pots received various combinations of a nutrient solution, with or without added N and with or without glucose. Bacterial, fungal, microarthropod, enchytraeid, nematode, and protozoan biomasses were determined after 398 days. After about 300 days, pine root and shoot biomass and respective N concentrations were determined. There was considerable stimulation of bacterial, yeast, and filamentous fungal growth by microbial grazing but only slight increases in fungal or bacterial feeders in the various treatments. Moreover, there was no increased plant growth that could be attributed to faunal grazing activity. The experiment had "humus" material with a somewhat low N content (C:N ratio of 44-1), such that N would be immobilized rapidly by the microflora.

In contrast to Ingham et al. (1984), Bååth et al. (1981) concluded that the fauna are not very important for nutrient return in the field. Presumably, the ubiquitous microflora have an advantage over the slower-growing and spatially limited plant root system of the pine, or other plants. There could be significant effects of mycorrhizae on nitrogen uptake, as shown by Melin and Nilsson (1952) for NH_4^+ uptake by ectomycorrhizae; this factor was not studied by Bååth et al. (1981). It would be useful to determine whether localized microsites exist which have a more narrow C:N ratio and to ascertain whether a small fraction of a total root system takes up the majority of the nutrients required by the plant during the growing season. Attention to this sort of spatial heterogeneity is considered next.

Microbial-faunal sites of activity. The spatial and temporal incidence of fauna-mediated immobilization and mineralization processes also are important for ascertaining their effect upon plant growth. Spatially, three regions exist where increased populations of microbes and fauna have either been shown to occur or have been postulated to occur: roots, litter, and aggregates. These regions (or microsites) exist at interfaces or zones where substrates accumulate (Marshall 1976, Berkeley et al. 1980, Bitton and Marshall 1980). Roots produce exudates such as carbohydrates, amino acids, exfoliates (sloughed cells) and mucigel (in conjunction with microbial activity) as the roots grow through soil (Coleman 1976, Rovira et al. 1979, Foster 1981). It is well known that microbes are more abundant in the rhizosphere (Barber and Lynch 1977, Van Vuurde and Schippers 1980, Bennett and Lynch 1981), and recent work by Ingham et al. (1984) has shown that microbial grazer populations also are considerably higher around roots.

Litter is another region of importance. Both within litter layers and in mineral soil near the litter, nutrients are available in greater quantities and microbial populations are higher than in mineral soil farther from roots or litter (Ulehlova 1980). Occasionally, direct cause-and-effect relationships have been demonstrated for enhanced decomposition of litter. Standen (1978) demonstrated this for moorland litter in the presence of bacterial-feeding enchytraeids, and Vossbrinck et al. (1979) showed enhanced decomposition of leaf litter in a shortgrass prairie when microbivorous microarthropods were present. The role of microarthropods in decomposition is extensively reviewed by

Seastedt (in press). These phenomena are discussed further in the section on field studies of microbial-faunal interactions.

The third microsite where C is likely to accumulate is on the surfaces of aggregates in illuviated soil horizons (Berkeley et al. 1980, Haska 1981, Kirchman and Mitchell 1982). These surfaces play a similar role to those of active charcoal filters in sewage treatment plants (Wimpenny 1981). Presently, there is little information on the extent or degree of the microbial or faunal activity at aggregate surfaces in soils. Organic matter, although principally plant and microbe derived, also is produced by certain members of the soil fauna. Thus macro-, and meso-fauna are well-known to produce fecal pellets, even enclosing them (collembola) in chitinous peritrophic membranes. Other fauna, such as protozoa, produce distinctive organic compounds which accumulate in soils. For example, using ^{31}P NMR analysis, nearly 10% of the organic P in certain cool, moist New Zealand soils was found to be composed of phosphonates (C-P bonded) versus the usual ester forms (C-O-P bonded) (Newman and Tate 1980, Tate and Newman 1982). These compounds, salts of 2-amino ethylphosphonic acid, are produced by ciliates. The relative turnover rate of phosphonates compared with that of orthophosphate monoesters, is unknown. Newman and Tate (1982) postulate that these compounds may provide a ready supply of "available P" in undisturbed natural ecosystems, which have P present in forms largely unavailable to plants.

Field studies of microbial-faunal interactions

Recently, a number of investigators have begun studies of selective removal or inhibition of certain target organisms in the field to determine microbial-faunal interactions (Santos and Whitford 1981, Santos et al. 1981, reviewed by Whitford et al. 1982). Whitford et al. (1982) applied fungicides, insecticides and nematicides, singly or in combination in 1 or 2 m^2 field plots in the Chihuahan Desert of southeastern New Mexico and then ascertained decomposition by various measurements including soil respiration. After

application of a general systemic fungicide (Benomyl® plus
Dexon) numbers of bacteria and bacterial-feeding (Cephalobid)
nematodes greatly increased with a shifting of decomposition
pathways through the bacteria-only system. In addition,
various nematode predators such as certain tydeid mites
became much more abundant. By using an acaricide, and thus
removing predatory mites, Santos et al. (1981) observed an
inhibition of decomposition via bacteria as a result of the
absence of mites predatory on the bacterivorous nematodes.

Recent work by Ingham et al. (1983) has followed up on
this interesting lead by applying various biocides in the
shortgrass prairie in northeastern Colorado. Six different
biocides were applied, including a bacteriocide, nematicide,
systemic fungicide, compounds specific for VA mycorrhizae, a
general insecticide and an acaricide. There was an increase
in NH_4^+ after 40 days, particularly when the bacterial
populations were stongly suppressed using streptomycin
compounds. These effects were translated into enhanced N
concentrations in blue grama shoots. The captan treatments
reduced total fungal standing crops, but there were no
attendant changes in soil inorganic N or nutrient uptake by
the grass.

The treatments receiving carbofuran showed significant
reductions in all nematodes, and slight increases in
bacterial-feeding protozoa. The possibility of compensatory
action by similar trophic "functional groups" will be worth
investigating further.

AGROECOSYSTEM STUDIES CONTRASTED WITH NATURAL ECOSYSTEMS
Field experiments
I will now examine biotic and abiotic aspects of
minimum tillage or no-tillage (Phillips 1980) and how it
affects microbial populations and associated nutrient
dynamics. General aspects of nutrient dynamics and
decomposition in agroecosystems is discussed in Chapter 3.
General aspects of microbial activity in relation to pH,
aeration, and soil management are reviewed by Greenland

(1981) and Parr and Papendick (1978). Considering more specific aspects of crop residue management, Doran (1980a,b) examined microbial changes associated with reduced tillage. In one study, Doran (1980a) applied corn stover, either by discing or with surface application at rates of 0, 7, and 14 tons/ha. He measured surface temperatures, microorganism populations, and manipulated soil pH by adding lime as well as using two kinds of herbicides. There were no effects of residues on pH. The residues generally increased all microflora including nitrifiers (<u>Nitrobacter</u> and <u>Nitrosomonas</u>) and denitrifiers.

Doran (1980b) also examined soil microbial and biochemical changes associated with reduced tillage in soybeans and corn in sites in Kentucky, Minnesota, West Virginia, Nebraska and Oregon. Doran found differences in microbial populations related principally to changes in soil water, organic C, N and pH when the reduced tillage regimes were imposed. There were generally higher C, N and water content (Fig. 10) in the surface soil under no till from 0-7.5 cm which is reflected by higher microbial populations and enzyme activities. These relations were reversed from 7.5 to 15 cm, probably because plowing places crop residues at lower depths.

In the top 7.5 cm of no-till soil there was from 20-100 kg more potentially mineralizable N (Stanford and Smith 1976) per hectare. This increased mineralizable N would be closely correlated to increased microbial biomass. It is unknown if the potential reserve of N is an integral component of microbial cells or microbial metabolites. The higher microbial populations in the surface 7.5 cm would act as a greater sink for immobilization of surface applied N fertilizer. Interestingly, increased numbers of microarthropods were found in zero tilled versus conventional tilled fields in Georgia (Stinner and Crossley 1980). Similar results have been found for several microarthropod groups by Moore <u>et al</u>. (in press). It will be important to ascertain whether these increases in some microarthropods

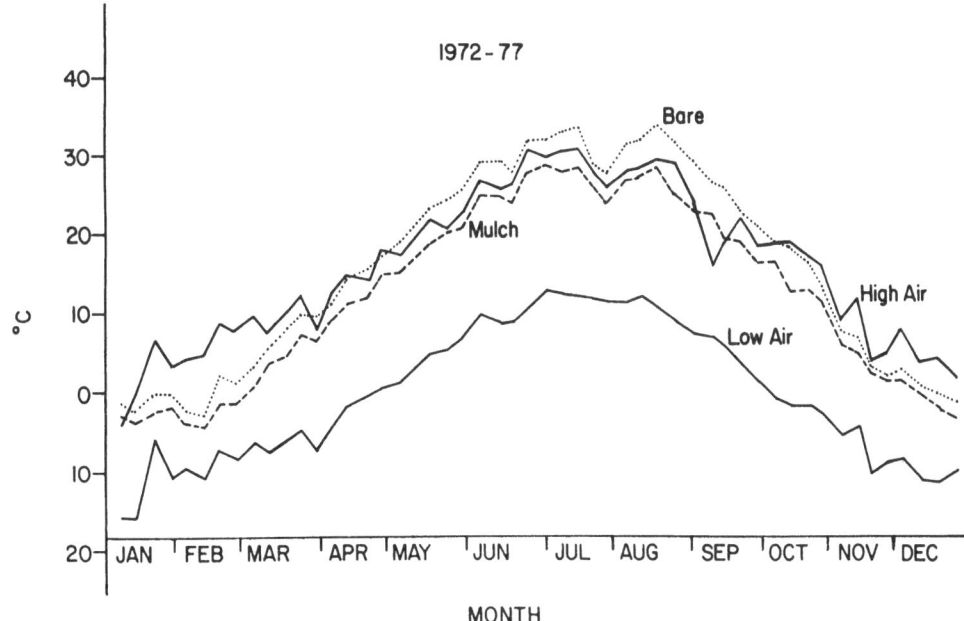

FIGURE 10. Mean maximum and minimum air temperatures and soil
temperatures at the 5-cm soil depth as affected by mulching
for the period 1972 through 1977 at the High Plains Agricultural
Laboratory, Sidney, Nebraska (Fenster and Peterson 1979).

and other micrograzers are due to the effects of tilling on
abiotic physical parameters or to increased food availability.

Doran (1980a) found 4 to 15 times the density of
anaerobes and denitrifiers with reduced tillage compared to
conventional tillage systems. These findings need to be
investigated in other field trials to see if there is more
denitrification and possible gaseous losses under these
field conditions. For further discussion on N dynamics
consult Chapter 6.

ENERGY AND NUTRIENT FLOWS AND THEORETICAL ASPECTS OF
MICROBIAL-FAUNAL INTERACTIONS

DeAngelis (1980) and Pimm (1982) have made a useful synthesis
of food chains and flows of energy in nutrients within food
webs. DeAngelis (1980) notes that recovery time increases
with mean transit time (Fig. 11) which would occur with longer
food chains. Thus, resilience, which is the reciprocal of
recovery time from a perturbation, is thought to actually

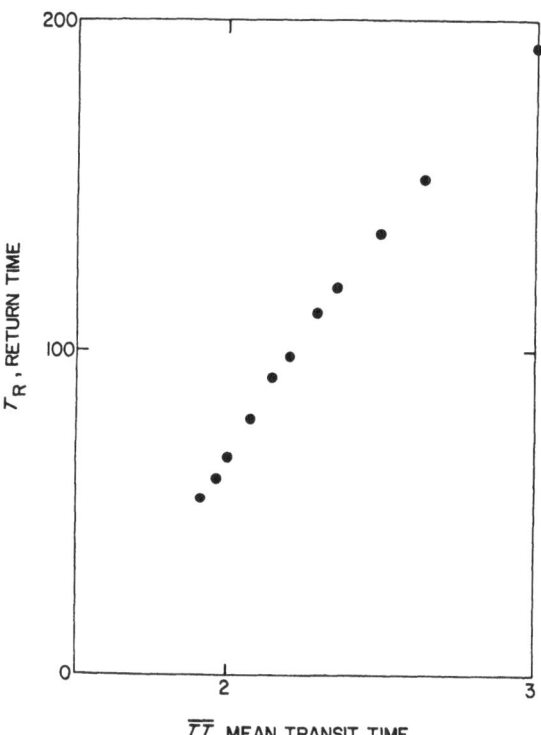

FIGURE 11. The return time, T_R, plotted as a function of the
mean time of transfer, $\overline{TT}$, for a Lotka-Volterra model of
a four trophic-level food chain. Points were generated
by a Monte Carlo sampling technique (De Angelis 1980).

decrease with greater amounts of recycling. It is important
to investigate the amount of interconnectedness of food chains,
food web linkages, and organism feeding specificity range
(omnivory to stenophagy) to ascertain whether these factors
are generally stabilizing or destabilizing. It seems very
likely that the highly stochastic nature of microbial and
faunal interactions, combined with certain life history
features, such as anabiosis or cryptobiosis (Demeure et al.
1979), may overwhelm or supersede the theoretical aspects of
food chain stability.

Certainly it is possible, from theoretical calculations,
to envision long food chains due to the very efficiency with
which some fauna can convert their microbial food source into
secondary production. For instance, Rogerson (1981) showed
that an amoeba (Amoeba proteus) could produce biomass with a
gross production efficiency (P/C) of up to 47% of the amount
of food ingested, while net production efficiencies (P/A)
ranged from 65 to 82%. For various microbial feeding protozoa,
other estimates of gross production efficiencies have gone
as high as 37 to 40% (Stout and Heal 1967, Coleman et al.
1978). In contrast, microbivorous nematode production
efficiencies vary between 8 and 20% (Coleman et al. 1978,
Sohlenius 1980). Certain microbivorous mites have gross
production efficiencies of 30-50%, depending on sex and life-
stage (Thomas 1979). These and other studies are ably
reviewed by Luxton (1982). On the basis of theoretical
calculations, therefore, one would expect to have food chains
that might be 6 or even 7 links long. The prevalence of
polyphagic feeding by various organisms, such as predators
feeding on several trophic levels cannot be discounted at
this point, however.

It seems important to use some colligative properties
of ecosystems, particularly some of those relating to
functional groups or guilds so that we can consider groups
of organisms that feed on similar types of substrates.
Nematodes and protozoa feeding on bacteria may have a more
related role in ecosystem function than various fungal

feeding nematodes and fungal feeding mites and collembola, which, in turn form their own "guilds" (Fig. 6). Feeding on symbiotic fungi is considered next.

The effects of fungal-feeding organisms on mycorrhizal populations on plant roots is also of considerable interest and has been reported on recently by several authors. In the presence of a fungal (including mycorrhizae) feeding collembolan, Folsomia candida, leeks (Allium porrum) under greenhouse conditions grew more when mycorrhizal feeders were absent (Warnock et al. 1982). Similar results of nematodes feeding on mycorrhizal fungi have been observed by Riffle (1975); but in an extensive field experiment, Hussey and Roncadori (1981) found no significant inhibition of plant growth or mycorrhizal growth with normal (in the field) amounts (ca. 1 individual / g soil) of Aphelenchus avenae feeding on mycorrhiza-infected cotton (Gossypium hirsutum) seedlings.

It is important to develop a wide array of laboratory and field experimental conditions for examining microbial and faunal interactions to better understand the diverse processes which function in the general framework of decomposition, immobilization and mineralization. These processes are key focal points in determining dynamics of terrestrial ecosystems.

ACKNOWLEDGMENTS

Support for preparation of this paper was provided by NSF grant DEB 80-04193A02 to Colorado State University. E. T. Elliott, E. R. Ingham, and H. W. Hunt provided helpful comments and suggestions on the manuscript.

REFERENCES

1. Abrams, B., and M. J. Mitchell. 1980. Role of nematode-bacterial interactions in heterotrophic systems with emphasis on sewage sludge decomposition. Oikos 35:404-410.

2. Anderson, D. W. 1979. Processes of humus formation and transformation in soils of the Canadian Great Plains. J. Soil Sci. 30:77-84.

3. Anderson, R. V., E. T. Elliott, J. F. McClellan, D. C. Coleman, C. V. Cole, and H. W. Hunt. 1978. Trophic interactions in soil as they affect energy and nutrient dynamics. III. Biotic interactions of bacteria, amoebae and nematodes. Microb. Ecol. 4:361-371.

4. Anderson, R. V., D. C. Coleman, C. V. Cole, E. T. Elliott, and J. F. McClellan. 1979. The use of soil microcosms in evaluating bacteriophagic nematode responses to other organisms and effects on nutrient cycling. Int. J. Environ. Stud. 13:175-182.

5. Anderson, R. V., D. C. Coleman, and C. V. Cole. 1981. Effects of saprotrophic grazing on net mineralization. In F. E. Clark and T. Rosswall, editors. Terrestrial nitrogen cycles. Ecol. Bull. (Stockholm) 33:201-216.

6. Anderson, R. V., D. C. Coleman, C. V. Cole, and E. T. Elliott. Effect of the nematodes Acrobeloides sp. and Mesodiplogaster lheritieri on substrate utilization and nitrogen and phosphorus mineralization in soil. Ecology 62:549-555.

7. Bååth, E., U. Lohm, B. Lundgren, T. Rosswall, B. Söderström, B. Sohlenius, and A. Wiren. 1978. The effect of nitrogen and carbon supply on the development of soil organism populations and pine seedlings: A microcosm study. Oikos 31:153-163.

8. Bååth, E., U. Lohm, B. Lundgren, T. Rosswall, B. Söderström, and B. Sohlenius. 1981. Impact of microbial-feeding animals on total soil activity and nitrogen dynamics: a soil microcosm experiment. Oikos 37:257-264.

9. Bennett, R. A., and J. M. Lynch. 1981. Bacterial growth and development in the rhizosphere of gnotobiotic cereal plants. J. Gen. Microbiol. 125:95-102.

10. Berkeley, R. C. W., J. M. Lynch, J. Melling, P. R. Rutter, and B. Vincent. 1980. Microbial Adhesion to Surfaces. John Wiley and Sons, New York.

11. Bitton, G., and K. C. Marshall. 1980. Adsorption of Microorganisms to Surfaces. John Wiley and Sons, New York.

12. Brady, N. C. 1978. The Nature and Properties of Soil. Macmillan, New York.

13. Breznak, J. A. 1982. Intestinal microbiota of termites and other xylophagous insects. Ann. Rev. Microbiol. 36:323-343.

14. Clarholm, M. 1981. Protozoan grazing of bacteria in soil. Microb. Ecol. 7:343-350.

15. Clarholm, M., and T. Rosswall. 1980. Biomass and turnover of bacteria in a forest soil and a peat. Soil Biol. Biochem. 12:49-57.

16. Cole, C. V., E. T. Elliott, H. W. Hunt, and D. C. Coleman. 1978. Trophic interactions in soils as they affect energy and nutrient dynamics. V. Phosphorus transformations. Microb. Ecol. 4:381-387.

17. Coleman, D. C. 1976. A review of root production processes and their influence on soil biota in terrestrial ecosystems. Pages 417-434 in J. M. Anderson and A. Macfadyen, editors. The role of terrestrial and aquatic organisms in decomposition processes. Blackwell, Oxford.

18. Coleman, D. C. 1983. The impacts of acid deposition on soil biota and C cycling. Environ. Exp. Bot. 23:225-233.

19. Coleman, D. C., R. V. Anderson, C. V. Cole, E. T. Elliott, L. Woods, and M. K. Campion. 1978. Trophic interactions in soils as they affect energy and nutrient dynamics. IV. Flows of metabolic and biomass carbon. Microb. Ecol. 4:373-380.

20. Coleman, D. C., C. P. P. Reid, and C. V. Cole. 1983. Biological strategies of nutrient cycling in soil systems. Adv. Ecol. Res. 13:1-55.

21. Cutler, D. W., L. M. Crump, and J. Sandon. 1922. A quantitative investigation of the bacterial and protozoan population of the soil. Phil. Trans. R. Soc. Lond. (B) Biol. Sci. 211:317-350.

22. Darbyshire, J. F. 1976. Effects of water suctions on the growth in soil of the ciliate Colpoda steinii and the bacterium Azotobacter chrococcum. J. Soil Sci. 27:369-376.

23. Darwin, C. 1837. On the formation of mould. Trans. Geol. Soc. London 3:505-510.

24. Darwin, C. 1881. The formation of vegetable mould through the action of worms, with observations on their habits. John Murray, London.

25. DeAngelis, D. L. 1980. Energy flow, nutrient cycling and ecosystem resilience. Ecology 61:764-771.

26. Demeure, Y., D. W. Freckman, and S. D. Van Gundy. 1979. Nahydrobiotic coiling of nematodes in soil. J. Nematol. 11:189-195.

27. de Soyza, K. 1973. Energetics of Aphelenchus avenae in monoxenic culture. Proc. Helmith. Soc. Wash. 40:1-10.

28. Doran, J. W. 1980a. Microbial changes associated with residue management with reduced tillage. Soil Sci. Soc. Am. J. 44:518-524.

29. Doran, J. W. 1980b. Soil microbial and biochemical changes associated with reduced tillage. Soil Sci. Soc. Am. J. 44:765-771.

30. Dyer, M. I., J. K. Detling, D. C. Coleman, and D. W. Hilbert. 1982. The role of herbivores in grasslands. Chapter 10 (pp. 255-295) in J. R. Estes, R. J. Tyrl, and J. N. Brunken, editors. Grasses and grasslands: Systematics and ecology. Univ. Oklahoma Press, Norman.

31. Edwards, C. A., and J. R. Lofty. 1978. The biology of earthworms. 2nd edn. Chapman & Hall, London.

32. Elliott, E. T., D. C. Coleman, and C. V. Cole. 1979.
     The influence of amoebae on the uptake of nitrogen by
     plants in gnotobiotic soil. Pages 223-229 in J. L.
     Harley and R. S. Russell, editors. The soil-root
     interface. Academic Press, London.
33. ·Elliott, E. T., R. V. Anderson, D. C. Coleman, and C.
     V. Cole. 1980a. Habitable pore space and microbial
     trophic interactions. Oikos 35:327-335.
34. Elliott, E. T., D. C. Coleman, R. V. Anderson, C. V.
     Cole, H. W. Hunt, L. E. Woods, W. D. Gould, and J. F.
     McClellan. 1980b. Microbial trophic structure and
     habitable pore space in soil. Microcosms in Ecological
     Research, Symposium Series, 52:1050-1070.
35. Fenster, C. R., and G. A. Peterson. 1979. Effects of
     no-tillage fallow as compared to conventional tillage
     in a wheat-fallow system. Res. Bull. 289, Agric. Exp.
     Sta., Inst. Agric. Natur. Res., Univ. Nebraska, Lincoln.
36. Greenland, D. J. 1981. Soil management and soil
     degradation. J. Soil Sci. 32:302-322.
37. Haska, G. 1981. Activity of bacteriolytic enzymes
     adsorbed to clays. Microb. Ecol. 7:331-341.
38. Hunt, H. W., and W. J. Parton. 1984. The role of
     mathematical models in research on microfloral and
     faunal interactions in natural- and agroecosystems.
     Chapter 12 in M. J. Mitchell, editor. Microfloral and
     faunal interactions in natural and agro-ecosystems.
39. Hussey, R. S., and R. W. Roncadori. Influence of
     Aphelenchus avenae on vesicular-arbuscular endomycorrhizal
     growth response in cotton. J. Nematol. 13:48-52.
40. Ingham, E. R., R. N. Ames, C. R. Morley, J. C. Moore,
     and D. C. Coleman. 1983. Field soil microbial and faunal
     population removal and effects on non-target soil
     populations and nutrient cycling by streptomycin, captan,
     carbofuran, PCNB, and cygon in a semi-arid grassland.
     Proc. 3rd Int. Sym. Microb. Ecol., East Lansing, Mich.
41. Ingham, R. E., J. A. Trofymow, R. V. Anderson, and D.
     C. Coleman. 1982. Relationships between soil type and
     soil nematodes in a shortgrass prairie. Pedobiologia
     24:139-144.
42. Ingham, R. E., J. A. Trofymow, E. R. Ingham, and D. C.
     Coleman. 1984. Interactions of bacteria, fungi, and
     their nematode grazers and effects on nutrient cycling
     and plant growth. Ecol. Monogr. (in press).
43. Jenny, H. 1941. Factors of soil formation. McGraw-
     Hill, New York.
44. Jones, F. G. W., and A. J. Thomasson. 1976. Bulk
     density as an indicator of pore space in soils usable
     by nematodes. Nematologica 22:133-137.
45. Jongerius, A. (ed.). 1964. Soil micromorphology.
     Elsevier, Amsterdam and New York.
46. Kirchman, D., and R. Mitchell. 1982. Contribution of
     particle-bound bacteria to total microheterotrophic
     activity in five ponds and two marshes. Appl. Environ.
     Microbiol. 43:200-209.

346

47. Kubiëna, W. L. 1938. Micropedology. Collegiate Press, Ames, Iowa.
48. Kubiëna, W. L. 1970. Micromorphological features of soil geography. Rutgers Univ. Press, New Brunswick, New Jersey.
49. Luxton, M. 1982. 7. Quantitative utilization of energy by the soil fauna. Oikos 39:342-354.
50. Marshall, K. C. 1976. Interfaces in microbial ecology. Harvard University Press, Cambridge.
51. Melin, E., and H. Nilsson. 1952. Transport of labelled nitrogen from an ammonium source to pine seedlings through mycorrhizal mycelium. Sven. Bot. Tidskr. 46:281-285.
52. Mitchell, M. J. Chapter 1, this volume.
53. Moore, J. C., R. J. Snider, and L. S. Robertson. Effects of different management practices on Collembola and Acarina in corn production systems. I. The effects of no-tillage and atrazine. Pedobiologia (in press).
54. Newman, R. H., and K. R. Tate. 1980. Soil phosphorus characterization by ^{31}P nuclear magnetic resonance. Commun. Soil Sci. Plant Anal. 11:835-842.
55. Odum, E. P. 1971. Fundamentals of ecology, 3rd ed. Saunders, Philadelphia.
56. Odum, E. P. 1983. Fundamentals of ecology, 4th ed. Saunders, Philadelphia.
57. Parkinson, D., S. Visser, and J. Whittaker. 1979. Effects of collembolan grazing on fungal colonization of leaf litter. Soil. Biol. Biochem. 11:529-536.
58. Parr, J. F., and R. I. Papendick. 1978. Factors affecting the decomposition of crop residues by microorganisms. Chapter 6, pp. 101-129 in W. R. Oschwald, editor. Crop residue management systems. Am. Soc. Agron. Spec. Pub. No. 31. American Society of Agronomy, Madison, Wis.
59. Persson, T., E. Bååth, M. Clarholm, H. Lundkvist, B. H. Söderström, and B. Sohlenius. 1980. In T. Persson, editor. Structure and function of Northern Coniferous forests--an ecosystem study. Ecol. Bull. (Stockholm) 32:419-459.
60. Petersen, H. 1982. The total soil fauna biomass and its composition. Oikos 39:330-339.
61. Phillips, R. E., R. L. Blevins, G. W. Thomas, W. W. Frye, and S. H. Phillips. 1980. No-tillage agriculture. Science 208:1108-1113.
62. Piearce, T. G., and M. J. Phillips 1980. The fate of ciliates in the earthworm gut: An in vitro study. Microb. Ecol. 5:313-319.
63. Pimm, S. L. 1982. Food webs. Chapman and Hall, New York and London.
64. Pomeroy, L. R. 1970. The strategy of mineral cycling in ecosystems. Annu. Rev. Ecol. Syst. 1:171-190.
65. Riffle, J. 1975. Two Aphelenchoides species suppress formation of Suillus granulatus ectomycorrhizae with Pinus ponderosa seedlings. Plant Dis. Rep. 59:951-955.

66. Rogerson, A. 1981. The ecological energetics of Amoeba proteus (Protozoa). Hydrobiologia 85:117-128.

67. Rovira, A. D., R. C. Foster, and J. K. Martin 1979. Note on terminology: Origin, nature, and nomenclature of the organic materials in the rhizosphere. Pages 1-4 in J. L. Harley and R. S. Russell, editors. The soil-root interface. Academic Press, London.

68. Santos, P. F., and W. G. Whitford 1981. The effects of microarthropods in litter decomposition in a Chihuahuan desert ecosystem. Ecology 62:664-669.

69. Santos, P. F., J. Phillips, and W. G. Whitford 1981. The role of mites and nematodes in early stages of buried litter decomposition in a desert. Ecology 62:654-663.

70. Seastedt, T. R. 1984. The role of microarthropods in decomposition and mineralization processes. Ann. Rev. Entomol. 29:25-46.

71. Sharpley, A. N., J. K. Syers and J. A. Springett. 1979. Effects of surface-casting earthworms on the transport of phosphorus and nitrogen in surface runoff from pasture. Soil Biol. Biochem. 11:459-462.

72. Sohlenius, B. 1980. Abundance, biomass, and contribution to energy flow by soil nematodes in terrestrial ecosystems. Oikos 34:186-194.

73. Standen, V. 1978. The influence of soil fauna on decomposition by micro-organisms in blanket bog litter. J. Anim. Ecol. 47:25-38.

74. Stanford, G., and S. J. Smith 1976. Estimating potentially mineralizable soil nitrogen from a chemical index of soil nitrogen availability. Soil Sci. 122:71-76.

75. Stewart, J. W. B., and R. B. McKercher 1982. Phosphorus cycle. Pages 221-238 in R. G. Burns and J. H. Slater, editors. Experimental microbial ecology. Blackwells, Oxford.

76. Stinner, B. R., and D. A. Crossley, Jr. 1980. Comparison of mineral element cycling under till and no-till practices: An experimental approach to agroecosystem analysis. Pages 280-288 in D. Dindal, editor. Soil biology as related to land use practices. U.S. EPA, Washington, D. C.

77. Stout, J. D., and O. W. Heal. 1967. Protozoa. Pages 149-211 in A. Burges and F. Raw, editors. Soil biology. Academic Press, New York.

78. Stout, J. D. 1980. The role of protozoa in nutrient cycling and energy flow. Adv. Microb. Ecol. 4:1-50.

79. Syers, J. K., A. N. Sharpley, and D. R. Kenney. 1979. Cycling of nitrogen by surface-casting earthworms in a pasture ecosystem. Soil Biol. Biochem. 11:181-185.

80. Tate, K. R., and R. H. Newman 1982. Phosphorus fractions of a climosequence of soils in New Zealand tussock grasslands. Soil Biol. Biochem. 14:191-196.

81. Thomas, J. O. M. 1979. An energy budget for a woodland population of oribatid mites. Pedobiologia 19:346-378.

348

82. Tisdall, J. M., and J. M. Oades. 1982. Organic matter and water-stable aggregates in soils. J. Soil Sci. 33:141-163.
83. Trofymow, J. A., and D. C. Coleman. 1982. The role of bacterivorous and fungivorous nematodes in cellulose and chitin decomposition in the context of a root/rhizosphere/soil conceptual model. Pages 117-138 in D. Freckman, editor. Nematodes in soil ecosystems. University of Texas Press, Austin.
84. Ulehlova, B. 1980. Contents, accumulation and release of energy in free, dead and decomposing plant materials in an upland grassland near Kamenicky, Czechoslovakia. Folia Microbiol. 25:162-167.
85. Usher, M. B., R. G. Booth, and K. E. Sparkes. 1982. A review of progress in understanding the organization of communities of soil arthropods. Pedobiologia 23:126-144.
86. van Vuurde, J. W. L., and B. Schippers. 1980. Bacterial colonization of seminal wheat roots. Soil Biol. Biochem. 12:559-565.
87. Vossbrinck, C. R., D. C. Coleman, and T. A. Woolley. 1979. Abiotic and biotic factors in litter decomposition in a semiarid grassland. Ecology 60:265-271.
88. Walker, T. W. 1965. The significance of phosphorus in pedogenesis. Pages 195-316 in E. G. Hallsworth and D. V. Crawford (editors). J. Exp. Ped. Butterworth, London.
89. Wallace, H. R. 1971. The movement of nematodes in the external environment. Pages 201-212 in A. M. Fallis, editor. Ecology and Physiology of Parasites. University of Toronto Press, Toronto, Canada.
90. Warnock, A. J., A. H. Fitter, and M. B. Usher. 1982. The influence of a springtail Folsomia candida (Insecta, Collembola) on the mycorrhizal association of leek Allium porrum and the vesicular-arbuscular mycorrhizal endophyte Glomus fasciculatus. New Phytol. 90:285-292.
91. Wasilewska, L., H. Jakubczyk, and E. Paplinska. 1975. Production of Aphelenchus avenae Bastian (Nematoda) and reduction of mycelium of saprophytic fungi by them. Pol. Ecol. Stud. 1:61-73.
92. Whitford, W. G., D. W. Freckman, P. F. Santos, N. Z. Elkins, and L. W. Parker. 1982. The role of nematodes in decomposition in desert ecosystems. Pages 98-116 in D. W. Freckman, editor. Nematodes in soil ecosystems. Univ. Texas Press, Austin.
93. Wilkinson, H. T., R. D. Miller, and R. L. Millar. 1981. Infiltration of fungal and bacterial propagules into soil. Soil Sci. Soc. Am. J. 45:1034-1039.
94. Wimpenny, J. W. T. 1981. Spatial order in microbial ecosystems. Biol. Rev. 56:295-342.
95. Yeates, G. W. 1981. Soil nematode populations depressed in the presence of earthworms. Pedobiologia 22:191-195.
96. Yeates, G. W., and D. C. Coleman. 1982. Nematodes and decomposition. Pages 55-80 in D. Freckman, editor. Nematodes in soil ecosystems. Univ. Texas Press, Austin.

EFFECTS OF MANAGEMENT ON SOIL DECOMPOSERS AND DECOMPOSITION PROCESSES IN GRASSLAND

J.P. CURRY*

INTRODUCTION

Decomposition processes in the world's grasslands and croplands are considerably influenced by management. Factors such as fertilizer use, stocking and cutting intensity, choice of crops, cultivation practices and use of pesticides directly influence the nutrient supply for the decomposer community by affecting net primary production and how the resultant fixed chemical energy is utilized. Net primary production in grassland sites studied during the International Biological Programme ranged from 239 g/m^2 dry mass in semi-arid areas to 4557 g/m^2 in subhumid tropics with temperate grasslands falling in the range 702 to 3470 g/m^2 (Coupland 1979). The ranges reflect inherent differences in soil fertility and climate, but are also very much influenced by management. As much as 75% of plant production may be returned to the soil in extensively grazed, semiarid, natural grassland and as little as 3-5% may be assimilated by cattle (Coleman et al. 1976). At the other extreme, under conditions of intensive sheep grazing in temperate grasslands up to 60% of annual shoot production may be assimilated by the sheep (Hutchinson and King 1980a) while in intensively fertilized, mown grassland over 90% of shoot production may be harvested (Andrzejewska and Gyllenberg 1980).

The degree to which management affects soil decomposer communities and decomposition processes increases from extensively managed natural and seminatural grasslands to intensively managed short term leys and is most pronounced in annual crops. Here the complex biotic communities of grassland are replaced by the simplified communities associated with the crop monoculture, and the system is maintained by extensive inputs of fertilizers, pesticides and mechanical cultivation. The absence of plant cover for a significant part of the year, the periodic disruption of the habitat by mechanical cultivation, and the frequent use of pesticides in

*Department of Agricultural Zoology and Genetics, University College, Belfield, Dublin 4, Ireland.

arable crops impose considerable stress on soil organisms exposed to fluctuating moisture, temperature and food resources as well as a variety of chemicals of varying toxicity.

This account briefly reviews some general features of the soil decomposer community before considering how the main components are influenced by management practices. A discussion of how the microflora and fauna interact to influence soil processes can be found in Chapter 8, while the influence of management on successional patterns in grassland decomposer communities is examined in Chapter 2.

THE DECOMPOSER COMMUNITY

The microflora constitute the largest proportion of the decomposer biomass of most soils with bacterial biomass generally ranging from 30 to 300 g/m^2, fungi accounting from 50 to 500 g and actinomycetes probably encompassing a similar range of biomass in many soils (Alexander 1977). Fungi tend to dominate microbial respiration in cultivated and uncultivated soils as well as in virgin grasslands (Anderson and Domsch 1975, Paul et al. 1979).

Soil invertebrate biomasses range from little more than 1 g/m^2 in arid soils to over 200 g/m^2 in temperate mull soils dominated by earthworms. In arid soils only the microfauna (mainly Protozoa and Nematoda) which inhabit the water film on soil particles can live, whereas in mesic climates the macrofauna which require capillary-free water dominate. Under acidic, raw humus conditions the macrofauna (particularly Lumbricidae) are poorly represented and the mesofauna (mainly Acari, Collembola and Enchytraeidae) dominate. Most groups of macrofauna are considerably scarcer in cultivated soils, and faunal biomass rarely exceeds 50 g/m^2.

Microorganisms account for most of the heterotrophic metabolism in soils. Under semiarid prairie conditions microbial respiration can account for 98% of non-root soil respiration (Paul et al. 1979) whereas estimates of the contribution of the soil fauna in non-cultivated ecosystems range from 4% in Swedish coniferous forest (Persson et al. 1980) to 20% in sheep-grazed Polish pasture (Kajak 1974). Invertebrates, particularly the macrofauna, are believed to accelerate decomposition processes through ingestion and comminution of plant litter, mixing of organic and mineral soil components, and improving soil structure and aeration by burrowing. The overriding importance of mechanical cultivation in determining soil

environmental conditions, and the relative scarcity of mesofauna and
macrofauna in cultivated soils, suggest a relatively minor role for
invertebrates in decomposition processes under those conditions. However,
most cultivated soils sustain high populations of protozoans and nematodes
which are associated with the plant rhizosphere and decaying organic
residues and feed on the microflora (Darbyshire and Greaves 1973,Wasilewska
1979). In addition some fungal-feeding Collembola and Acari can also be
locally very abundant in decaying plant residues (Andrén and Lagerlöf
1983a, b; Emmanuel et al. 1984). The effects of microbial feeding invertebrates on microbial activity and organic matter mineralization rates
may be important in nutrient cycling (Coleman et al. 1978, Clarholm 1981,
1983; see also Chapter 8).

Preliminary data from Swedish cropping systems indicate that the fauna
account for 11 to 12% of heterotrophic soil metabolism, with Protozoa
accounting for over 50% of faunal respiration (Paustian 1983). The fauna
account for 19 to 36% of net N mineralization, and has an increasingly
important role in this process as the system becomes poorer in N.

GRAZING BY VERTEBRATE HERBIVORES

Grazing has been one of the major factors in the evolution of grasslands
and is important in regulating the structure and dynamics of much of the
world's semi-natural grasslands in tropical and temperate regions. The
grassland community is affected by grazing intensity, and once grazing
pressure is decreased or eliminated, the community undergoes rapid changes.
This is well exemplified by chalk grassland in England where a rich
herbaceous and grass flora is maintained by rabbit and sheep grazing.
When grazing pressure was reduced due to a decline in the rabbit population
colonization by coarse grasses and scrub occurred with a concomitant
decrease in floral diversity (Wells 1971).

Sown grass-legume leys tend to consist of a small range of cultivars
with the capacity to respond to high levels of fertilizers and frequent
defoliation by grazing or cutting; these swards invariably have lower
species diversity than old grassland. Intensively managed old pastures
also generally tend to have fewer plant species than lightly grazed,
unfertilized ones, but this is not invariably the case (Harper 1969).
Hutchinson and King (1980a) consider that floristic changes in intensively
grazed grasslands depend on the resistance of plants to grazing, selectivity

by grazing animals, frequency of defoliation and the level of plant nutrients and climate. Uncontrolled grazing allows selective grazing of more palatable species and this promotes structural heterogeneity in the sward with heavily grazed areas interspersed with clumps of ungrazed vegetation. This tussock type formation has an important effect on the distribution and abundance of the surface-dwelling invertebrate fauna (Luff 1966). Where grazing is controlled with periods of intensive grazing followed by recovery periods, the result is a greater degree of uniformity within the sward.

Increasing intensity of grazing is usually accompanied by a decrease in the amount of standing dead plant material and surface litter in grass-land; this decrease ranged from 10 to 20% in studies cited by Clark and Paul (1970). Intensity of grazing strongly influences the form and amount of decaying plant material entering the soil decomposer system. This is illustrated by a study at Armidale, N.S.W. Australia, where the effects of sheep stocking levels on the soil invertebrate community and on the partitioning of energy from shoot production among sheep, invertebrate herbivores and decomposers were assessed (Table 1). In this example sheep consumed 59% of all shoot production utilized by animals at a low stocking density of 10 sheep/ha and 94% at the highest stocking density. The amount of unconsumed shoot production entering the decomposer system declined from 27% at 10 sheep/ha to little more than 2% at 30 sheep/ha.

Table 1. Energy budgets ($GJ \cdot ha^{-1} \cdot yr^{-1}$) for sheep and invertebrates at three stocking levels in temperate sown grassland in Australia (after Hutchinson and King 1980b).

Sheep/ha	Ingested by invertebrate herbivores	Ingested by sheep Assimilated	Feces	Plant litter ingested by invertebrates decomposers
10	21	53	32	29
20	21	102	60	21
30	7.4	104	62	3.3

There was a concomitant increase in the proportion of plant material returned to the soil via sheep feces. Thus, with increasing grazing pressure the relative importance of plant litter decomposition declined and that of feces decomposition increased. Dung on the soil surface provides an important, if ephemeral, substrate which is utilized first by

specialized coprophilous species and later by more general saprophages.
The dung community is considered further when organic manures are discussed.

Grazing intensity has an important influence on the rate of nutrient
cycling in grassland. Where grazing intensity is low, the principal route
for nutrient transfer is through the return of dead plant material to the
soil, its incorporation into the soil and its resultant decomposition and
mineralization. This is a relatively slow process: for example, Floate
(1981) reported that only 19 to 35% of N in grass litter confined in litter
bags was lost in 115 days under acid hill pasture conditions in Britain.
By contrast, when herbage is ingested by ruminants the initial stages of
digestion take place in the rumen. About 60% of the N in herbage of 60%
digestibility is voided as urine by sheep and this is either immediately
available or is rapidly converted into plant available form (Floate 1981).
Rapid incorporation of dung into the soil is important for N conservation:
Gillard (1967) found that under Australian conditions 80% of the N is lost
by volatilization from dung which remains on the surface compared with
only 15% loss when coprophagous beetles are active.

The relationships between grazing intensity and rates of nutrient cycling
are further illustrated by data from Armidale (Hutchinson and King 1980a).
In experiments using the radionuclide ^{35}S an increase in sheep numbers
from 10 to 20/ha resulted in an increased uptake of S by the sward from
19 to 25 kg ha^{-1} yr^{-1}, an increase which reflected the overall increase
in primary productivity of the sward. The mean residence times for S
were calculated to be 280 and 170 days for the two stocking treatments,
respectively. The increased cycling rate for S in the former treatment
was attributed to consumption and excretal return by sheep. Relationships
between management and rates of decomposition and nutrient cycling are
considered more fully in Chapter 3.

An important influence of grazing on the distribution of nutrients
within a pasture can arise from the behaviour of gregarious domestic animals
which may congregate in corners of fields and thus cause local concentrations
of excreta. This 'camping' behaviour by sheep in Australian pastures was
responsible for a drift of nutrients away from the major portions of the
pastures and subsequent high localized concentrations of nutrients which
are more subject to volatilization and leaching (Hilder 1966).

Some of the main effects of sheep stocking density on the grassland
habitat at Armidale are summarized in Table 2. Major factors which affect

the soil community include type and amount of food (i.e. herbage), living
space (i.e. soil porosity) and microclimate (i.e. temperature). However
the relatively minor differences in temperature ranges (based on 9 a.m.
temperatures) are of little significance for soil organisms compared with
the marked effects on diurnal temperature ranges resulting from differential
removal of vegetation by grazing. Thus, Davidson et al. (1980) recorded
a maximum soil temperature in lightly grazed pasture in N.S.W. of $28^{\circ}C$
compared with $42^{\circ}C$ in heavily grazed areas. This greater temperature

Table 2. Effects of sheep numbers on the grassland habitat at Armidale
(Hutchinson and King 1980a).

Sheep nos./ha	Green herbage kg dry mass/ha	Dead Herbage kg dry mass/ha	Roots kg dry mass/ha	Soil temp. annual range $^{\circ}C$	Soil Porosity (Vol.%)
10	3125	4225	6800	7-18	52
20	1959	1238	4263	7-19	49
30	620	104	2516	7-21	44

variation in heavily grazed swards was responsible for high mortality in
scarab larvae and, presumably, other soil invertebrates.

There appears to be a general relationship between plant and invertebrate
biomass in grassland (Andrzejewska 1979), and most studies indicate a
decline in population of herbage layer and surface-dwelling invertebrates
as stocking rates increase (Morris 1971, Andrzejewska 1979, Purvis and
Curry 1981, East and Pottinger 1983). This decline has been attributed
to reductions in the amounts of herbage and litter, reduction in pore
space, and changes in soil microclimate. Exceptions include certain
Hymenoptera and Acrididae which prefer the sparsely covered ground conditions
under intensive grazing (Knutson and Campbell 1976, Hutchinson and King
1980b).

Adverse effects of grazing on soil organisms are likely to be more
pronounced in the surface layers and this is reflected in marked reductions
in populations of hemiedaphic invertebrates in moderately to heavily
grazed pastures compared with ungrazed or lightly grazed areas (Morris
1968, King and Hutchinson 1976, King et al. 1977, Hutchinson and King
1980b). Deeper soil dwelling (euedaphic) invertebrates usually are less
adversely affected, and some groups benefit from moderate levels of grazing.
Hutchinson and King (1980b) reported maximum numbers of root feeding

Scarabaeidae and large Oligochaetae in Australian pasture at an intermediate stocking density which coincided with maximum primary productivity. Other studies also show that highest earthworm populations occur in more fertile pastures (Waters 1955, Cotton and Curry 1980a, b), indicating that these animals are adapted to conditions of fast energy and nutrient cycling in contrast to many groups of saprophagous invertebrates which are most abundant when litter decomposition is slow.

Microbial activity appears to be little influenced by stocking density (Hutchinson and King 1982). Microbes are likely to be less sensitive to reduction in pore volume than are invertebrates, and adverse effects of moisture and temperature fluctuations may be counterbalanced by increased inputs of readily metabolised animal excreta in heavily grazed pasture.

MOWING

Mowing is similar to grazing in that defoliation prolongs the vegetative phase of grass growth and increases primary production up to a point. It differs in being non-selective and in that most of the shoot production of agricultural swards is removed with little return of litter to the soil. The effects of mowing on the grassland community depend very much on the frequency and timing of mowing. Grasslands intensively utilized for hay or silage production tend to have less diverse flora than occasionally mown, semi-natural meadows, but on the other hand judiciously timed cutting to simulate grazing can be an effective means of maintaining high floral and faunal diversity in areas of conservation interest (Morris 1971, Wells 1971).

Table 3 illustrates some of the main differences in production and utilization of plant biomass among three Polish grassland sites under different management regimes. In heavily fertilized meadows mown two or three times during the season over 90% of above ground production was

Table 3. Shoot production and utilization in three Polish grasslands (after Andrzejewska and Gyllenberg 1980).

	Unused meadow	Mown meadow	Sheep grazed pasture
Shoot production (g dry mass$\cdot$m$^{-2}\cdot$yr^{-1}	472	909	592
Harvested grass ''		864	
Consumed by sheep ''			435
Returned as feces ''			360

removed as harvested grass whereas in grazed pasture up to 80% of plant production was consumed but up to 60% was returned to the soil as feces. In the unused meadow almost all plant production was eventually returned to the soil as litter.

Some features of the response of the sward to mowing were illustrated by Jankowska (1971) who compared part of a meadow unmown during a three year period with a portion mown twice annually. In the unmown section, there was a decrease in plant production and in the proportion of green shoots, while the proportion of monocotyledons in the sward increased and dead organic matter accumulated on the surface.

Invertebrate communities in mown swards generally have lower species diversity than in unmown (Morris and Lakhani 1979). Biomass of phytophagous species able to exploit the more favorable food supply in mown swards is higher (Southwood and Jepson 1962, Breymeyer 1971, Andrzejewska 1979) while polyphagous predators, particularly spiders, are adversely affected by the reduction in structural diversity associated with cutting (Kajak 1971). Curry and Tuohy (1978) concluded that cutting returned the sward community to an earlier successional stage characterized by mobile species which have short life cycles, rapid developmental time and the ability to rapidly recolonize the swards after cutting, while species with longer life cycles and slower developmental rates become less prevalent.

As in the case of grazing, the greater productivity of swards managed for mowing is probably reflected in accelerated rates of decomposition and nutrient cycling although there is little experimental evidence on this point. Jakubczyk (1971) compared decomposition rates of dead plant material and of cellulose in unmanaged Polish meadow on moist, boggy soil with those on mown, fertile meadows on an alluvial soil. Dead plant material confined in litter bags disappeared at the rate of 4.8 $mg \cdot g^{-1} \cdot day^{-1}$ during the growing season in the unmanaged meadow compared with 8.7 - 14$mg \cdot g^{-1} \cdot day^{-1}$ in the mown meadows. The corresponding figures for cellulose were 15.2 and 27.0 $mg \cdot g^{-1} \cdot day^{-1}$. The lower decomposition rates in the unmanaged meadow were reflected in low densities of bacteria, fungi and actinomycetes. In this instance it is not possible to conclude whether the differences in decomposition rates were due to inherent differences in soil fertility and microclimate or to differences in management.

It is likely that the differences in herbage utilization between mown and unmown grassland are reflected in significant differences in the soil

decomposer communities, but there are no adequate data available from long
term experiments to test this hypothesis. However, reasonable inferences
can be made from comparisons between sites where the only major difference
is in method of utilization. One such comparison can be made between two
sets of grassland sites in Ireland on similar, medium textured, mineral
soils. Two of the sites, at Lyons Hill, Co. Kildare and Phoenix Park,
Co. Dublin were long established parkland which were intermittently grazed
by cattle and deer (Curry 1969, Cotton and Curry 1982) while the other
sites - at Johnstown Castle, Co, Wexford - were cut three times yearly
(Curry et al. 1980). The parklands had a well developed litter layer
reflecting the low level of utilization while the mown swards had no
surface litter. The most striking differences in the soil decomposer
communities were seen when comparing density and biomass of the microarthropods and earthworms (Table 4). This comparison suggests that

Table 4. Microarthropod and earthworm fauna in sites receiving contrasting
management.

	Parklands intermittently grazed			Mown swards		
	Indivi- duals /m^2	Biomass g/m^2	No. of spp.	Indivi- duals/m^2	Biomass g/m^2	No. of spp.
Microarthropods	220,000	2.2	200	12,000	0.18	73
Earthworms	143	88	12	413	162	13

increased intensity of utilization is accompanied by a decline in species
diversity and abundance of microarthropods and an increase in the dominance
of earthworms, notably the larger species such as Lumbricus terrestris.
Microarthropod communities tend to reach their greatest development in
infertile sites where decomposition is slow and litter accumulates (see
Chapter 2); their decline in significance in mown grassland probably
reflects increased decomposition rates and the absence of a litter layer.
Earthworm dominance seems to increase in spite of decreasing litter input
from above ground, suggesting that the extra food required to sustain the
increased population comes from root debris derived from enhanced root
turnover.

FERTILIZERS

Heavy fertilizer use is an integral part of intensive crop production and grassland utilization. Fertilizers affect the soil ecosystem through nutrient enrichment and in the case of organic fertilizers, the organic fraction provides an additional food source for the decomposer community.

Mineral Fertilizers

The most obvious effect of mineral fertilizers is an increase in net primary production. Traczyk et al. (1976) reported a fourfold increase in herbage dry mass harvested from intensively fertilized grass plots compared with unfertilized plots in Poland. Fertilizers bring about a drastic change in sward composition, with a decrease in species diversity, a decline in the importance of many dicotyledon species, and an increase in the proportion of some grass species in the sward (Rorison 1971, Van der Maarel 1971, Traczyk et al. 1976). Increased fertilizer application increases the nutrient content of the vegetation (Mochnacka-Lawacz 1978) thereby improving its quality as food for herbivorous invertebrates (Andrzejewska 1976, Prestidge 1982).

Some studies have indicated that fertilizers increase litter decomposition rates and microbial activity in grassland, while others have shown little or no effect. Jakubczyk (1976a) reported an average rate of litter disappearance of 16 mg g^{-1} day^{-1} in soil fertilized with 360 kg N, 120kg P and 200 kg K/ha compared with 13.5 $mg \cdot g^{-1} \cdot day^{-1}$ from unfertilized control plots. Numbers of ammonifying bacteria, microorganisms utilizing mineral N, fungi and actinomycetes were greater in litter from the fertilized plots after four weeks. Root decomposition rates appeared to be unaffected by fertilizer level. Results of cellulose decomposition experiments were variable: decomposition rates did increase with increasing doses of fertilizers in the laboratory when moisture was adequate but in the field there were no consistent relationships (Jakubczyk 1976a, b). She concluded that the response of decomposition rates to fertilizers was very much influenced by soil conditions. Kubicka (1978) reported enhanced soil respiration in a fertilized meadow and attributed this to increased microbial activity in response to elevated N, Mg and P concentrations in the dead plant material. Úlehlová (1979) cites U.S.S.R. data which also indicates a very pronounced response by microbial populations to fertilizers in steppe soil. Microbial biomass in unfertilized sites was 100-160 g/m^2 compared with 310-480 g/m^2 in sites fertilized with P.

Microbial populations in arable soils are often C limited and response

to added fertilizers is frequently more dramatic than that seen in grassland. This is often attributed to increased litter input, but decomposer
activity may also benefit from lower soil temperatures due to increased
shading by the crop canopy in crops fertilized with N (Hay et al. 1978).
Thus Rosswall and Paustian (1984) reported marked increases in microbial
density in fertilized cereal plots in Sweden compared with unfertilized
plots. This was most dramatically expressed in a 40 to 50 fold increase
in protozoan density which was probably a reflection of increased bacterial
production. Lynch and Panting (1982) recorded an increase in microbial
biomass in response to N fertilizers in direct drilled but not in plowed
plots. They suggest this might be because the added N provides a priming
action to induce the breakdown of the organic mat in the direct drilled
plots, whereas with lower organic matter levels in plowed plots this
response did not occur since C was probably limiting.

Available data indicate variable responses by invertebrate decomposers
to mineral fertilizers. Population increases can probably be attributed
mainly to increased quantity and quality of litter in fertilized swards
and crops, and population declines may be due to temporarily toxic
concentrations of NH_3 and nitrogenous salts following high rates of
fertilizer applications. Increased emergence of saprophagous Diptera in
intensively fertilized Polish meadows was reported by Olechowicz (1976a).
Apterygota increased by 36% under the same conditions (Zyromska-Rudzka
1976) but acarine numbers declined. Further analysis of the data for one
of the most severely affected acarine groups, the Cryptostigmata, revealed
that fecundity and population turnover rates were in fact greater in the
fertilized swards, but this was offset by increased mortality, resulting
in a lower standing crop. Edwards and Lofty (1975a) also reported
moderate to severe reductions in populations of Acari, Collembola, Diptera,
Coleoptera and Myriapoda in permanent pasture receiving 144 kg $N \cdot ha^{-1} \cdot yr^{-1}$
in England. Andrén and Lagerlöf (1983a) reported generally beneficial
effects of inorganic fertilizers on soil microarthropods, enchytraeids
and nematodes in Swedish cropping systems, but recorded negative effects
in some instances in dry weather. Variable responses of nematodes to
fertilizers have been reported. Wasilewska (1976) found no response to
mineral N, P and K, whereas Boström and Sohlenius (1983) reported marked
increases in root feeders and bacterial feeders in N fertilized plots.

Earthworm populations benefit from moderate applications of nitrogenous fertilizers but may be depressed by higher levels. Zajonc (1975) reported maximum populations in meadows receiving 100 kg N/ha, with lower numbers at higher application rates. Nowak (1976) reported that earthworm populations were depressed in grass plots receiving 300 kg N/ha in Poland, and Edwards and Lofty (1975a) reported marked earthworm depression at much lower levels of N in English grassland. On the other hand, these authors reported a positive correlation between earthworm numbers and rates of N application within the range 0 to 192 kg/ha in cereal plots (Edwards and Lofty 1982a). Gerard and Hay (1979) likewise recorded an increase in earthworm biomass with increasing rates of N from 0 to 150 kg/ha in arable plots, and Lofs-Holmin (1983b) noted positive responses of earthworm populations to inorganic fertilizers in Swedish cultivated soils. She occasionally observed harmful effects on light-textured soils. Edwards and Lofty (1982a) found more marked responses of earthworms to fertilizers in arable soils than in grassland, and attributed these to a greater impact on food supply. They reported a strong positive correlation between the C content of the soil and the amount of N that had been added annually in a long-term fertilizer experiment on continuous wheat, supporting the hypothesis that N influences earthworms by increasing plant production and the amount of dead plant material returned to the soil.

Large doses of ammonium sulfate have sometimes been linked with toxicity to earthworms and other soil animals (Satchell 1955, Edwards 1977); this effect may be related to increased soil acidity and is more likely to occur in acid soils (Edwards 1983).

Organic Manures

When herbivore dung is deposited on the soil surface it is rapidly colonized by a specialized coprophilous microflora and fauna. Several groups of fungi are represented (Cooke 1979); in many cases the spores are ingested by the animal and passage through the digestive tract hastens germination. Harper and Webster (1964) described a succession of fungi on rabbit dung consisting of an initial phase of phycomycetes such as Mucor, Pilaira and Pilobolus spp. followed by discomycetes including Rhyparobius and Ascobolus spp. and then pyrenomycetes such as Sordaria and Podospora spp. Basidiomycetes such as Coprinus spp. appeared late in the succession. The phycomycetes can only utilize the simple sugars present in fresh dung, discomycetes and pyrenomycetes can break down cellulose, while basidiomycetes are largely responsible for lignin decomposition. Herbivore dung is a rich source of energy and nutrients and the initial rate of decomposition attributable to the resident microflora is high compared with that of other kinds of organic detritus (Angel and Wicklow 1974).

The early invertebrate colonizers of dung are mainly coprophagous Diptera of the Families Muscidae, Scathophagidae, Sepsidae, Borboridae, and dung beetles of the Family Scarabaeidae (Laurence 1954, Valiela 1974, Breymeyer 1974, Olechowicz 1976b). The activities of the large burrowing Scarabaeidae, notably Aphodius and Sphaeridium, are particularly important for subsequent community development: their tunnels riddle the dung and are used by air-breathing dipterous larvae to reach the interior (Valiela 1974). As decomposition proceeds an increasingly complex food web develops as the dung is invaded by many general litter-feeding species. For example Curry (1979) recorded a total of 144 arthropod species from cattle manure on the soil surface. Aspects of the interactions between coprophagous arthropods and microorganisms in sheep dung were studied in the laboratory by Breymeyer et al. (1975). They found that Diptera and scarabaeid activity decreased fungal activity and increased the numbers of ammonifying bacteria, indicating an enhancement of mineralization.

The importance of the coprophagous fauna in dung decomposition is illustrated by many examples from Australia where the indigenous fauna is not adapted to cope with large quantities of ungulate dung which is produced in intensively grazed pastures and it thus becomes concentrated on the surface. This accumulation causes pasture fowling, sward deterioration, immobilization of plant nutrients in the feces and the nuisance problem of dung-breeding flies. Coprophagous beetles have been introduced experimentally in attempts to alleviate these conditions (Gillard 1967, Ferrar 1973).

The role of dipterous larvae in the initial stages of dung decomposition was stressed by Olechowicz (1976b) who calculated that they consumed about 16% of the total dung produced by grazing sheep in a Polish mountain pasture. Holter (1979) concluded that the main agents in dung decomposition in Denmark were earthworms, notably Aporrectodea longa. They accounted for 50% of the disappearance from the surface of experimental cowpats and Aphodius larvae accounted for another 14-20%. The proportion of dung actually metabolised by dung beetles was negligible; their main role was in mixing dung into the soil and in stimulating microbial activity.

Grassland soil microbial activity only appears to respond to high inputs of animal manure. Thus, Jakubczyk (1974) noted little response to dung deposited during normal sheep grazing in a Polish mountain pasture, but in an area where sheep had been folded during the previous season and large amounts of dung had been deposited microbial activity was significantly affected. Initially ammonification, proteolysis and cellulose decomposition increased. Subsequently nitrification was particularly marked, actinomycete populations increased, but the development of fungi was poor. Eight to ten months after sheep folding nitrification and cellulose decomposition were inhibited as the supply of N became exhausted and fungal biomass increased again due to the supply of dead plant material (Czerwinski et al. 1974). Actinomycetes are generally favored by the addition of organic matter including animal manures (Alexander 1977). They have a significant role in the decomposition of resistant organic components and in the process of humus formation. Davidson (1979) reported a five to ten fold increase in respiration when soil was amended with manure. Hutchinson and King (1982) reported only small differences in microbial respiration between pastures stocked with 10, 20 and 30 sheep/ha, but respiration was substantially higher in areas where sheep gathered at night and which, therefore, received high inputs of dung and urine.

Microbial activity is usually considerably greater in arable land fertilized with organic manure than in land receiving inorganic fertilizers only. Rosswall and Paustian (1984) recorded highest microbial density in long term field experimental plots fertilized with farmyard manure: this was correlated with higher soil organic C content. MacGregor and Naylor (1982) reported enhanced respiratory activity in cropland receiving farmyard manure, and a considerably greater effect when municipal sludge was applied. Doran (1980a) reported a 2 to 6 fold increase in populations of bacteria, actinomycetes and fungi in cornfields mulched with corn residues and related this to increased soil moisture and lower surface temperatures. Very pronounced responses of microbial activity have been reported when organic materials were added to arable soil in the absence of a crop. For example, Mitchell et al. (1978) reported a 40 to 70 fold increase in viable bacteria in sludge-amended soil after 139 days, followed by a decline to background levels over the following 120 days as utilizable substrate became depleted.

Large quantities of manure are produced in intensive animal production enterprises and are often spread on land as semi-liquid slurry with less than 10% dry mass, often at rates considerably in excess of crop nutrient requirements. Adverse effects of slurry on the soil atmosphere and aeration have been reported (Stevens and Cornforth 1974, Burford 1976), although Delpui (1980) recorded no effects of heavy pig slurry applications on the microbial population of a grass ley in France. None of the main physiological groups of soil microorganisms (ammonifiers, nitrifiers, denitrifiers, cellulolytic aerobic bacteria, sulfate reducing bacteria) was affected. Nutrient-rich organic residues from milk and meat processing plants can create local anaerobic conditions when spray-irrigated onto farmland (Stout et al. 1976). Under those conditions the normal decomposer community can be suppressed and be replaced by a community consisting of anaerobes, yeasts and protozoa. The normal soil population can usually cope with moderate applications of more dilute wastes, and an increase in metabolic activity may be recorded (Stout et al. 1976).

Soil invertebrate populations often benefit from the increased food supply resulting from organic amendments, but adverse effects may also occur. These have been variously attributed to NH_3, benzoic acid, phenols, salts of N and S, decomposition products such as butyric acid, and oxygen depletion (Gisiger 1961, Moursi 1962, Curry 1976, Fortuner and Jacq 1976, Huhta et al. 1979, Guiran et al. 1980). Generally earthworms benefit from light to moderate rates of application of farmyard manure, animal slurry or organic wastes (Zajonc 1975, Curry 1976, Dindal et al. 1977, Andersen 1980, Cotton and Curry 1980a, b, Edwards and Lofty 1982a). The increase in earthworm density is usually modest in grassland (0-50%), but Edwards and Lofty (1982a) recorded an increase in the surface feeding species L. terrestris of 184% in response to farmyard manure. Earthworm responses to farmyard manure are more pronounced in arable land than in grassland (Edwards and Lofty 1982a, Lofs-Holmin 1983a) with litter feeding Lumbricus spp. being favored by surface spreading and soil ingesting species such as Aporrectodea caliginosa benefiting more from plowed-in dung. High rates of slurry application are toxic and can drastically reduce populations, probably because of high levels of NH_3 and soluble salts (Curry 1976, Curry and Cotton 1980, Andersen 1980). Species such as L. terrestris and A. longa with well defined burrow openings are particularly vulnerable (Andersen 1980). However, this is a transitory effect and populations generally recover within a year or so, sometimes reaching levels as much as five times greater than those in unaffected areas (Nowak 1975). Mineral soil species such as Allolobophora spp., Aporrectodea spp., Octolasion spp. and L. terrestris are slowest to recover from heavy slurry doses. In soils subject to repeated contamination a discrete organic layer develops on the surface and earthworm species characteristic of raw humus (e.g. Dendrobaena spp., L. rubellus, L. castaneus) tend to predominate (Table 5).

Table 5. Earthworm density (individuals/m^2) as affected by pig slurry contamination (Curry and Cotton 1980).

| | distance from source (m) | | | | | |
	0	5	15	35	55	75
Allolobophora chlorotica (Sav.)	38	22	29	74	130	82
Aporrectodea caliginosa (Sav.)	2	2	17	24	23	29
A. rosea (Sav.)	1			3		11
A. longa (Sav.)		1	2	3	4	4
Allolobophora/Aporrectodea immatures	16	16	38	54	61	132
Dendrobaena mammalis (Sav.)	26	26	7	7	8	39
Dendrobaena rubida (Sav.)	30	26	12			
Dendrobaena immatures	15	24	5	7	12	4
Eisenia foetida (Sav.)	65	42	14		4	
Lumbricus terrestris (L.)		1	2	17	8	14
L. festivus (Sav.)	61	32	50	17	38	4
L. castaneus (Sav.)	82	42	14	10	15	18
L. rubellus Hoff.	19	26	62	17	8	11
Lumbricus immatures	155	155	34	94	119	82
Lumbricus eiseni Lev.	3	6				
Eiseniella tetraedra (Friend)	1					
Octolasion cyaneum (Sav.)						7
Octolasion immatures			5			
Total	514	422	291	327	430	427

Czerwinski et al. (1974) suggest that enhanced earthworm activity in dung-enriched soil may contribute significantly to humus formation. They estimated that earthworm casts in mountain sheep pasture receiving high inputs of sheep manure and supporting 523 earthworms/m^2 added 145 g/m^2 of humus to the soil per season, compared with 28 g/m^2 in pasture which only supported 63 worms/m^2. They also noted that earthworm casts contained 4 to 10 times as much available P as soil and also had a higher content of the cations Ca, Mg and K; hence increased earthworm activity could enhance the cycling of these nutrients in nutrient-poor soils.

Varying responses of other soil fauna to organic amendments have been

described in a variety of studies (Marshall 1977). Arthropod populations
are usually higher in arable land fertilized with moderate amounts of
farmyard manure (Raw 1967, Andrén and Lagerlöf 1983a). Weil and Kroontje
(1979) reported high populations of acarid and macrochelid mites in arable
land in the U.S.A. which had received poultry manure at the rate of 10 kg
dry mass$\cdot$m$^{-2}\cdot$yr^{-1}. These were added with the manure, but manuring also
greatly increased populations of some indigenous arthropod groups although
species diversity was greatly reduced. Dindal (1978) reported that mite
populations were suppressed by spray irrigation of municipal wastes while
Collembola showed periodic increases. Bolger and Curry (1980) found that
most groups were drastically reduced initially by heavy applications of
cattle slurry. Substantial recolonization by hemiedaphic species occurred
within nine months, but many of the euedaphic species remained scarce after
this time. Responses to moderate cattle slurry applications were generally
favorable, particularly in the cases of surface dwelling collembolans and

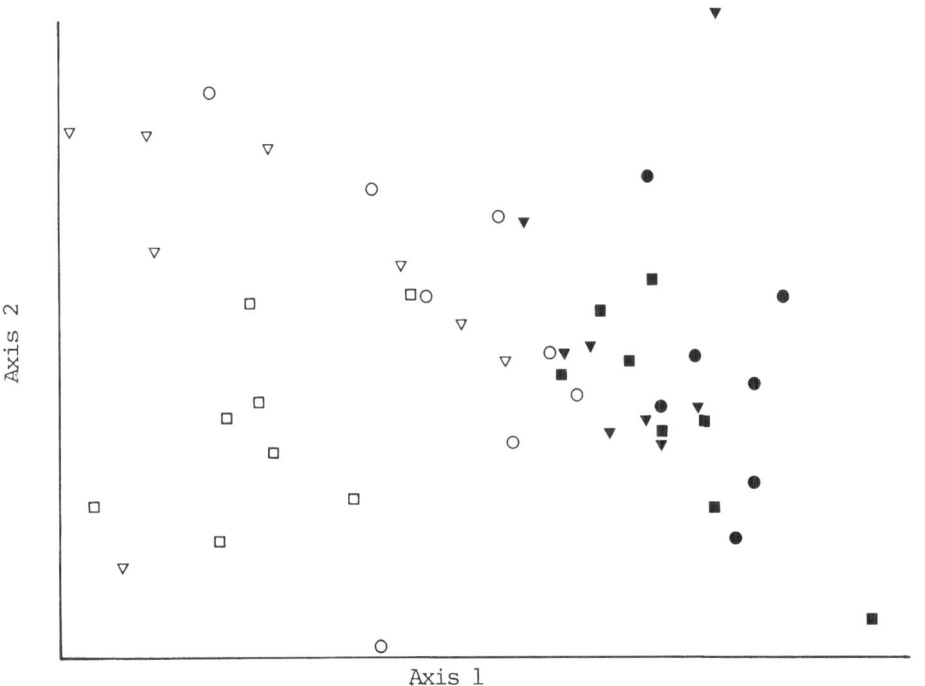

Figure 1. Soil microarthropod samples along a gradient of pig slurry
contamination ordinated by Detrended Correspondence Analysis (based on
data in Bolger and Curry 1984).
  ▽ 5m   □ 15m   ○ 35m   ▼ 55m   ■ 75m   ● 95m from source.

dipterans normally associated with decaying plant litter. Soil microarthropods showed only minor responses to pig slurry applied at rates of
6.6 to 44.5 kg·m^{-2}· yr^{-1}, but marked effects on the structure of the
microarthropod community were detected in a site which had been heavily
contaminated (Bolger and Curry 1984). Detrended Correspondence Analysis
separated samples near the source of contamination from those farther
away, indicating the existence of a coprophilous community in the most
heavily contaminated soil (Fig.1). No effects of sludge amendment on
Acari or Collembola populations were noted by Mitchell et al. (1978) in a
glasshouse experiment. Different ecological groups of nematodes responded
differently to heavy doses of sheep dung in Polish pastures (Wasilewska
1974): microbivores and fungivores increased while omnivores decreased.
Guiran et al. (1980) reported a decrease in populations of phytophagous
nematodes and an increase in populations of microphytophagous forms in
French grass leys heavily fertilized with pig slurry. Mitchell et al.
(1978) reported a rapid increase in numbers of nematodes, particularly
bacterial feeders in sludge amended glasshouse soil, while enchytraeid
numbers increased at a slower rate.

BURNING

Fire is a factor of great importance in the maintenance of much of the
earth's grasslands. In temperate latitudes burning is often carried out
to improve the quality of rough upland pasture, and in tropical areas it
is a key factor in the maintenance of seral grasslands. Fire is widely
used in the management of savannah grasslands where its effects include
the elimination of woody plants, the removal of inedible, dried vegetation
and the stimulation of nutrient cycling and grass growth (Spence and Angus
1971, Singh and Joshi 1979). As much as 53% of primary production in
African tall grass savannah is consumed annually by fire (Sinclair (1975).
Straw burning after harvest is widely practiced in intensive cereal growing
areas of Europe: in England about 50% of all cereal straw is disposed of
in this way (Hughes 1979).

The effects of fire on soil decomposers and decomposition processes
depend on the intensity and duration of the fire. During hot fires in
coniferous forest the surface temperatures may reach 1100°C, consuming all
the litter and litter dwelling organisms (Daubenmire 1968). In African
grasslands surface temperatures exceeding 700°C have been recorded, but

for such brief periods that the direct effect on the soil may be negligible. Generally grassland fires are of low intensity and rarely does the surface temperature exceed 100°C for more than a few minutes (Daubenmire 1968). Under these conditions the litter layer is only partially burned and adverse effects on soil organisms are limited.

The immediate effects of fire on the habitat may include elevated pH, higher concentrations of available bases, elevated surface temperatures, and reduced water availability in the surface horizons (Woodmansee and Wallach 1981). Some loss of plant nutrients occurs through volatilization and leaching, but natural ecosystems appear to be able to retain nutrients following fire and to quickly replace those lost. Mechanisms for nutrient retention include adsorbtion of ions leached out of ash on soil organic matter and clay colloids, immobilization by microbial and root activity, and uptake by rapidly growing pioneer plants. N lost as a result of fire may be replaced quickly by enhanced microbial mineralization and biological N fixation (Woodmansee and Wallach 1981). These effects are usually much greater in forest ecosystems than in grasslands.

Typically, fire stimulates the activity of soil microorganisms, particularly following high-intensity fires where above ground vegetation and roots are killed and a large amount of C becomes available to deomposers (Woodmansee and Wallach 1981). However, Pochon and Bacvarov (1973) reported reduced fungal and bacterial populations while actinomycetes increased in burned Ivory Coast savannah. Results from temperate arable soils are variable: Lynch and Panting (1980b) reported that microbial biomass was greater in the presence of straw from a preceding crop than when it was burned, but subsequent experiments did not confirm this result (Lynch and Panting 1982). However, it appears likely that long term straw burning could deplete soil organic matter and the soil decomposer community in arable land.

Fire promotes earlier and more vigorous grass regrowth and burned savannah supports a denser invertebrate phytophagous fauna during the growing season than unburned (Andrzejewska 1979). Many surface-dwelling arthropods are able to find refuge from fire under stones and in other sheltered situations (Gillon, D. 1971), and groups such as grasshoppers benefit from the rejuvenated habitat following fire (Gillon, Y. 1971). The total biomass of litter-dwelling and hemiedaphic soil invertebrates is often significantly reduced by burning: this is partly a direct

368

consequence of the fire and partly results from the destruction of the
litter layer and the changed microclimatic conditions once the insulating
effect is removed. Lamotte (1976) reports that, on average, faunal density
and biomass were reduced by about 60% one month after fire in Ivory Coast
savannah. Couteaux (1980) found that fire destroyed the testate amoebae
in savannah topsoil since these organisms are incapable of rapid migration
to refuges. She found that Testacea in burned savannah are virtually
confined to the roots of grass tussocks which are little affected by fire.
Athias (1976) reported considerably fewer soil microarthropods (mainly Acari
and Collembola) in burned savannah at the same site. After burning, marked
reductions of normal seasonal changes in density of invertebrates from
litter and surface soil have been recorded in the U.S.A. (Bulan and Barrett
1971, Metz and Dindal 1975) and in England (Edwards and Lofty 1979). Majer
(1984) reported a variety of responses by soil and litter invertebrates to
a cool autumn burn in Eucalyptus forest in Western Australia. Some groups
were unaffected while others were stimulated by fire, but the densities of
many groups were reduced either immediately or after a time lag.

The adverse effects of fire are largely confined to litter and surface
dwellers; euedaphic species are little affected in uncultivated soils and
may even benefit from the greater productivity. Lamotte (1976) gives
data from a study by Lavelle which showed that smaller, litter-dwelling
earthworm species were scarce in burned savannah but the total earthworm
biomass was greater than in unburned areas. Adverse effects are more
likely to be seen in arable soil as soil organic matter is progressively
depleted by cultivation and repeated burning: Edwards and Lofty (1979)
noted that numbers of the litter feeding L. terristris declined after 3
or 4 seasons of straw burning.

IRRIGATION AND DRAINAGE

Biological processes are severely limited over much of the earth by
unfavorable moisture conditions. Macroarthropod decomposers are dependent
on free soil moisture and are virtually absent from arid grasslands; under
those conditions only the microfauna which dwell in water film such as
Protozoa and Nematoda are abundant (Smolik 1974, Elliott and Coleman 1977).
The microbial biomass may be substantial (Paul et al. 1979) but is largely
inactive during periods of drought. In tropical soils which have a
pronounced dry season, the activity of decomposers is largely confined to

the wet season (Dwivedi 1979). Even in moist temperate grasslands where moisture conditions are mainly favorable temporary shortages are reflected in reduced activity and in some cases high mortality in susceptible groups such as Oligochaetae (Nielsen 1955, Gerard 1967, Martin 1978).

Irrigation is a major tool for increasing food production in many of the world's drylands, and even in areas favorable for agriculture where summer moisture deficits occur it is likely that crop yields could be considerably increased by irrigation (Frissel and Van Veen 1981). Irrigation benefits phytophagous species by prolonging the growing season (Andrzejewska 1979) and the removal of moisture restrictions promotes soil decomposition processes. Soil microfloral populations were usually higher in irrigated pastures in New Zealand (Ruscoe 1973) and protozoan activity increased dramatically when arid grasslands were watered (Elliott and Coleman 1977, Dash and Guru 1980). Clarholm (1983) linked increased protozoan biomass with increased availability of microbial food when dry soil was wetted. Data from microcosm studies indicated that microbial grazing by Protozoa significantly increased mineralization of N; this topic is considered further in Chapter 8. Dodd and Lauenroth (1979) reported a four-fold increase in soil microarthropod abundance in watered semiarid prairie. On the other hand Steinberger et al. (1984) did not record any response by nematodes to watering under desert conditions. Population densities of several microarthropod taxa increased in response to watering only when soil litter levels were high, indicating that water has less effect than adequate organic matter on population growth of desert soil nematodes and microarthropods. There are several examples of earthworms becoming established following irrigation of arid areas (Barley and Kleing 1964, Ghilarov and Mamajev 1966, Reinecke and Visser 1980).

Increasing salinity in the root zone can be a major problem in irrigation agriculture, restricting plant growth and soil decomposition processes (Allison 1964). Nitrate production is commonly retarded in soils with high salt concentrations, as nitrifiers have a low tolerance for salinity (Alexander 1977). High salt levels can also be a factor in limiting earthworm populations (El Duweini and Ghabbour 1965).

Waterlogged soils are characterized by a scarcity of soil macro-invertebrate decomposers, slow decomposition rates, accumulation of raw organic matter and peat formation in extreme cases. Reclamation including drainage and liming are followed by a marked increase in biological activity

(Skoropanov 1961) and a gradual increase in earthworm populations (Curry and Cotton 1983). Dowding (1981) considers some aspects of the effects of drainage on N cycling in peatland. He concludes that excessive drainage of eutrophic peat can result in large losses of soil through respiration and the production of large amounts of nitrate which may be lost unless a perennial crop is grown to take it up. In oligotrophic peats on the other hand biological processes are often limited by available P, and rates of mineralization and nitrification may only be increased slightly by drainage.

CULTIVATION

Grassland Reseeding

In the past improving the productivity of old grasslands often involved plowing, cultivation and reseeding, but currently the option of suppressing the sward with herbicides and surface seeding is available. Both methods, and particularly the former, involve severe perturbation of the established ecosystem. Removal of the plant cover and mechanical disturbance of the soil affect the soil microclimate drastically, causing greater fluctuations in moisture and temperature. Plowing and cultivation enhance soil mixing and aeration and promote organic matter oxidation, mineralization and nitrification (Power 1981). This latter effect is seen in a pronounced decrease in soil organic matter when virgin prairie is tilled. Cultivation of old grassland also disrupts the closed nutrient cycles characteristic of such systems, and major losses of labile nutrients such as N may occur (Khanna 1981).

Microbial populations appear to benefit from cultivation of virgin grassland, at least initially. Burges (1958) states that bacterial densities increased twenty to thirty fold in a few days following plowing, maintained a high level for about two months and then fell rapidly. Actinomycetes showed a smaller increase (2 to 3 fold) for a few weeks while isolates of fungi increased 2 to 3 fold for a few months and then declined less rapidly than bacteria.

Most studies indicate fairly marked reductions in numbers of arthropods in the soil following cultivation of old pasture (Sheals 1956, Curry 1970, Edwards and Lofty 1975b). Edwards and Lofty (1975b) reported an overall reduction of over 50% in arthropod density six months after plowing and reseeding, hemiedaphic Collembola and cryptostigmatic mites being most affected. Curry (1970) reported short term reductions of 20 to 50% in

populations of most groups in the surface soil (0-7.5 cm) following
cultivation, but these reductions were offset by higher densities deeper
in the soil (7.5-15 cm), presumably associated with the decaying sward.
Once the new sward became established populations in the surface soil
recovered rapidly.

Earthworms appear to benefit from plowing and reseeding once the initial
stress has passed. Edwards and Lofty (1975b) found that earthworm numbers
increased for two years after treatment: presumably this was a response
to the increased food supply arising from the incorporated sward and litter.
Some accounts indicate that the fauna is less affected by direct drilling
than by plowing and cultivation (Edwards 1975, Wilkinson 1977), but Curry
(1970) found that many soil arthropod groups were reduced equally by both
methods.

Arable Cropping

Repeated cultivations associated with continuous arable cropping affect
many of the primary variables which influence soil organisms, including
aeration, temperation, organic matter content, pH and distribution of
inorganic nutrients (Alexander 1977). Microorganisms in particular benefit
from improved aeration and mixing, while adverse effects of cultivation
are usually ascribed to depletion of soil organic matter, mechanical
damage during cultivation and the absence of an insulating vegetation
layer for part of the year. Bacterial biomass tends to be greater in
cultivated soils than in virgin soils reflecting more oxidative conditions
and more active root growth (Burges 1958, Alexander 1977, Power 1981).
Root exudates and detritus derived from actively growing roots are probably
a major source of C for bacterial populations in cultivated soils, but the
amounts involved are difficult to quantify (Lynch 1979; see also Chapter 5).
Many studies indicate greater overall microbial biomass in managed grassland
than in cultivated soils associated with greater root density and soil
organic matter content (Alexander 1977, Lynch and Panting 1980b), although
Schnürer et al. (1983) report no significant differences between barley
crops and grass leys in Sweden in regard to bacterial, fungal or protozoan
populations. Lynch and Panting (1980a, 1982) reported a strong relationship between microbial biomass and the phenology of the cereal crop, with
biomass being relatively constant between autumn and early spring when
plant roots were scarce or absent and increasing to a maximum in late

spring and early summer as the availability of substrates derived from plant roots increased. These authors (1982) reported a 40% drop in microbial biomass in the spring following rotary cultivation the previous autumn.

Nematode populations in managed grassland and cultivated land tend to be similar in density and composition (Wasilewska 1979, Boström and Sohlenius 1983), but phytophagous forms tend to be less abundant and omnivores and predators more abundant in natural grasslands than in cultivated soils. Boström and Sohlenius (1983) recorded a pronounced increase with time in the density of root feeding nematodes in lucerne as compared with barley or grass leys.

The mesofauna and macrofauna are more adversely affected by cultivation than the microflora and microfauna and tend to be impoverished in most cultivated soils. Table 6 shows how arthropod population density and species richness in an arable field increased rapidly during a three year ley break and declined equally rapidly when tillage was resumed. Andrén and Lagerlöf (1983a) in a survey of Swedish cropping systems reported generally higher abundance of mesofauna in grass leys than in arable land, but some arable sites can have unexpectedly high populations of Acari and Collembola (Andrén and Lagerlöf 1983a, Emmanuel et al. 1984). These consist mainly of microbial feeding Collembola of the Family Isotomidae and Acari of the Families Acaridae, Tarsonemidae, Pyemotidae and Scutacaridae whose life cycles are adapted to exploit the flush of microbial activity associated with decaying crop residues.

Earthworm biomass in cultivated soils is usually less than 50 g/m^2 while grasslands of similar soil type often support 150 to 200 g/m^2 (Gerard and Hay 1979, Andersen 1980, Edwards 1983). Their role in biological processes is correspondingly smaller: Andersen (1983) calculated that the annual N turnover attributable directly to earthworm metabolism in Danish arable plots was at most 14 kg/ha while Keogh (1979) gives a corresponding estimate of 109 to 147 kg/ha for productive New Zealand pasture and Lee (1983) suggests that this can exceed 300 $kg \cdot ha^{-1} \cdot yr^{-1}$ in soils where earthworms are abundant. Deep burrowing species such as L. terrestris and A. longa which have relatively permanent burrows and which feed mainly on surface litter are most adversely affected, while species such as A. caliginosa and A. chlorotica which do not have permanent burrows and which benefit from plowed in organic matter are less affected and usually become dominant in cultivated soils (Edwards 1983).

Crop production systems involving no or minimum tillage conserve soil organic matter, reduce soil erosion and improve water infiltration and

Table 6. Soil arthropod species richness (R) and density (D) in a field at Celbridge, Co. Kildare under grass and arable crops (data from Purvis and Curry 1981 and Curry and Purvis 1982). Standard errors are given in parentheses.

Crop	Acari		Collembola		Other arthropods	
	R	D	R	D	R	D
1974 Barley	–		–		–	
1975 Grass ley	21	1,285(344)	11	891(216)	9	140(55)
1976 Grass ley	54	22,270(5,348)	15	11,370(2,024)	19	2,165(1,081)
1977 Grass ley	75	61,917(10,135)	19	41,253(5,768)	24	11,243(4,771)
1978 Barley	–		–		–	
1979 Sugar beet	38	11,867(3,304)	15	4,320(648)	12	1,740(394)

storage (Power 1981, Stanford 1981). Generally the less the disturbance the less the soil population is affected. Doran (1980b) reported higher microbial populations and greater enzyme activity associated with higher C, N and water content in the surface layers of a "no till" soil, while Lynch and Panting (1980a, 1982) found greater microbial biomass associated with the greater abundance of plant roots in direct drilled cereals. Greater microbial activity results in greater immobilization of N in the surface zone of minimum tillage crops; this reduces N available for plant uptake and can reduce yields if N is limited (Stanford 1981).

Earthworm populations are less affected by light cultivation such as disk harrowing than by plowing (Zicsi 1969), and are least affected by direct drilling (Gerard and Hay 1979, Edwards and Lofty 1982b). This is particularly true of the deep burrowing species: Edwards and Lofty (1982b) reported that L. terrestris and A.longa were 17.5 and 37.3 times more numerous, respectively, in plots direct drilled for eight years than in plowed plots. Most groups of soil invertebrates appear to be favored by minimum cultivation as compared with conventional tillage, but not invariably so. Gers (1982) found greater density and diversity of Collembola and Edwards (1975) reported higher densities of most soil animals in direct drilled than in plowed plots. On the other hand, Emmanuel et al. (1984) reported highest acarine density in a plowed plot while Edwards and Lofty (1975b) and Loring et al. (1981) concluded that populations of euedaphic Collembola may be stimulated by conventional cultivation. Faunal activity in incorporating organic matter into the soil and in improving soil aeration and drainage is important in maintaining the fertility of minimum tillage soils. Edwards and Lofty (1978, 1980) showed that the presence of soil fauna, particularly, earthworms, improved root growth and yields of cereals in soil that had been direct drilled for several years. A negative aspect of minimum tillage is that there are often greater soil pest problems associated with it and consequently greater pesticide use may be required (Speight 1983).

Crop residue management is an important variable affecting soil organic matter content and decomposer activity. In drier climates, surface mulching with crop residues reduces water and wind erosion, conserves soil moisture, and leads to a build up of soil C and organic N near the surface (Power 1981). The advantages of surface residues are less apparent under the humid conditions of Northern Europe, and poor crop establishment occurs when straw from the preceding crop is left in situ (Ellis 1979). Plowed in residues are decomposed and mineralized much more quickly than surface residues (Power 1981), and Doran (1980b) recorded higher C, N and water content and greater microbial activity at a depth of 7.5 to 15 cm in the soil when crop residues were plowed in than when they were left on the surface. On the other hand, Ellis (1979) concluded that plowing in of straw had little effect on soil fertility under British conditions, except in soils with poor structure and soils low in organic matter. However, there is considerable evidence that long term complete removal of crop residues depletes soil C and organic N reservoirs and results in eventual loss of productivity (Power 1981).

Soil invertebrates tend to be most abundant in cropping systems where large amounts of organic residues are returned to the soil. Microarthropods, enchytraeids, nematodes and earthworms are most abundant in Swedish cropping systems where crop residues were plowed in annually (Andrén and Lagerlöf 1983a, Lofs-Holmin 1983b). Edwards (1980, 1983) concludes that earthworm density tends to build up under continuous cereal cropping where there is a significant annual return of organic matter to the soil in the form of stubble, dead roots and straw whereas they tend to be depleted under root crops where most of the plant production is removed. Cultivated soils under orchards and fruit bushes receive a large annual input of leaf litter and support a higher earthworm biomass than annual crops (Edwards 1983).

Chemical and physical properties of plant residues such as C:N ratios, lignin and polyphenol content and the nature of leaf surfaces have an important influence on palatability to decomposers and on decomposition rates (Swift et al. 1979). Heath et al. (1966) found that leaves of kale, beet, lettuce and beans were readily decomposed and disappeared within two months in the field whereas oak and birch leaves took 14 months. Curry (1973) reported differences in the decomposition rates of some grass and broad leaved weeds of pasture with Dactylis glomerata decomposing most slowly. Barley straw decomposes fairly slowly compared with other crop residues: only 40% mass loss occurred in the course of one year in Swedish arable land (Andrén and Lagerlöf 1983b), whereas wheat straw may lose over 80% of its mass in the same period (Lynch 1979). A comparison of decomposition rates over a two month period (April to June) revealed only 13% mass loss for barley straw compared with 26% for meadow fescue and 66% for lucerne in central Sweden (Andrén personal communication). Substrate quality is considered further in relation to decomposition and humification in Chapter 4.

PESTICIDES

A feature of intensive crop production is the degree to which pesticides are used for the control of weeds, pathogens and pests. Pesticides are less widely used in grassland, but heavy applications of insecticides for the control of soil pests can have potentially serious effects on litter incorporation in the absence of mechanical cultivation.

Hundreds of chemicals used for plant protection enter the environment every year; these give rise to vast numbers of breakdown products and metabolites which can be more or less toxic to non target organisms (Menzie 1972, Crosby 1973, Edwards 1973). Soil decomposer populations are most affected by large doses of pesticides applied directly to the soil, but a large proportion of the pesticides applied to the crop canopy also runs off onto the soil (Metcalf 1975). Invertebrates such as L. terrestris can also ingest considerable amounts of pesticides when feeding on contaminated crop residues. Although the environmental effects of pesticides are as yet incompletely understood, a considerable literature has accumulated in this field. Some of the main conclusions relating to soil decomposers are summarized below; more detailed accounts can be found in the reviews cited.

Several of the older organic herbicides have been reported to inhibit soil microbial activity, but currently used materials have only minor effects on microbial processes at normal rates of application (Parr 1974, Grossbard 1976, Wainwright 1978). Often there is an initial reduction in microbial activity, followed by a surge related to increased substrate availability. Ammonification appears to be little affected but nitrification is more pesticide sensitive and can be inhibited by some materials such as amitrole, diallate and dalapon (Parr 1974). However, Domsch and Paul (1974) reported only negligible effects on nitrification of thirty five herbicides applied at field rates. Transient effects of some herbicides on soil respiration and enzymatic activity have been reported. Schinner et al. (1983) reported an initial decrease in CO_2 evolution, dehydrogenase, xylanase and urease activity in soils treated with dinoseb, paraquat, 24D, 245T and simazine followed by an increase, and concluded that practical dose rates will influence these processes for 3 to 4 months. Malanchuk and Joyce (1983) observed a 40 to 50% decrease in CO_2 evolution from soil treated with 24D, but a second application seventy days later elicited no response, suggesting adaptation of the microflora.

Herbicides indirectly affect the soil fauna by altering plant cover and food supply (Fox 1964, Curry 1970, Andrén and Lagerlöf 1983a, Purvis and Curry 1984), but most are probably not directly toxic to soil animals to any great extent. Eijsackers and Van der Drift (1976) reviewed the literature and concluded that the phenoxy acids have little effect, but the triazine-, urea-, and phenol-containing herbicides often have short lived inhibitory effects.

Soil microbial populations and processes can be drastically affected by large doses of contact, broad spectrum fungicides such as verdasan, captan and thiram (Parr 1974, Wainwright 1978, Anderson et al. 1981, Goring and Laskowski 1982). There is usually a temporary reduction in the number of fungi with a concomitant rise in the number of heterotrophic bacteria. Nitrogen transformations are inhibited, particularly nitrification, resulting in an accumulation of ammonium N. Soil respiration may be markedly depressed for a number of weeks, followed by resurgence to a higher level. Typically, the greater the initial biocidal effect the greater the substrate available for surviving heterotrophic microorganisms and the greater the resurgence. Domsch (1970) reported that captan completely inhibited cellulose degradation for six weeks and markedly inhibited chitin and cutin breakdown. The more recent systemic fungicides have only marginal effects on soil microbial processes (Parr 1974, Wainwright 1978, Anderson et al. 1981).

Some of the broad spectrum fungicides have insecticidal and acaricidal properties (Morgan et al. 1958, Fungo and Curry 1983), but few of the currently used materials adversely affect the soil fauna. Routine use of mercury-based fungicides on golf greens and copper-based materials in orchards have resulted in a surface mat of undecayed litter which has been attributed to suppression of earthworm populations and a reduction in cellulolytic fungi (Raw 1962, Pugh and Williams 1971, Van Rhee 1977); these materials are no longer widely used. The greatest risk to earthworms at present arises from the extensive use of substituted benzimidazoles such as benomyl, carbendazim and thiophanate-methyl for the control of fungal pathogens in orchards and arable crops. These materials are highly toxic to earthworms and have been shown to depress populations in the field (Keogh and Whitehead 1965, Stringer and Wright 1976, Wright 1977), although toxicity very much depends on rates of application and soil type (Lofs-Holmin 1981). L. terrestris is particularly susceptible because of its surface-feeding habits.

Fumigants such as DD, chloropicrin, methyl bromide and carbon disulphide are partial soil sterilants which drastically affect soil decomposition processes. Fungi are particularly susceptible, while bacteria and actinomycetes are tolerant and may increase in numbers as they utilize the dead fungal biomass (Brown 1977). Some fumigants stimulate N mineralization (Jenkinson and Powlson 1970), but most inhibit N transformations, particularly nitrification, and depress soil respiration in the short term (Parr 1974, Wainwright 1978, Goring and Laskowski 1982), followed by a resurgence later. Most soil invertebrates are highly susceptible to these fumigants and densities may remain depleted for up to two years (Edwards and Thompson 1973). However, these materials are not extensively used in agricultural soils.

Most insecticides affect soil invertebrate decomposers to a greater or lesser degree depending on many variables including rate and method of application, formulation, and soil type. Table 7 summarizes the responses of the main groups of soil invertebrates to some of the more widely used insecticides.

Table 7. Responses of soil invertebrates to some commonly used insecticides.
= Density reduced by more than 50%,   - Density reduced by less than 50%,
+ Density increased,   +/- Variable response.

	Acari	Collembola	Carabidae	Pauropoda	Diplopoda	Symphyla	Nematoda	Lumbricidae	Enchytraeidae	Mollusca
ORGANOCHLORINES										
Aldrin	-	-	=	=	=	=	+/-			-
Chlordane	-	-	=	-				=		
Dieldrin	-	-	=		=					
DDT	=	+	=	=	=	=	+/-	-		
Endrin	-	-		-				=		
Heptachlor	-	-	=				+/-	-	-	
HCH	-	=	=		=	=	+/-			
Isobenzan	=	=						-		
ORGANOPHOSPHATES										
Chlorfenvinphos	-	+		=						
Chlorpyrifos	-	-						-	-	
Demeton		-								
Diazinon	+/-	+	=	=	-	-	-			
Disulfoton	+/-	+/-	=	=	-					
Fenitrothion		-								
Fensulfothion		-	-				=	=		
Fenthion		-								
Fonofos		-								
Malathion							-	=	=	
Menazan		-								
Parathion	-	+/-	=	=	=	=	-			-
Phorate	-	-	=	=		-	-	=	+	-
Thionazin	+/-	+/-					-			
Trichlorfon		+					-	-		
CARBAMATES										
Aldicarb	=	=					=	-	=	
Carbaryl	=	=	=		=		-	=		-
Carbofuran	-	-					=	=		-
Methiocarb										=

References
Sheals 1956, Raw 1965, Edwards 1965, 1974, 1977, 1980, 1983, Edwards, Dennis and Empson 1967, Edwards, Thompson and Lofty 1967, Fox 1967, Griffiths et al. 1967, Davis 1968, Edwards et al. 1968, Kelsey and Arlidge 1968, Voronova 1968, Way and Scopes 1968, Thompson 1971, Thompson and Gore 1972, Edwards and Thompson 1973, 1975, Wibo 1973, Tomlin and Gore 1974, Martin 1975, Brown 1977, Breslin 1979, Broadbent and Tomlin 1980, Hoy 1980.

The use of organochlorine insecticides is now restricted in many countries because of their long persistence in the environment and the readiness with which they accumulate in food chains. D.D.T. and dieldrin persist longest in soil, with estimated 95% disappearance times ranging from 4 to 30 years, while the corresponding times for the least persistent organochlorines such as aldrin and heptachlor are 1 to 6 years (Edwards 1973). Moderate to severe depressions in numbers of several groups of soil invertebrates have been attributed to organochlorine insecticides (Table 7); these effects may persist for several years. Some studies have reported reductions in densities of susceptible groups exceeding 50%, but usually the effects at normal rates of application are less severe (Edwards and Thompson 1973, Edwards 1977). Collembola are not affected by D.D.T. and often increase in density in treated soils following depletion of their acarine predators (Sheals 1956, Edwards et al. 1967), although they are susceptible to some other organochlorines. Earthworms are not usually affected by normal rates of organochlorines except chlordane, heptachlor and to some extent endrin (Edwards 1980), but residues are readily accumulated in their tissues creating hazards for vertebrates further up the food chain. Edwards (1973) cites concentrations of D.D.T. in earthworm bodies up to 19 times greater than those in the soil.

Organophosphorous insecticides with few exceptions such as chlorfenvinphos and fonofos persist for only a few weeks in soil (Edwards 1973). They are less injurious to soil invertebrates than organochlorines; generally adverse effects are short lived and only excessive doses depress densities to the same extent as normal doses of organochlorines (Edwards and Thompson 1973, Brown 1977). Carbamate insecticides are also non persistent, but can be more toxic to many soil animals than organochlorines or organophosphates (Edwards 1977). Most carbamates including carbaryl, carbofuran and methiocarb are relatively toxic to earthworms (Edwards 1980, 1983). Aldicarb can be taken up by earthworms in large quantities in wet soils, causing irritation which brings them to the surface where they may be eaten by birds. Other insecticides currently being used include members of the synthetic pyrethroid group. These have not been long enough in use to permit adequate evaluation, but there is no evidence to date that they present major environmental hazards.

Insecticides have little effect on soil microbial processes at normal rates of application, and microbial activity plays an important part in the biodegradation of many insecticides in the soil (Parr 1974, Brown 1977, Wainwright 1978). Tu (1970) reported increases in ammonium-N and oxygen consumption proportional to the concentrations of organophosphate insecticides applied to soil, and he suggested that microorganisms utilized these insecticides as substrates. A number of insecticides have been reported to stimulate N mineralization, probably reflecting release of mineral N from the tissues of dead soil animals (Goring and Laskowski 1982), and Davidson (1979) cites studies which indicate that some insecticides may stimulate nitrification and N fixation. These effects may contribute to the dramatic response in grass production which often follows insecticidal application.

Relationships between pesticide effects on fauna and decomposition rates are often difficult to assess. Severe disruption of decomposition processes resulting in the accumulation of a surface layer of undecomposed organic matter can occur in grassland following heavy doses of pesticides which kill soil fauna and particularly earthworms (Kelsey and Arlidge 1968, Clements and Henderson 1976, Keogh 1979), but at moderate rates of application the relationships can be more complex. Malone (1969) reported a drastic reduction initially in microarthropod density in an old field which had been sprayed with diazinon, but the rate of disappearance of organic detritus was greater in treated than in untreated areas. This was attributed to stimulation of microfloral activity by the insecticide. In the first post-treatment growing season species diversity, density and production of herb species was greater in the treated area than in the untreated. This might have been a response to decreased herbivore pressure on seeds and seedlings and enhanced nutrient availability resulting from microbial stimulation. Such changes in the flora would undoubtedly influence the soil microclimate and the nature and quantity of litter returned to the soil decomposer system.

The use of biocides to manipulate soil decomposer communities can provide valuable information on the role of various organisms in decomposition processes. Thus, Malone and Reichle (1973) used formalin, phorate and sodium chlorate to selectively depress earthworms, arthropods and microbes in a fescue meadow soil. The results of litter bag studies and studies with ^{134}Cs labelled litter suggested that faunal activity had little effect on the decomposition of fescue shoots but that fragmentation by the fauna was an important first step in the decomposition of tougher root tissue. Also, when animals were suppressed loss of ^{134}Cs was retarded due to proliferation of microorganisms and microbial immobilization of the radionuclide. Over 90% of the radionuclide remained in the top cm of the soil where the fauna was suppressed compared with only 60% in untreated soil, indicating the important role of animals in element redistribution within soil. The use of pesticides for manipulating the soil system is explored further in Chapter 10.

While the evidence does not suggest that decomposition processes are severely inhibited by normal pesticide usage, nevertheless the fact that there can be some disruption lends weight to arguments based on pest natural enemy, wildlife and human health considerations for reducing pesticide use. Metcalf (1975) suggested ways in which this could be achieved. He pointed out that 50 to 75% of insecticides applied to crops either fall to the ground or drift away, and less than 1% actually hit the pest target. He estimated that a policy of spraying when necessary rather than routinely could reduce present insecticide use in the U.S. by 50%. There are many instances in which pest management strategies integrating biological, cultural and chemical approaches could reduce pesticide use to a fraction of present levels. Possibilities for reducing environmental impact of pesticides include the use of non persistent chemicals coupled with better timing of applications based on life table data and use of lower dosage rates and more precise methods of application. Long chain polymer formulations offer prospects for controlled release of biodegradable pesticides in the correct amounts over appropriate time periods (Allan et al. 1971).

HEAVY METALS

Many industrial processes, particularly those involving extraction and smelting of metals, result in the emission of small particles containing heavy metals in various chemical forms which may be deposited over a wide area. The rate of deposition is greatest near the pollution source, and often results in marked accumulation of heavy metals in the litter and organic layers of soil (Hughes et al. 1980). Heavy metals may also be introduced into agricultural land through the spreading of sewage sludge containing high levels of potentially toxic metals such as Cu, Zn, Ni, Cd (CAST 1976) and farm wastes such as pig slurry which may contain fairly high levels of Cu and other metals added to the food as growth promoters (Curry and Cotton 1980).

The impact of heavy metals on the soil ecosystem is complex and poorly understood (Hughes et al. 1980), with factors such as the chemical forms of the metals, soil organic matter content, the presence of chelating agents, the nature of the clay minerals and soil pH affecting their biological activity. Many laboratory and field studies have shown that heavy metals can affect soil microbial processes, particularly in acid soils. Strojan (1978a,b) noted marked accumulation of litter in a forest site 1 km from a zinc smelter where very high metal concentrations occurred in the litter (26000 ppm Zn, 10000 ppm Fe, 2300 ppm Pb, 900 ppm Cd, 34000 ppm Cu). He found that mass loss and CO_2 evolution from litter samples decreased as proximity to the zinc smelter increased. Tyler 1975 (a,b) likewise reported litter accumulation in a spruce forest near an old brassworks and showed that organic matter decomposition rate, N and P mineralization rates and soil enzymatic activity were negatively related to Cu plus Zn concentrations. Nitrogen mineralization was depressed at Cu levels as low as 50 ppm in acid soil. Microbial populations are frequently severely depressed in heavily contaminated soils but populations of tolerant or resistant species are often reported (Jordan and Lechevalier 1975, Jensen 1977). Severe depression of litter microarthropod populations, especially oribatid mites, were reported by Strojan in his study, and by Watson et al. (1976) in the region of a lead mining smelter. On the other hand, Williamson and Evans (1973) reported high densities of Collembola and Acari in overgrown spoil heaps of disused lead mines in England.

Heavy metals in sewage sludge appear to be relatively stable and are not absorbed by plants to the same extent as inorganic salts (Chang et al. 1983). They do not appear to inhibit decomposition; in fact sludge amendment provides a readily available C source and usually considerably enhances soil respiration (Mitchell et al. 1978, MacGregor and Naylor (1982).

Earthworms are particularly at risk since they may ingest large quantities of contaminated soil and organic material. Metals in ionic form pose a greater risk than those which are organically bound. Malecki et al. (1982) reported significant adverse effects of some Cd salts on growth and reproduction of the compost worm Eisenia foetida at concentrations down to 50 ppm, but this species can tolerate relatively high levels (several hundred ppm) of several heavy metals in sludge without adverse

effects (Hartenstein et al. 1980). Earthworm mortality in the field has
been attributed to high levels of Cu derived from pig slurry and fungicides
(Van Rhee 1977, Curry and Cotton 1980). Hunter and Johnson (1982) reported
low populations of earthworms near a refinery where soil Cu levels were
2000 to 3000 ppm, but earthworms were abundant some distance away from
the refinery where Cu levels were 235 to 255 ppm, considerably above
normal soil levels. Ireland (1983) reviewed the literature and concluded
that earthworms can tolerate relatively high levels of most heavy metals
without adverse effects. They can accumulate high concentrations of Cd
and other metals within their bodies, and these may constitute a hazard
for earthworm-feeding vertebrates.

CONCLUSIONS

Table 8 summarizes the effects of management on the main soil biotic
and abiotic parameters affecting decomposition. Management practices
aimed at increasing grassland productivity induce changes in the decomposer community which in many ways parallel those occurring in the aboveground herbage. Enhanced primary production and accelerated root and
shoot turnover result in an increased supply of litter of higher nutrient
status. More rapid decomposition and mineralization result from the
enhanced activities of soil microorganisms and macrofauna resulting in the
disappearance of the surface litter layer. The highly-diversified
invertebrate community associated with the litter of undisturbed grassland
cannot persist; it is replaced by a simpler community adapted to the new
conditions. These changes primarily reflect changes in rates of energy
and nutrient cycling but are accentuated by periodic stresses imposed by
grazing, mowing, fertilizers, fire etc.

Arable cropping involves drastic simplification of the vegetation,
repeated mechanical cultivations, depletion of soil organic matter, frequent
pesticide applications and the absence of plant cover for part of the year
and thus cropping imposes severe and repeated stress on the deomposer
community. Large invertebrate decomposers are most affected and their
role in physically working the soil is largely subsumed by mechanical
cultivation. Protozoans, nematodes and some small arthropods are less
affected, and their activities as microbial grazers may be of considerable
significance in mineralization processes in arable land. The less the
mechanical disturbance, the less soil decomposers are affected and direct-

Table 8. General responses of some soil biotic and abiotic parameters to management.

+ positive   - negative   o neutral   +/- variable   () slight effect

	Temperature	Moisture content	Porosity/aeration	Organic matter	Microbial Activity	Microfaunal Activity	Mesofaunal Activity	Macrofaunal Activity	Decomposition Rate
GRASSLAND									
Mowing	+	-	o	-	+	o	-	+	+
Grazing	+	-	-	-	+	o	-	+	+
Inorganic fertilizers	o	o	o	-	+	(+)	(-)	+/-	+
Organic fertilizers	-	+	o	+	+	+	+	+	+
Burning	+	-	o	-	+	-	-	-	+
Irrigation	-	+	o	o	+	+	+	+	+
Drainage	+	-	+	-	+	+	+	+	+
ARABLE LAND									
Plowing	+	-	+	-	+	+/-	-	-	+
Minimum cultivation	(+)	(-)	-	(-)	(+)	o	(-)	(-)	(+)
Inorganic fertilizers	o	o	o	-	+	+	(+)	(+)	+
Organic fertilizers	-	+	o	+	+	+	+	+	+
Surface crop residues	--	+	o	+	+	(+)	(+)	+	(+)
Incorporated residues	(-)	(+)	o	+	+	+	(+)	(+)	+
Residue removal	+	-	o	-	-	-	-	-	-
Herbicides	o	o	o	o	(-)	o	o	o	(-)
Fungicides	o	o	o	o	-	o	o	(-)	-
Fumigants	o	o	o	o	-	-	-	-	-
Insecticides	o	o	o	o	o	--	--	-	-
Heavy metals	o	o	o	o	-	-	-	-	-

drilled crops support a greater microbial and faunal biomass than plowed systems. Decomposers benefit from organic amendments including crop residues and organic manures although heavy applications of raw organic wastes can be toxic to sensitive groups. Moderate applications of mineral fertilizers are beneficial through increasing plant production and litter return, but high applications can have detrimental effects. Many pesticides and heavy metal pollutants at high levels can drastically affect susceptible decomposers, but normal doses of most pesticides and moderate heavy metal loads do not usually adversely affect decomposition and mineralization processes. Information on the long term effects of repeated applications is lacking but risks should be minimized by curtailing land spreading of sludges with high metal content on agricultural land and by restricting the use of the more toxic pesticides.

REFERENCES

1. Alexander, M. 1977. Introduction to soil microbiology, 2nd edn. Wiley, New York.
2. Allan, G.C., C.S. Chopra, A.N. Neogi, and R.M. Wilkins. 1971. Design and synthesis of controlled release pesticide-polymer combinations, Nature (Lond.) 234:349-351.
3. Allison, L.E. 1964. Salinity in relation to irrigation. Adv. Agron. 16:139-180.
4. Andersen, C. 1980. The influence of farmyard manure and slurry on the earthworm population (Lumbricidae) in arable soil, pages 325-335, in D.L. Dindal, editor. Soil biology as related to land use practices. EPA, Washington.
5. ———. 1983. Nitrogen turnover by earthworms in animal manured plots, pages 139-150, in J.E. Satchell, editor. Earthworm ecology, Chapman and Hall, London.
6. Anderson, J.P.E., R.A. Armstrong, and S.N. Smith. 1981. Methods to evaluate pesticide damage to the biomass of the soil microflora. Soil Biol. Biochem. 13:149-153.
7. Anderson, J.P.E., and K.H. Domsch. 1975. Measurement of bacterial and fungal contribution to respiration of selected agricultural and forest soils. Can. J. Microbiol. 21:314-322.
8. Andrén, O., and J. Lagerlöf. 1983a. Soil fauna (Microarthropods, Enchytraeids, Nematodes) in Swedish agricultural cropping systems. Acta agric. Scand. 33:33-52.
9. ———. 1983b. Succession of soil microarthropods in decomposing barley straw, pages 644-646, in Ph. Lebrun, H.M. André, A. de Medts, C. Grégoire-Wibo, G. Wauthy, editors. New trends in soil biology. Imprimeur Dieu-Brichart, Ottignies - Louvain-la-Neuve.
10. Andrzejewska, L. 1976. The influence of mineral fertilization on the meadow phytophagous fauna. Pol. ecol. Stud. 2:93-109.
11. ———. 1979. Herbivorous fauna and its role in the economy of grassland ecosystems. 1. Herbivores in natural and managed meadows. Pol. ecol. Stud. 5:5-44.

386

12. ———, and G. Gyllenberg. 1980. Small herbivore subsystem, pages 201-267, in A.I. Breymeyer and G.M. Van Dyne, editors. Grasslands, systems analysis and man. Cambridge University Press, Cambridge.

13. Angel, K., and D.T. Wicklow. 1974. Decomposition of rabbit faeces: an indication of the significance of the coprophilous microflora in energy flow schemes. J. Ecol. 62:429-437.

14. Athias, F. 1976. Recherche sur les microarthropods du sol de la savane de Lamto (Côte d'Ivoire). Ann. Univ. Adidjan sér. E (Ecol.) IX:193-271.

15. Barley, K.P., and C.R. Kleing. 1964. The occupation of newly irrigated lands by earthworms. Aust. J. Sci. 26:290-291.

16. Bolger, T., and J.P. Curry. 1980. Effects of cattle slurry on soil arthropods in grassland. Pedobiologia 20:246-253.

17. ———. 1984. Influences of pig slurry on soil microarthropods in grassland. Rev. Ecol. Biol. Sol 21:269-281.

18. Boström, S., and B. Sohlenius. 1983. Abundance, biomass and metabolic activity of nematodes in different cropping systems, pages 58-77, in T. Rosswall, editor. Ecology of arable land. The role of organisms in nitrogen cycling. Progress report 1982. Swedish University of Agricultural Sciences, Uppsala.

19. Breslin, L.J. 1979. Aspects of the biology and chemical control of Collembola and Acari associated with the root systems of sugar beet seedlings in Co. Carlow. Unpublished M. Agr. Sc. thesis, National University of Ireland.

20. Breymeyer, A. 1971. Productivity investigation of two types of meadows in the Vistula Valley. XIII. Some regularities in structure and function of the ecosystem. Ekol. Pol. 19:249-261.

21. ———. 1974. Analysis of a sheep pasture in the Pieniny Mountains (the Carpathians). XI. The role of coprophagous beetles (Coleoptera, Scarabaeidae) in the utilization of sheep dung. Ekol. Pol. 22:617-34.

22. ———, E. Olechowicz, and H. Jakubczyk. 1975. Influence of coprophagous arthropods on microorganisms in sheep faeces -laboratory investigation. Bul. Acad. Pol. Sci., cl. II, 23:257-62.

23. Broadbent, A.B., and A.D. Tomlin. 1980. Effects of carbofuran on the soil microarthropod community in a cornfield, pages 82-87, in D.L. Dindal, editor. Soil biology as related to land use practices. EPA, Washington.

24. Brown, A.W.A. 1977. Ecology of pesticides. Wiley, New York.

25. Bulan, C.A., and G.W. Barrett. 1971. The effects of two acute stresses on the arthropod component of an experimental grassland ecosystem. Ecology 52:597-605.

26. Burford, J.R. 1976. Effects of the application of cow slurry to grass-land on the composition of the soil atmosphere. J. Sci. Food Agric. 27:115-126.

27. Burges, A. 1958. Microorganisms in the soil. Hutchinson, London.

28. Chang, A.C., A.L. Page, J.E. Warneke, M.R. Resketo, and T.E. Jones. 1983. Accumulation of cadmium and zinc in barley grown on sludge-treated soils: a long-term study. J. Environ. Qual. 12:391-397.

29. Clarholm, M. 1981. Protozoan grazing of bacteria in soil - impact and importance. Microb. Ecol. 7:343-350.

30. ———. 1983. Dynamics of soil bacteria in relation to plants, protozoa and inorganic notrogen. Report no. 17, Department of Microbiology, Swedish University of Agricultural Sciences, Uppsala.

31. Clark, F.E., and E.A. Paul. 1970. The microflora of grassland. Adv. Agron. 22:375-435.

32. Clements, R.O., and I.F. Henderson. 1976. Some consequences of a

prolonged absence of invertebrates from a perennial ryegrass sward, pages 129-134, in Proc. XIIIth Int. Grassl. Congr., Leipzig.

33. Coleman, D.C., R.V. Anderson, C.V. Cole, E.T. Elliott, L. Woods, and M.K. Campion. 1978. Trophic interactions in soils as they affect energy and nutrient dynamics. IV. Flows of metabolic and biomass carbon. Microb. Ecol. 4:373-380.

34. Coleman, D.C., R. Andrews, J.E. Ellis, and J.S. Singh. 1976. Energy flow and partitioning in selected man-managed and natural ecosystems. Agro-Ecosystems 3:45-54.

35. Cooke, N.B. 1979. The ecology of fungi. C R C Press Inc., Boca Raton, Florida.

36. Cotton, D.C.F., and J.P. Curry. 1980a. The effects of cattle and pig slurry fertilizers on earthworms (Oligochaeta, Lumbricidae) in grassland managed for silage production. Pedobiologia 20:181-188.

37. ————. 1980b. The response of earthworm populations (Oligochaeta, Lumbricidae) to high applications of pig slurry. Pedobiologia 20:189-196.

38. ————. 1982. Earthworm distribution and abundance along a mineral-peat soil transect. Soil Biol. Biochem. 14:211-214.

39. Council for Agricultural Science and Technology 1976. Application of sewage sludge to cropland: Appraisal of potential hazards of the heavy metals to plants and animals. Report no. 64, CAST, Ames, Iowa.

40. Coupland, R.T. 1979. Conclusion, pages 335-355, in R.T. Coupland, editor. Grassland ecosystems of the world: analysis of grasslands and their use. Cambridge University Press, Cambridge.

41. Couteaux, M.M. 1980. Effects of the annual burnings on testacea of two kinds of savannah in Ivory Coast, pages 466-471, in D.L. Dindal, editor. Soil biology as related to land use practices. EPA, Washington.

42. Crosby, D.G. 1973. The fate of pesticides in the environment. Annu. Rev. Plant Physiol. 24:467-492.

43. Curry, J.P. 1969. The qualitative and quantitative composition of the fauna of an old grassland site at Celbridge, Co. Kildare. Soil Biol. Biochem. 1:219-227.

44. ————. 1970. The effects of different methods of new sward establishment and the effects of the herbicides paraquat and dalapon on the soil fauna. Pedobiologia 10:329-361.

45. ————. 1973. The arthropods associated with the decomposition of some common grass and weed species in the soil. Soil Biol. Biochem. 5:645-657.

46. ————. 1976. Some effects of animal manures on earthworms in grassland. Pedobiologia 16:425-438.

47. ————. 1979. The arthropod fauna associated with cattle manure applied as slurry to grassland. Proc. R. Ir. Acad. 79B:15-27.

48. ————, and D.C.F. Cotton. 1980. Effects of heavy pig slurry contamination on earthworms in grassland, pages 336-343, in D.L. Dindal, editor. Soil biology as related to land use practices. EPA, Washington.

49. ————. 1983. Earthworms and land reclamation, pages 215-228, in J.E. Satchell, editor. Earthworm ecology. Chapman and Hall, London.

50. ————, T. Bolger, and V. O'Brien. 1980. Effects of landspread animal manures on the fauna of grassland, pages 314-325, in J.K.R. Gasser, editor. Effluents from livestock. Applied Science Publishers Ltd., London.

51. Curry, J.P., and G. Purvis. 1982. Studies on the influence of weeds and farmyard manure on the arthropod fauna of sugar beet. J. Life Sci. R. Dublin Soc. 3:397-408.

52. Curry, J.P., and C.F. Tuohy. 1978. Studies on the epigeal microarthropod fauna of grassland swards managed for silage production. J. appl.

388

Ecol. 15:727-741.

53. Czerwinski, A., H. Jakubczyk, and E. Nowak. 1974. Analysis of a sheep pasture ecosystem in the Pieniny Mountains (the Carpathians). XII. The effect of earthworms on the pasture soil. Ekol. Pol. 22:635-650.

54. Czerwinski, A., H. Jakubczyk, A. Tatur, and T. Traczyk. 1974. Analysis of a sheep pasture ecosystem in the Pieniny Mountains (the Carpathians). VII. The effects of penning up sheep on soil, microflora and vegetation. Ekol. Pol. 22:547-558.

55. Darbyshire, J.F., and M.P. Greaves. 1973. Bacteria and protozoa in the rhizosphere. Pestic. Sci. 4:349-360.

56. Dash, M.C., and B.C. Guru. 1980. Distribution and seasonal variation in numbers of Testacea (Protozoa) in some Indian soils. Pedobiologia 20:325-342.

57. Daubenmire, R. 1968. Ecology of fire in grasslands. Adv. ecol. Res. 5:209-266.

58. Davidson, R.C. 1979. Micro organisms, pages 267-276, in R.T. Coupland, editor. Grassland ecosystems of the world. Cambridge University Press, Cambridge.

59. Davidson, R.L., J.A. Hilditch, J.R. Wiseman, and V.J. Wolfe. 1980. Growth, fecundity and mortality responses of scarabs contributing to population increases in improved pastures, pages 66-70, in T.K. Crosby and R.P. Pottinger, editors. Proc. 2nd Australas. Conf. Grassl. Invertebr. Ecol. Government Printers, Wellington.

60. Davis, B.N.K. 1968. The soil macrofauna and organochlorine insecticide residues at twelve agricultural sites near Huntingdon. Ann. appl. Biol. 61:29-45.

61. Delpui, C. 1980. Landspreading of liquid pig manure V. Effect of soil microflora, pages 120-138, in J.K.R. Gasser, editor. Effluents from livestock. Applied Science Publishers Ltd., London.

62. Dindal, D.L. 1978. Soil organisms and stabilizing wastes. Compost Science/Land Utilization. J. Waste Recycling 19:8-11.

63. ———, D.P. Schwert, J.P. Moreau, and L. Theoret. 1977. Earthworm communities and soil nutrient levels as affected by municipal waste-water irrigation, pages 284-290, in U. Lohm and T. Persson, editors. Soil organisms as components of ecosystems. Ecol. Bull. no. 25, Stockholm.

64. Dodd, J.L., and U.K. Lauenroth. 1979. Analysis of the response of a grassland ecosystem to stress, pages 45-58, in N. French, editor. Perspectives in grassland ecology. Springer-Verlag, New York.

65. Domsch, K.H. 1970. Effects of fungicides on microbial population in soil. Symposium on pesticides in soil: ecology, degradation, movement. Michigan State University, East Lansing, Michigan.

66. ———, and W. Paul. 1974. Simulation and experimental analysis of the influence of herbicides on soil nitrification. Arch. Microbiol. 97:283-301.

67. Doran, J.W. 1980a. Microbial changes associated with residue management with reduced tillage. Soil Sci. Soc. Amer. J. 44:518-524.

68. ———. 1980b. Soil microbial and biochemical changes associated with reduced tillage. Soil Sci. Soc. Amer. J. 44:765-771.

69. Dowding, P. 1981. The effects of artificial drainage on nitrogen cycle processes, pages 615-625, in F.E. Clark and T. Rosswall, editors. Terrestrial nitrogen cycles. Ecol. Bull. no. 33, Stockholm.

70. Dwivedi, R.S. 1979. Micro-organisms, pages 227-230, in R.T. Coupland, editor. Grassland ecosystems of the world. Cambridge University Press, Cambridge.

71. East, R., and R.P. Pottinger. 1983. Use of grazing animals to control insect pests of pastures. N.Z. Entomol. 7:352-359.

72. Edwards, C.A. 1965. Some side effects resulting from the use of persistent insecticides. Ann. appl. Biol. 55:329-331.

73. ———. 1973. Persistent pesticides in the environment. 2nd edn. CRC Press, Cleveland, Ohio.

74. ———. 1974. Some effects of insecticides on myriapod populations, pages 645-655, in J.G. Blower, editor. Symp. Zool. Soc. Lond. no. 32. Myriapoda.

75. ———. 1975. Effects of direct drilling on the soil fauna. Outlook Agric. 8:243-244.

76. ———. 1977. Investigations into the influence of agricultural practice on soil invertebrates. Ann. appl. Bio. 87:515-520.

77. ———. 1980. Interactions between agricultural practice and earthworms, pages 3-12, in D.L. Dindal, editor. Soil biology as related to land use practices. EPA, Washington.

78. ———. 1983. Earthworm ecology in cultivated soils, pages 123-137, in J.E. Satchell, editor. Earthworm ecology. Chapman and Hall, London.

79. ———, E.B. Dennis, and D.W. Empson. 1967. Pesticides and the soil fauna:effects of aldrin and DDT in an arable field. Ann. appl. Biol. 60:11-22.

80. Edwards, C.A., and J.R. Lofty. 1975a. The invertebrate fauna of the Park Grass Plots. 1. Soil fauna, pages 133-154, in Rothamsted Exp. Stn. Rep. for 1974, part 2.

81. ———. 1975b. The influence of cultivations on soil animal populations, pages 399-407, in J. Vanek, editor. Progress in soil zoology. Academia, Prague.

82. ———. 1978. The influence of arthropods and earthworms upon root growth of direct drilled cereals. J. appl. Ecol. 15:789-795.

83. ———. 1979. The effects of straw residues and their disposal on the soil fauna, pages 37-44, in E. Grossbard, editor. Straw decay and its effect on disposal and utilization. Wiley, Chichester.

84. ———. 1980. Effects of earthworm inoculation upon the root growth of direct drilled cereals. J. appl. Ecol. 17:533-543.

85. ———. 1982a. Nitrogenous fertilizers and earthworm populations in agricultural soils. Soil Biol. Biochem. 14:515-521.

86. ———. 1982b. The effect of direct drilling and minimal cultivation on earthworm populations. J. appl. Ecol. 19:723-734.

87. Edwards, C.A., and A.R. Thompson. 1973. Pesticides and the soil fauna. Residue Rev. 45:1-79.

88. ———. 1975. Some effects of insecticides on predatory beetles. Ann. appl. Biol. 80:132-135.

89. ———, and K.I. Beynon. 1968. Some effects of chlorfenvinphos, an organophosphorous insecticide on populations of soil animals. Rev. Ecol. Biol. Sol 5:199-224.

90. Edwards, C.A., A.R. Thompson, and J.R. Lofty. 1967. Changes in soil invertebrate populations caused by some organophosphorous insecticides. Proc. 4th Br. Insectic. Fungic. Conf. 1:48-55.

91. Eijsackers, H., and J. Van der Drift. 1976. Effects on the soil fauna, pages 148-174, in L.J. Audus, editor. Herbicides. Physiology, Biochemistry, Ecology, Vol. 2, 2nd Edn. Academic Press, London.

92. El Duweini, A.K., and S.I. Ghabbour.1965. Population density and biomass of earthworms in different types of Egyptian soils. J. appl. Ecol. 2:271-287.

93. Elliott, E.T., and D.C. Coleman. 1977. Soil protozoan dynamics of a

390

Colorado short grass prairie. Soil Biol. Biochem. 9:113-18.

94. Ellis, F.B. 1979. Agronomic problems from straw residues with particular reference to reduced cultivation and direct drilling in Britain, pages 11-20, in E. Grossbard, editor. Straw decay and its effect on disposal and utilization. Wiley, Chichester.

95. Emmanuel, N., J.P. Curry, and G.O. Evans. 1984. The soil acarine fauna of barley plots (in review).

96. Ferrar, P. 1973. CSIRO dung beetle project. Wool Technol. Sheep Breed. 20:73-75.

97. Floate, M.I.S. 1981. Effects of grazing by large herbivores on nitrogen cycling in agricultural ecosystems, pages 585-601, in F.E. Clark and T. Rosswall, editors. Terrestrial nitrogen cycles. Ecol. Bull. no. 33, Stockholm.

98. Fortuner, R., and V.A. Jacq. 1976. In vitro study of toxicity of soluble sulfides to three nematodes parasitic on rice in Senegal. Nematologia 22:343-351.

99. Fox, C.J.S. 1964. The effects of five herbicides on the numbers of certain invertebrate animals in grassland soil. Can. J. Plant Sci. 44:405-409.

100. ———. 1967. Effects of several chlorinated hydrocarbon insecticides on the springtails and mites of grassland soil. J. econ. Entomol. 60:77-79.

101. Frissel, M.J., and J.A. Van Veen. 1981. Some aspects of irrigation relevant to the terrestrial nitrogen cycle, pages 603-614, in F.E. Clark and T. Rosswall, editors. Terrestrial nitrogen cycles. Ecol. Bull. no. 33. Stockholm.

102. Fungo, N.K., and J.P. Curry. 1983. The effects of eight fungicides on the glasshouse spider mite, Tetranychus urticae (Koch). J. Life Sci. R. Dublin Soc. 4:175-181.

103. Gerard, B.M. 1967. Factors affecting earthworms in pastures. J. Anim. Ecol. 36:235-252.

104. ———, and R.K.M. Hay. 1979. The effect on earthworms of ploughing, tined cultivation, direct drilling and nitrogen in a barley monoculture system. J. agric. Sci. 93:147-155.

105. Gers, G. 1982. Incidence de la simplification de travail du sol sur la microfaune edaphique hivernale: données préliminaires. Rev. Ecol. Biol. Sol 19:593-604.

106. Ghilarov, M.S., and B.M. Mamajev. 1966. Über die Ansiedlun von Regenwürmern in den artesisch bewässerten Oasen der Wüste Kysyl-Kum. Pedobiologia 6:197-218.

107. Gillard, O. 1967. Coprophagous beetles in pasture ecosystems. J. Aust. Inst. agric. Sci. 33:30-74.

108. Gillon, D. 1971. The effect of bush fire on the principal pentatomid bugs (Hemiptera) of an Ivory Coast savanna, pages 377-417, in Proc. Tall Timbers Fire Ecol. Conf. no. 11.

109. Gillon, Y. 1971. The effect of bush fire on the principal Acridid species of an Ivory Coast savanna, pages 419-471, in Proc. Tall Timbers Fire Ecol. Conf. no. 11.

110. Gisiger, L. 1961. Neue Erkentnisse über die Bereitung der Gülle und ihre zweckmässige Anwendung, Seiten 103-122, in Bund. Vers. Anst. Alpenland. Land W. Gumpenstein, Ber. 3. Arbeitstag. Fragen der Güllerei (1960).

111. Goring, C.A.I., and D.A. Laskowski. 1982. The effects of pesticides on nitrogen transformations in soils, pages 689-720, in F.J. Stevenson, editor. Nitrogen in agricultural soils. ASA-CSSA-SSSA, Madison, Wisconsin.

112. Griffiths, D.C., F. Raw, and J.R. Lofty. 1967. The effects on soil fauna of insecticides tested against wireworms (Agriotes spp.) in wheat. Ann. appl. Biol. 60:479-490.
113. Grossbard, E. 1976. Effects on the soil microflora, pages 99-147, in L.J. Audus, editor. Herbicides. Physiology, biochemistry ecology. Vol. 2, 2nd Edn. Academic Press, London.
114. Guiran, G. de, L. Bonnel, and M. Abirached. 1980. Landspreading of pig manures IV. Effect on soil nematodes, pages 109-119, in J.K.R. Gasser, editor. Effluents from livestock. Applied Science Publishers Ltd., London.
115. Harper, J.E., and Webster, J. 1964. An experimental analysis of the coprophilous fungus succession. Trans. Brit. mycol. Soc. 47:511-530.
116. Harper, J.L. 1969. The role of predation in vegetational diversity, pages 48-62, in Brookhaven Symp. Biol. no. 22.
117. Hartenstein, R., E.F. Neuhauser, and J. Collier. 1980. Accumulation of heavy metals in the earthworm Eisenia foetida. J. Environ. Qual. 9:23-26.
118. Hay, R.K.M., J.C. Holmes, and E.A. Hunter. 1978. The effects of tillage, direct drilling and nitrogen fertilizers on soil temperature under a barley crop. J. Soil Sci. 29:174-183.
119. Heath, G.W., M.K. Arnold, and C.A. Edwards. 1966. Studies in leaf litter breakdown. 1. Breakdown rates of leaves of different species. Pedobiologia 6:1-12.
120. Hilder, E.J. 1966. Distribution of excreta by sheep at pasture, pages 977-981, **in Proc.** Xth Int. Grassl. Congr., Helsinki, Finland. Valtio-neuvostóston Kirjapaino, Helsinki.
121. Holter, P. 1979. Effect of dung-beetles (Aphodius spp.) and earthworms on the disappearance of cattle dung. Oikos 32:393-402.
122. Hoy, J.B. 1980. Effects of lindane, carbaryl and chlorpyrifos on non-target soil arthropod communities, pages 71-81, in D.L. Dindal, editor. Soil biology as related to land use practices. EPA, Washington.
123. Hughes, M.K., N.W. Lepp, and D.A. Phippi. 1980. Aerial heavy metal pollution and terrestrial ecosystems. Adv. ecol. Res. 11: 218-327.
124. Hughes, R.G. 1979. Arable farmer's problems with straw, pages 3-10, in E. Grossbard, editor. Straw decay and its effect on disposal and utilization. Wiley, Chichester.
125. Huhta, V., E. Ikonen, and P. Vilkamaa. 1979. Succession of invertebrate populations in artificial soil made of sewage sludge and crushed bark. Ann. Zool. Fenn. 16:223-270.
126. Hunter, B.A., and M.S. Johnson. 1982. Food chain relationships of copper and coal mining in continental grassland ecosystems. Oikos 38:108-117.
127. Hutchinson, K.J., and K.L. King. 1980a. Management impacts on structure and function of sown grasslands, pages 823-852, in A.I. Breymeyer and G.M. Van Dyne, editors. Grasslands, systems analysis and man. Cambridge University Press, Cambridge.
128. ———. 1980b. The effects of sheep stocking level on invertebrate abundance, biomass and energy utilization in a temperate, sown grassland. J. appl. Ecol. 17:369-387.
129. ———. 1982. Microbial respiration in a temperate sown grassland grazed by sheep. J. appl. Ecol. 19:821-833.
130. Ireland, M.P. 1983. Earthworms and heavy metals, pages 247-265, in J.E. Satchell, editor. Earthworm ecology. Chapman and Hall, London.
131. Jakubczyk, H. 1971. Productivity investigation of two types of meadows in the Vistula Valley. III. Decomposition rate of organic

matter and microbiological activity. Ekol. Pol. 19:121-128.

132. ———. 1974. Analysis of a sheep pasture ecosystem in the Pieniny Mountains (the Carpathians). IX. Decomposition processes and development of microflora in the soil. Ekol. Pol. 22:569-588.

133. ———. 1976a. The dependence of the rate of plant material decomposition in a meadow upon mineral fertilization and environmental factors. Pol. ecol. Stud. 2(4):259-286.

134. ———. 1976b. The changes in numbers of meadow soil microflora under the influence of mineral fertilization. Pol. ecol. Stud. 2(4):231-258.

135. Jankowska, K. 1971. Net primary production during a three-year succession on an unmowed meadow of the Arrhenatheretum elatioris plant association. Bull. Acad. Pol. Sci. cl. II, 19:789-794.

136. Jenkinson, D.S., and D.S. Powlson. 1970. Residual effects of soil fumigation on soil respiration and mineralization. Soil Biol. Biochem. 2:99-108.

137. Jensen, V. 1977. Effects of lead on biodegradation of hydrocarbons in soil. Oikos 28:220-224.

138. Jordan, M.J., and M.P. Lechevalier. 1975. Effects of zinc smelter emissions on forest soil microflora. Can. J. Microbiol. 21:1855-1865.

139. Kajak, A. 1971. Productivity investigations of two types of meadows in the Vistula Valley. IX. Production and consumption of field layer spiders. Ekol. Pol. 19:197-211.

140. ———. 1974. Analysis of a sheep pasture ecosystem in the Pieniny Mountains (the Carpathians). XVII. Analysis of the transfer of carbon. Ekol. Pol. 22:711-732.

141. Kelsey, J.M., and E.Z. Arlidge. 1968. Effects of isobenzan on soil fauna and soil structure. N.Z. J. agric. Res. 11:245-260.

142. Keogh, R.G. 1979. Lumbricid earthworm activities and nutrient cycling in pasture ecosystems, pages 49-51, in T.K. Crosby and R.P. Pottinger, editors. Proc. 2nd Australas. Conf. Grassl. Invertebr. Ecol. Government Printer, Wellington.

143. ———, and P.H. Whitehead. 1965. Observations on some effects of pasture spraying with benomyl and carbendazim on earthworm activity and litter removal from pasture. N.Z. J. exp. Agric. 3:103-104.

144. Khanna, P.K. 1981. Leaching of nitrogen from terrestrial ecosystems - patterns, mechanisms and ecosystem responses, pages 343-352, in F.E. Clark and T. Rosswall, editors. Terrestrial nitrogen cycles. Ecol. Bull. no. 33., Stockholm.

145. King, K.L., and K.J. Hutchinson. 1976. The effects of sheep stocking intensity on the abundance and distribution of microfauna in pasture. J. appl. Ecol. 13:41-55.

146. ———, and P. Greenslade. 1977. The effects of sheep numbers on associations of collembola in sown pastures. J. appl. Ecol. 14:731-739.

147. Knutson, H., and J.B. Campbell. 1976. Relationships of grasshoppers (Acrididae) to burning, grazing and range sites of native tallgrass prairie in Kansas, pages 107-120, in Proc. Tall Timber Conf. Ecol. Anim. Control Habitat Manag. no. 6, 1974. Gainsville, Florida.

148. Kubicka, H. 1978. The effect of mineral fertilization on CO_2 evolution in meadow soil. Part II. Pol. ecol. Stud. 4:167-178.

149. Lamotte, M. 1976. The structure and function of a tropical savannah ecosystem, pages 179-222, in F.B. Golley and E. Medina, editors. Ecological Studies II, Tropical ecological system: analysis and synthesis. Springer- Verlag, New York.

150. Laurence, B.R. 1954. The larval inhabitants of cow pats. J. Anim. Ecol. 23:234-260.

151. Lee, K.E. 1983. The influence of earthworms and termites on soil nitrogen cycling, pages 35-48, in P.H. Lebrun, H.M. André, A. de Medts, C. Gregoire-Wibo, and G. Wauthy, editors. New trends in soil biology. Imprimeur Dieu-Brichart, Ottignies-Louvain-la-Neuve.

152. Lofs-Holmin, A. 1981. Influence in field experiments of benomyl and carbendazim on earthworms (Lumbricidae) in relation to soil texture. Swedish J. agric. Res. 11:141-147.

153. ————. 1983a. Earthworm population dynamics in different agricultural rotations, pages 151-160, in J.E. Satchell, editor. Earthworm ecology. Chapman and Hall, London.

154. ————. 1983b. Influence of agricultural practices on earthworms. Acta Agric. Scand. 33:225-256.

155. Loring, S.J., R.J. Snider, and L.S. Robertson. 1981. The effects of three tillage practices on Collembola and Acarine populations. Pedobiologia 22:172-184.

156. Luff, M.L. 1966. The abundance and diversity of the beetle fauna of grass tussocks. J. Anim. Ecol. 35:189-208.

157. Lynch, J.M. 1979. Straw residues as substrates for growth and product formation by soil micro-organisms, pages 47-56, in E. Grossbard, editor. Straw decay and its effects on disposal and utilization. Wiley, Chichester.

158. ————, and L.M. Panting. 1980a. Cultivation and the soil biomass. Soil Biol. Biochem. 12:29-33.

159. ————. 1980b. Variations in the size of the soil biomass. Soil Biol. Biochem. 12:547-550.

160. ————. 1982. Effects of season, cultivation and nitrogen fertilizer on the size of the soil microbial biomass. J. Sci. Food Agric. 33: 249-252.

161. MacGregor, A.N., and L.M. Naylor. 1982. Effect of municipal sludge on the respiratory activity of a cropland soil. Plant Soil 65:149-152.

162. Majer, J.D. 1984. Short term responses of soil and litter invertebrates to a cool autumn burn in Jarrah (Eucalyptus marginata) forest in Western Australia. Pedobiologia 26:229-247.

163. Malanchuk, J.C., and K. Joyce. 1983. Effects of 2, 4-D on nitrogen fixation and carbon dioxide evolution in a soil microcosm. Water Air Soil Pollut. 20:181-189.

164. Malecki, M.R., E.F. Neuhauser, and R. Loehr. 1982. The effect of metals on the growth and reproduction of Eisenia foetida (Oligochaeta, Lumbricidae). Pedobiologia 24:129-137.

165. Malone, C. 1969. Effects of diazinon contamination on an old-field ecosystem. Am. Midl. Nat. 82:1-27.

166. Malone, C.R., and D.E. Reichle. 1973. Chemical manipulation of soil biota in a fescue meadow. Soil Biol. Biochem. 5:629-639.

167. Marshall, V.G. 1977. Effects of manures and fertilizers on soil fauna: A review. Commonw. Bur. Soils Spec. Publ. no. 3. Commonwealth Agricultural Bureau, London.

168. Martin, N.A. 1975. Effect of four insecticides on the pasture ecosystem II. The fauna collected in pit-traps. N.Z. J. agric. Res. 18:179-182.

169. ————. 1978. Earthworms in New Zealand agriculture, pages 176-180, in Proc. 31st Weed Pest Contr. Conf.

170. Menzie, C.M. 1972. Fate of pesticides in the environment. Annu. Rev. Entomol. 17:199-222.

171. Metcalf, R.L. 1975. Insecticides in pest management, pages 235-274, in R.L. Metcalf and W.H. Luckmann, editors. Introduction to insect pest management. Wiley, New York.

172. Metz., L.J., and D.L. Dindal. 1975. Collembola populations and prescribed burning. Environ. Entomol. 4:583-587.
173. Mitchell, M.J., R. Hartenstein, B.L. Swift, E.F. Neuhauser, B.I. Abrams, R.M. Mulligan, B.A. Brown, D.Craig, and D. Kaplan. 1978. Effects of different sewage sludges on some chemical and biological characteristics of soil. J. environ. Qual. 7:551-559.
174. Mochnacka-Lawacz, H. 1978. Mineral NPK fertilization and fluctuations of some elements in the meadow vegetation . Pol. ecol. Stud. 4(1): 123-133.
175. Morgan, C.V.G., N.H. Anderson, and J.E. Swales. 1958. Influence of some fungicides on orchard mites in British Columbia. Can. J. Plant Sci. 38:94-105.
176. Morris, M.G. 1968. Differences between the invertebrate faunas of grazed and ungrazed chalk grassland. II. The faunas of sample turves. J. appl. Ecol. 5:601-611.
177. ———. 1971. The management of grassland for the conservation of invertebrate animals, pages 527-552, in E. Duffey and A.S. Watt, editors. The scientific management of animals and plant communities : for conservation. Blackwell, Oxford.
178. ———, and K.H. Lakhani. 1979. Response of grassland invertebrates to management by cutting. 1. Species diversity of Heniptera. J. appl. Ecol. 16:77-98.
179. Moursi, A. 1962. The lethal doses of CO_2, N_2 and H_2S for soil arthropods. Pedobiologia 2:9-14.
180. Nielsen, C.O. 1955. Studies on Enchytraeidae. 5. Factors causing seasonal fluctuations in numbers. Oikos 6:153-169.
181. Nowak, E. 1975. Population density of earthworms and some elements of their production in several grassland environments. Ekol. Pol. 23:459-491.
182. ———. 1976. The effect of fertilization on earthworms and other soil microfauna. Pol. ecol. Stud. 2(4): 195-207.
183. Olechowicz, E. 1976a. The effect of mineral fertilization on insect community of the herbage in a meadow. Pol. ecol. Stud. 2(4);129-136.
184. ———. 1976b. The role of coprophagous dipterans in a mountain pasture ecosystem. Ekol. Pol. 24:125-165.
185. Parr, J.F. 1974. Effects of pesticides on microorganisms in soil and water, pages 315-340, in W.D. Guenzi, editor. Pesticides in soil and water. Soil Science Society of America Inc., Madison, Wisconsin.
186. Paul, E.A., F.E. Clark, and V.O. Biederbeck. 1979. Microorganisms, pages 87-96, in R.T. Coupland, editor. Grassland ecosystems of the world. Cambridge University Press, Cambridge.
187. Paustian, K. 1983. Simulation modelling of carbon and nitrogen cycling, pages 252-279, in T. Rosswall, editor. Ecology of arable land. The role of organisms in nitrogen cycling. Progress report 1982. Swedish University of Agricultural Sciences, Uppsala.
188. Persson, T., E. Bååth, M. Clarholm, H. Lundkvist, and B.E. Soderström. 1980. Trophic structure, biomass dynamics and carbon metabolism of soil organisms in a Scots pine forest, pages 419-459, in T. Persson, editor. Structure and function of northern coniferous forests - an ecosystem study. Ecol. Bull. no. 32, Stockholm.
189. Pochon, J., and I. Bacarov. 1973. Données préliminaires sur l'activite microbiologique des sols de la savane de Lamto (Côte d'Ivoire). Rev. Ecol. Biol. Sol 10:35-45.
190. Power, J.F. 1981. Nitrogen in the cultivated ecosystem, pages 529-546, in F.E. Clark and T. Rosswall, editors. Terrestrial nitrogen cycles.

Ecol. Bull. no. 33, Stockholm.

191. Prestidge, R.A. 1982. The influence of nitrogenous fertilizer on the grassland Auchenorrhyncha (Homoptera). J. appl. Ecol. 19:735-749.

192. Pugh, G.J.F., and J.I. Williams. 1971. Effect of an organomercury fungicide on saprophytic fungi and on litter decomposition. Trans. Br. Mycol. Soc. 57:164-166.

193. Purvis, G. and J.P. Curry. 1981. The influence of sward management on foliage arthropod communities in a ley grassland. J. appl. Ecol 18:711-725.

194. ———. 1984. The influence of weeds and farmyard manure on the activity of Carabidae and other ground-dwelling arthropods in a sugar beet crop. J. appl. Ecol. 21:271-284.

195. Raw, F. 1962. Studies of earthworm populations in orchards 1. Leaf burial in apple orchards. Ann. appl. Biol. 50:389-404.

196. ———. 1965. Current work on side effects of soil-applied organophosphorus insecticides. Ann. appl. Biol. 55:342-343.

197. ———. 1967. Arthropoda (except Acari and Collembola), pages 327-362, in A. Burges and F. Raw, editors. Soil biology. Academic Press, London.

198. Reinecke, A.J., and F.A. Visser. 1980. The influence of agricultural land use practices on the population densities of Allolobophora trapizoides and Eisenia rosea (Oligochaeta) in Southern Africa, pages 310-324, in D.L. Dindal, editor. Soil biology as related to land use practices. EPA, Washington.

199. Rorison, I.H. 1971. The use of nutrients in the control of the floristic composition of grassland, pages 65-77, in E. Duffey and A.S. Watt, editors. The scientific management of animal and plant communities for conservation. Blackwell, Oxford.

200. Rosswall, T. and K. Paustian. 1984. Cycling of nitrogen in modern agricultural systems. Plant Soil 76:3-21.

201. Ruscoe, Q.W. 1973. Changes in the mycofloras of pasture soils after longterm irrigation. N.Z. J. Sci. 16:9-20.

202. Satchell, J.E. 1955. Some aspects of earthworm ecology, pages 180-201, in D.K. McE Kevan, editor. Soil zoology. Butterworths, London.

203. Schinner, F., H. Bayer, and M. Mitterer. 1983. The influence of herbicides on microbial activity in soil materials. Bodenkultur 34:22-30.

204. Schnürer, J., M. Clarholm, and T. Rosswall. 1983. Fluctuations in bacterial biomass, protozoan numbers and FDA-active hyphal lengths in a fertilized barley and grass ley, pages 79-87, in T. Rosswall, editor. Ecology of arable land. The role of organisms in nitrogen cycling. Progress Report 1982. Swedish University of Agricultural Sciences, Uppsala.

205. Sheals, J.G. 1956. Soil populations studies. 1. The effects of cultivation and treatment with insecticides. Bull. entomol. Res, 47:803-822.

206. Sinclair, A.R.E. 1975. The resource limitation of trophic levels in tropical grassland ecosystems. J. Anim. Ecol. 44:497-520.

207. Singh, J.S., and M.C. Joshi. 1979. Primary production, pages 197-218, in R.T. Coupland, editor. Grassland ecosystems of the world. Cambridge University Press, Cambridge.

208. Skoropanov, S.G. 1961. Reclamation and cultivation of peat-bog soils. Israel Program for Scientific Translations, Jerusalem 1968.

209. Smolik, J.D. 1974. Nematode studies at the Cottonwood site. US/IBP Grassland Biome Technical Report No. 251. Colorado State University, Fort Collins, Colorado.

210. Southwood, T.R.E., and W.F. Jepson. 1962. The productivity of grass-
    land in England for Oscinella frit and other stem-boring Diptera.
    Bull. entomol. Res. 53:395-407.
211. Speight, M.R. 1983. The potential of ecosystem management for pest
    control. Agric. Ecosystems Environ. 10:183-199.
212. Spence, D.H.N., and A. Angus. 1971. African grassland management-
    burning and grazing in Murchison Falls National Park, Uganda, pages
    319-332, in E. Duffey and A.S. Watt, editors. The scientific management
    of animal and plant communities for conservation. Blackwell, Oxford.
213. Stanford, G. 1981. Nitrogen in the cultivated ecosystem, pages 547-550,
    in F.E. Clark and T. Rosswall, editors. Terrestrial nitrogen cycles.
    Ecol. Bull. no. 33, Stockholm.
214. Steinberger, Y., D.W. Freckman, L.W. Parker, and W.G. Whitford. 1984.
    Effects of simulated rainfall and litter quantities on desert soil
    biota:nematodes and microarthropods. Pedobiologia 26:267-274.
215. Stevens, R.J., and I.S. Cornforth. 1974. The effect of pig slurry
    applied to a soil surface on the composition of the soil atmosphere.
    J. Sci. Food Agric. 25:1263-1272.
216. Stout, J.D., K.R. Tate, and L.F. Molloy. 1976. Decomposition processes
    in New Zealand soils with particular respect to rates and pathways
    of plant degradation, pages 97-144, in J.M. Anderson and A. Macfadyen,
    editors. The role of terrestrial and aquatic organisms in decomposition
    processes. Blackwell, Oxford.
217. Stringer, A., and M.A. Wright. 1976. The toxicity of benomyl and
    some related 2-substituted benzimidazoles to the earthworm Lumbricus
    terrestris. Pesticide Sci. 7:459-464.
218. Strojan, C.L. 1978a. Forest leaf litter decomposition in the vicinity
    of a zinc smelter. Oecologia 32:203-212.
219. ———. 1978b. The impact of zinc smelter emissions on forest litter
    arthropods. Oikos 31:41-46.
220. Swift, M.J., O.W. Heal, and J.M. Anderson. 1979. Decomposition in
    terrestrial ecosystems. Blackwell, London.
221. Thompson, A.R. 1971. Effects of nine insecticides on the numbers and
    biomass of earthworms in pasture. Bull. environ. Contam. Toxicol.
    5:577-586.
222. ———, and F.L. Gore. 1972. Toxicity of twenty-nine insecticides
    to Folsomia candida: Laboratory studies. J. econ. Entomol. 65:
    1255-1260.
223. Tomlin, A.D. and F.L. Gore. 1974. Effects of six insecticides and a
    fungicide on the numbers and biomass of earthworms in pasture. Bull.
    environ. Contam. Toxicol. 12:487-492.
224. Traczyk, T., H. Traczyk, and D. Pasternak. 1976. The influence of
    intensive mineral fertilization on the yield and floral composition
    of meadows. Pol. ecol. Stud. 2(4):39-47.
225. Tu, C.M. 1970. Effect of four organophosphorous insecticides on
    microbial activites in soil. Appl. Microbiol. 19:479-484.
226. Tyler, G. 1975a. Heavy metal pollution and mineralisation of nitrogen
    in forest soils. Nature (Lond.) 255:701-702.
227. ———. 1975b. Effect of heavy metal pollution on decomposition and
    mineralisation rates in forest soils, pages 217-226, in Proc. Int.
    Conf. on Heavy Metals in the Environment, Toronto. Agricultural
    Institute of Canada, Ottowa.
228. Ülehlová, B. 1979. Micro-organisms in meadows, pages 155-164, in R.T.
    Coupland, editor. Grassland ecosystems of the world. Cambridge
    University Press, Cambridge.

229. Valiela, I. 1974. Composition, food webs and population limitation in dung arthropod communities during invasion and succession. Am. Midl. Nat. 92:370-385.

230. Van der Maarel, E. 1971. Plant species diversity in relation to management, pages 45-64, in E. Duffey and A.S. Watt, editors. The scientific management of animal and plant communities for conservation. Blackwell, Oxford.

231. Van Rhee, J.A. 1977. Effects of soil pollution on earthworms. Pedobiologia 17:201-208.

232. Voronova, L.D. 1968. The effect of some pesticides on the soil invertebrate fauna in the South Taiga Zone in the Perm region (U.S.S.R.). Pedobiologia 8:507-525.

233. Wainwright, M. 1978. A review of the effects of pesticides on microbial activity in soils. J. Soil Sci. 29:287-298.

234. Wasilewska, L. 1974. Analysis of a sheep pasture ecosystem in the Pieniny Mountains (the Carpathians). XIII. Quantitive distribution, respiratory metabolism and some suggestions on predation of nematodes. Ekol. Pol. 22:651-668.

235. ———. 1976. The effect of intensive fertilization on the structure and productivity of meadow ecosystems. 13. The role of nematodes in the ecosystem of a meadow in Warsaw environs. Pol. ecol. Stud. 2(4):137-156.

236. ———. 1979. The structure and function of soil nematode communities in natural ecosystems and agrocenoses. Pol. ecol. Stud. 5:97-145.

237. Waters, R.A.S. 1955. Numbers and weights of earthworms under a highly productive pasture. N.Z. J. Sci. Technol. 36:516-525.

238. Watson, A.P., R.I. Van Hook, D.R. Jackson and D.E. Reichle. 1976. Impact of a lead mining-smelting complex on the forest floor litter arthropod fauna in the new lead belt region of Southwest Missouri. ORNL/NSF/EATC-30, Oakridge National Laboratory, Tennessee.

239. Way, M.J., and N.E.A. Scopes. 1968. Studies on the persistence and effects on soil fauna of some soil-applied systemic insecticides. Ann. appl. Biol. 62:199-214.

240. Weil, R.R., and W. Kroontje. 1979. Effects of manuring on the arthropod community in an arable soil. Soil Biol. Biochem. 11:669-679.

241. Wells, T.C.E. 1971. A comparison of the effects of sheep grazing and mechanical cutting on the structure and botanical composition of chalk grassland, pages 497-515, in E. Duffey and A.S. Watt, editors. The scientific management of animal and plant communities for conservation. Blackwell, Oxford.

242. Wibo, C. 1973. Etude de l'action d'un insecticide organophosphore sur quelques populations de microarthropodes edaphiques. Pedobiologia 13:150-163.

243. Williamson, P., and P.R. Evans. 1973. A preliminary study of the effects of high levels of inorganic lead on soil fauna. Pedobiologia 13:16-21.

244. Wilkinson, W. 1977. Effects of direct drilling on soil micro-arthropods. Ann. appl. Biol. 87:520.

245. Woodmansee, R.G., and L.S. Wallach. 1981. Effects of fire regimes on biogeochemical cycles, pages 649-669, in F.E. Clark and T. Rosswall, editors. Terrestrial nitrogen cycles. Ecol. Bull. no. 33 Stockholm.

246. Wright, M.A. 1977. Effects of benomyl and some other systemic fungicides on earthworms. Ann. appl. Biol. 87:520-524.

247. Zajonc, I. 1975. Variations in meadow associations caused by the influence of nitrogen fertilizers and liquid-manure irrigation,

pages 497-503, in J. Vanek, editor. Progress in soil zoology. Academia Praque.

248. Zicsi, A.W. 1969. Über die Auswirkung der Nachfrucht und Bodenbearbeitung auf die Aktivitat der Regenwürmer. Pedobiologia 9:141-146.

249 Zyromska-Rudzka, H. 1976. The effect of mineral fertilization of a meadow on the oribatid mites and other soil mezofauna. Pol. ecol. Stud. 2(4):157-182.

RECENT ADVANCES IN QUANTITATIVE SOIL BIOLOGY

D. W. FRECKMAN[1], K. CROMACK, JR.[2], and J. A. WALLWORK[3]

INTRODUCTION

In addressing the question of recent advances in quantitative soil biology, it is most important to ask, recent? Since when? We have chosen to define this question even further, by dividing the question into three parts, i.e., 1) are there recent advances in the measurement of microbial populations and biomass, 2) are there recent advances or modifications of old established methods in the measurement of population density, biomass and contributions of the soil fauna to energy and nutrient fluxes, and 3) are there recent advances in techniques to determine the influence of the soil biota on soil decomposition and mineralization pathways. In each of these questions, the term "recent" will take on a different meaning. In the question concerning soil fauna, advances shall be designated as recent since Petersen and Luxton's (1982) excellent critique of procedures used during the International Biological Program to determine density and biomass for soil animals of the detritus food web. Swift, Heal and Anderson (1979) emphasized unique advances in methodology following Phillipson's (1971) IBP Handbook of Methods. There have been, of course, many specialized reviews of population sampling techniques, extraction techniques, statistical analyses and activity measurements for each group of the biota, some of which we will note here, along with limitations. The contributions of agriculturalists to this

---

1) Dept. of Nematology, Univ. of California, Riverside.
2) Dept. of Forest Science, Oregon State Univ., Corvallis.
3) Dept. of Zoology, Queen Mary College, London, England.

area of quantitative soil biology has increased in recent
years due to the need to determine economic thresholds for
those pests in soil causing yield loss in crops.
Techniques, such as microcosms in field and laboratory, are
continuing to prove a reliable method for determining the
effects of soil decomposition and nutrient pathways. It
should be emphasized that many of the "recent" advances in
quantitative soil biology to be discussed in this chapter
are actually refinements of excellent methods established in
the 1960-1970's.

DIRECT METHODS FOR ESTIMATING MICROBIAL POPULATIONS

Direct estimates of microbial biomass have long been of
interest due to the importance of microbial processes in
energy flow and nutrient cycling (Waksman 1932, Lindemann
1942, MacFadyen 1963, Odum 1971, Parkinson et al. 1971).
The value of obtaining better quantitative estimates of
microbial biomass and activity rates has been reinforced in
recent years by the realization that in many terrestrial
ecosystems >50% of net primary production is channeled into
production of coarse and fine roots and microorganisms
(Fogel and Hunt 1979, Coleman et al. 1980, Harris et al.
1980). A large variety of invertebrate microfauna and
macrofauna and even a number of vertebrates depend upon
these organisms to supply a substantial portion of their
nutrition. There is therefore a rich variety of food web
relationships in which microbes play an integral part.

Direct observation of the soil system in situ was
pioneered by Kubiena (1938), with the objective of observing
directly the spatial arrangement of soil particles, soil
pores, and soil organisms. Gray (1967) showed that it was
possible to examine microbes directly in situ utilizing
scanning electron microscopy. Invention of resin embedding
techniques has helped to facilitate such work (Anderson and
Healey 1970, Pande and Berthet 1973). Utilizing
transmission electron microscopy, direct in situ examination
has permitted initial inferences to be drawn concerning food

storage products within individual bacterial cells in a soil matrix (Foster 1981). These techniques are, however, time-consuming and perhaps are still best reserved for qualitative information in taxonomic or ecological survey work and for experiments which selectively manipulate portions of the soil flora and fauna (Satchell 1974).

Direct microscopic counts of soil microorganisms from soil or litter samples were brought to a reasonable state of statistical reliability by Jones and Mollison (1948), including refinements for estimating fungal hyphal length (Olson 1950). Ingham and Klein (1984a) improved the Jones and Mollison (1948) method for fungal hyphae by using fresh slide preparations, appropriate stains and phase contrast microscopy. Other recent studies using the fluorescent antibody (FA) technique for known species of fungi in the rhizosphere (Chao and Holland 1970), or litter (Frankland et al. 1981), or bacteria in soil (Bohlool and Schmidt 1968) have facilitated quantitative studies of selected microorganisms in selected habitats. Use of metabolic stains, such as fluorescein diacetate (FDA), have facilitated quantitation of physiologically active fungal hyphae vs. total fungal hyphae (Ingham and Klein 1984b) and also developed a correlation between FDA staining and CO_2 evolution which should prove useful in assessing fungal contributions to carbon transformations. Direct methods have worked well for types of protozoans such as testate forms which are better able to withstand the rigors of slide preparation (Lousier and Parkinson 1984).

An indirect method of estimating viable microbial populations is the most probable numbers method (MPN), in which populations of the particular organisms or organismal groups are isolated into selective media at the highest dilution which will provide acceptable statistical replication (Cutler 1920). This method is still used for microorganisms such as protozoans (Clarholm 1981, Parker et al. 1984), especially for forms such as naked amoebae which are too delicate to examine by most direct techniques (Clarholm 1981).

Use of sampling for direct observations in situ or for making population estimates of microorganisms on slides or selective media has a variety of interesting applications including: microbial biomass turnover to energy flow and nutrient cycling, microbial physiology under a variety of natural conditions, habitat relationships, and food chain analyses.

Accurate information concerning carbon turnover by different microbial components such as bacteria, fungi and protozoans is important if we are to assess the functional contributions of these groups in various ecosystems. For example, studies by Clarholm and Rosswall (1981) and Clarholm (1981) showed the importance of bacterial turnover and protozoan predation in acid forest soils in which dominance of fungi had been previously assumed. Parker et al. (1984) studied the effects of rapidly changing environmental conditions in desert soil upon bacteria, fungi, and protozoans. Forest ecosystems which are fungal dominated (Fogel and Hunt 1979), particularly with ectomycorrhizal roots, may have a substantial turnover of fungal tissue associated with fine root turnover.

Investigations of microbial physiology, based upon in situ observations, prepared slides, or most probable numbers (MPN) cultures in selective media, focus on determinations such as the proportion of the microbial cells which are viable and what enzyme systems are operative. Ingham and Klein (1982) were able to obtain good correlations between FDA stained, and therefore viable fungal hyphae, and CO_2 evolution. Except in young, rapidly growing cultures, total hyphae (which included both FDA stained and unstained hyphae) did not correlate significantly with CO_2 evolution. Ingham and Klein's (1984b) FDA work presumably could be extended to include selected microbial species, particularly dominates, with species-specific antigen-antibody reactions, as in the FA technique (Frankland et al. 1981), or in the even more sensitive technique of enzyme-linked immunosorbent assay (ELISA) (Fichter and Stephen 1981). The ELISA method is approximately 1000 times more sensitive than the classic precipitin test used in immunological studies (Boreham 1979). Other physiological studies, such as those by Foster (1981) or Graustein et al. (1979) and Cromack et al. (1979) have been concerned with questions regarding internal physiological state of the microorganisms, such as Foster's (1981) pioneering study of bacterial food reserves in situ in soil, or with observations concerning microbial products such as oxalate production by fungi in soil and litter (Graustein et al. 1979). Data on the macroscopic scale of ectomycorrhizal fungal mats were combined with scanning electron microscope observations to obtain information on fungal soil colonization and soil weathering (Cromack et al. 1979).

Habitat studies of microorganisms are of interest because of the great diversity of physical and chemical substrate colonization by microbes. Direct observations of such habitats provide evidence of the kinds of microbes present and provide a basis for studies of the relative contribution of different functional groups to ecosystem processes. Inferences concerning habitat associations between microorganisms and larger organisms, such as microarthropods, or other kinds of invertebrates, can be determined with information concerning digestive tract analyses of microbial materials using the ELISA technique as previously adapted for insect predator-prey work by Fichter and Stephen (1981). This technique would extend the type of work done by Mitchell and Parkinson (1976), which utilized information concerning kinds of fungal hyphae present in oribatid mite guts to infer feeding on these types of hyphae in forest floor litter, and would permit conclusions to be drawn based on dominant fungi present in litter and soil substrates.

A variety of indirect methods have been developed to assay microbial activities or biomass (Jenkinson and Powlson

1976, Anderson and Domsch 1978). Such techniques afford the
advantage of being more rapid and easier than direct
observation methods. Recent work evaluating the ATP method
(Fairbanks et al. 1984) cautions that such methods are not
always a shortcut to useful data concerning microbial
activities. In general, however, indirect techniques should
be viewed as complementary to direct methods (Nannipieri
1984, Paul and Voroney 1984). In many ecosystem studies, it
may be useful to combine both approaches.

MEASUREMENT OF POPULATION DENSITY AND BIOMASS OF SOIL FAUNA

Inherent in estimations of population density and biomass
of soil fauna is the definition of the sampling scheme. Not
only must the investigator have knowledge of the animal and
its biology, but there must be a clearly defined concept of
the habitat to be sampled. The unit area sampled, whether a
cm^3 for direct observation of a specimen, or several cores
that will be mixed, are often extrapolated to a much larger
area. The data from this unit area will eventually become
the basis for hypotheses and principles detailing an aspect
of the interaction of a particular animal in the soil system
and its contribution to functioning in an ecosystem.
Sampling techniques can basically be divided into two types,
direct and indirect, direct being observation or extraction
of the total population within a defined area, and indirect
relying on baits or attractants. Sampling and sampling
designs, therefore, are important considerations in the
estimation of soil animal populations (O'Connor 1957, Murphy
1962, MacFadyen 1962, Satchell 1969, Petrusewicz and
MacFadyen 1970, Wallwork 1976, Ayoub 1980, Barker and
Campbell 1981, Freckman 1982a, Petersen and Luxton 1982).

It is imperative that the investigator compare extraction
methods prior to the beginning of the study. Extraction
methods for a particular group of fauna vary with habitat,
soil type, and in their ability to extract different species
and life stages equally. An ideal extraction method would
extract all individuals regardless of size, life stage, or

species. Since this ideal is rarely obtained, different extraction methods should be tested and evaluated and the investigation should report the efficiency of the technique. The determination of extraction efficiency is a difficult task. Seeding of the soil, litter or lichens, using mark-recapture techniques, handsorting, or extraction of the fauna continually from the soil until no further animals are recovered, have been used to estimate efficiency. There have been many general reviews of extraction techniques for soil fauna (Burges and Raw 1967, Wallwork 1970, Phillipson 1971, Petersen and Luxton 1982). In addition, more specific information on individual taxa are also available for tardigrades and rotifers (Sohlenius 1979); nematodes (Barker 1978, Minagawa 1979, Goodell 1982, Viglierchio and Schmitt 1983, Barker et al. 1985); enchytraeids (O'Connor 1957, Willard 1972, Abrahamsen 1972, Huhta and Koskennemi 1975, Lundvist 1978, 1982, Standen 1973, 1982); microarthropods (Edwards and Fletcher 1971, Pande and Berthet 1973, Tamura 1976, Petersen 1978, Takeda 1979, Van Straalen and Rijninks 1982); and earthworms (Satchell 1971, Axelsson et al. 1971, Piearce and Piearce 1979, Terhivuo 1980, Springett 1981, Bouche and Gardner 1983).

The purpose of estimating the abundance or population density of microfauna is often to provide a measure of the importance of a particular taxon in the soil. One estimate of this importance is an estimate of number of individuals. Whereas agriculturalists have traditionally referred to faunal density as numbers per mass or volume of soil, ecologists have expressed density on an area basis. However, measurement of importance based on density does not accurately reflect the metabolic role of individual species. More useful measurements including biomass and secondary production necessitates accurate determinations of population density estimates and the mass of specific stages. This is a formidable problem because the small size of the micro- and mesofauna limits accurate assessment of density and prevents accurate weighing of the individuals.

This limitation has been examined for all the soil fauna, and biomass is still considered to be a reliable method for estimating biological activity when comparing similar fauna. Mass determinations for adult and immature mites can be determined indirectly by use of regression equations developed from actual determinations on the relationships between mass, length and width (Van der Drift 1951, MacFadyen 1952, Englemann 1961, O'Connor 1971). Direct measures of mass using very accurate microbalances are also employed and then the true mass to size relationships determined and used in studies measuring only size (Berthet 1971, Lebrun 1971, Luxton 1975, Mitchell 1979, Thomas 1979). The mass of nematodes is generally calculated from their volume, as measured by microscopy (Andrassy 1956, Warwick and Price 1979). The development of digital image analyzers interfaced with a computer has improved the volume measurements (Robinson 1984). The usual procedure for determining biomass is to multiply the individual/m^2 by average individual masses representing species, life stage, trophic or taxonomic grouping. For a discussion of the sources of error when termining biomass for different habitats, seasons or by different extraction techniques, see Thomas (1979), Freckman (1982b) and Petersen and Luxton (1982).

All of these procedures, sampling, extraction, identification and calculation of density and biomass are of equal importance and should be carefully considered and evaluated in preliminary tests for each habitat studied before beginning with actual field observations (Englemann 1961, Burges and Raw 1967, Berthet 1971, Phillipson 1971, Yeates 1979, Freckman 1982a). Each of the above procedures has sources of error, which can over-estimate or underestimate the importance of a particular taxon in a habitat.

Sampling

There are several factors to consider in developing a sampling plan. One of the most important, the biology of the animal, will determine the place, time, or season, and depth of the sample. Variations in soil texture, moisture, pH, temperature, oxygen, litter accumulation or root association all contribute to determining whether the animal is distributed through the soil in a homogeneous or clumped pattern. Seastedt (1984) reviewed studies on microarthropod densities and noted that on a global scale, they were substrate limited and their densities were not correlated with litterfall or plant productivity. However, microarthropod densities within a given environment were influenced by both amounts and quality of detritus. A study by Goodell and Ferris (1980) found the distribution of several plant parasitic nematode species in a smaller habitat, an alfalfa field, to be best represented by a negative binomial distribution, which is indicative of clumping or aggregation. Plant parasitic nematodes from individual soil cores were taken on a 6 m grid pattern in a 7 ha alfalfa field. Although the host was the same for all nematode species, the differences in the population distribution could best be correlated with soil texture and the biology of the individual species, i.e., whether the nematode was migratory or sedentary. The negative binomial distribution has been confirmed for other plant parasitic nematode species (Anscombe 1950, Merny and Dejardin 1970, Proctor and Marks 1975), microarthropods (Elliott 1971, Southwood 1975), and other soil fauna (Petersen and Luxton 1982). However, negative binomial distributions can be construed from factors other than aggregation. Information of the type of distribution of soil animals being studied is useful, because the number of soil samples required to estimate the density with a certain level of precision can then be calculated (Elliott 1971, Southwood 1975).

Seasonal variation in demographic characteristics will obviously affect population estimates. Therefore, knowledge of the population fluctuation and age structure throughout the year is necessary for calculating biomass and other energetic parameters. In ectothermic animals, such as arthropods and nematodes, where physiologic activity is dependent on ambient temperature conditions, sampling intervals may be defined according to a physiological time sequence (e.g., degree days above a threshold), to allow for increased sampling intensity when the animals are physiologically active (Baskerville and Emin 1969).

Spatial distribution of the fauna can be affected by season, abiotic factors, life history and physiological activity. Seastedt (1984) noted that reports of microarthropod densities were probably underestimated because lower soil horizons are often not sampled. Many microarthropods and nematodes migrate to deeper soil horizons in response to abiotic factors such as temperature and moisture gradients. The sampling design finally chosen by the investigator is to a large degree dependent on the objectives of the study. If the objective is to find presence or absence of a species, then intense sampling should occur at specific periods when the particular life stage needed for taxonomic identification is known to be abundant. If the objective is to determine how the population distribution is affected by a treatment over time, the sampling design should incorporate more frequent sampling intervals as well as habitat subsampling to ascertain horizontal and vertical distribution. Studies in which the goal is to estimate production or the role of fauna in nutrient cycling should be sampled every two weeks, or a minimum of once a month, with particular attention to age, structure, and seasonal distribution (Berthet 1971, Healey 1971). Petersen and Luxton (1982) noted that populations of all the fauna are probably subject to fluctuations in numbers and biomass both seasonally and annually. Most studies, however, have based conclusions only on a single year which may have been atypical. They suggest that less frequent samplings over many years such as Bouche's (1975) study of earthworms may provide a more useful estimate of density and biomass than a one year intensive study.

A major emphasis should be placed on the precision of the results to be obtained from the study. What is to be considered accurate estimates of the population, and what will be the sampling error? Should the variance be examined rather than the mean? Increased replication (Hurlbert 1984) will reduce sampling error but the number of replicates which can be taken and processed is often limited by various constraints. Alternatively, it may be possible to increase the sample size, or to increase the number of units of soil comprising one sample. For example, Goodell (1982), found increasing the sample size by bulking a greater number of soil cores was less expensive and more accurately estimated the nematode population than increasing the number of samples. O'Connor (1971) also recommended 10 cores per sample. However, Zicsi (1958) noted that increasing the size of the sample decreased the number of earthworms/m^2 recovered by handsorting. The decrease was attributed to an increase in the tediousness of the handsorting technique. For further discussions of sample sizes, see Andrassy (1962), Berthet (1971), Phillipson (1971) and Petersen and Luxton (1982). In most cases, a preliminary sampling protocol should be tested to determine labor requirements, effectiveness of sampling without biasing the population and the importance of spatial and temporal variations.

A final factor important in soil sampling is common sense. Collecting and processing soil samples requires considerable effort. The soil biologist can increase accuracy of any sampling procedure by proper handling and storage of the samples. Too often, investigators forget they are dealing with fragile animals in a fragile environment. Winfield and Cooke (1974) noted that soil samples in plastic bags, thrown or dropped into the back of a truck, resulted in a 98% kill of the nematode, Paratrichodorus. Storage temperature and length of storage time for soil samples will affect hatch of eggs, survival of various life stages, etc., and should be monitored carefully (Edwards and Fletcher 1971, Goodell 1982).

Direct methods. Direct observation is rarely achieved for the purpose of quantitatively estimating density of soil animals, since the animals are quite small. The techniques that are used for direct observation actually involve assessing density by sampling the habitat and enumerating the removed animals (Anderson and Healey 1970, Pande and Berthet 1973). This can be done by three procedures depending on the size of the animals. Soil sections and soil cores are used for microscopic animals; and quadrats, the removal of a designated amount of soil within a certain larger area, are used for larger animals.

Soil cores. Soil cores are appropriate for numbers of the soil meso- and micro- fauna including nematodes, microarthropods and enchytraeids. Generally, small soil cores (less than 10 cm diam) are not used for larger animals such as earthworms. However, Springett (1981) successfully used a larger 16 cm diameter core for earthworms. Investigators have adapted soil corers for both habitats and biota, from the split soil corer for enchytraeids (O'Connor 1957) to a soil corer with serrated edges used for nematodes and enchytraeids in the Arctic taiga forest (Freckman et al. 1977, MacLean et al. 1977).

Quadrats. This method is used for earthworms and isopods and is based on determining density of fauna within a randomly placed quadrat to a given depth. The size of the quadrat is determined by the size of the animals to be extracted. For example, many workers examining earthworms use 0.25 or 0.5 m^2 quadrats to a depth of 40-50 cm (Satchell 1971), whereas for woodlice a quadrat of 10 cm^2 is more appropriate (Paris and Pitekla 1962).

Embedding soil samples. The purpose of this technique is to observe the biota in situ. A core of soil or litter is impregnated with gelatin or other materials, sectioned and examined under a dissecting microscope (Anderson and Healey 1970, Richards 1974, Satchell 1974, Seastedt et al. 1980). This is a most useful technique for determining function and vertical distributions of microarthropods, fungi, and annelids, but is very time-consuming and little used for enumerating microarthropod populations. However, Anderson and Healey (1970) have demonstrated its potential in estimating populations of oribatids in woodland soils.

Indirect methods. Indirect methods utilize behavorial characteristics of the animals, i.e., motility, attractants, or temperature gradients. These methods are frequently used for habitats with known low densities of animals or for animals that cannot be reliably estimated by a more direct method.

Bait and trapping. This procedure is usually used for seasonal or comparative studies of large oligochaetes, and relies on the positive attractive properties of the substance to the animals. Bait, usually fresh dung, is placed at intersections of a grid, and the number of animals attracted are counted over a short time span of about 14 days. Limitations of the method are: 1) dung has been noted to be attractive to only certain species of earthworms, and 2) the results may reflect activity of a certain age group or a seasonal or diurnal pattern and thus misrepresent the density (Whitford et al. 1981). Pitfall traps (Southwood 1975) for collecting surface active collembola, isopods, insects and spiders is a commonly used method which has the disadvantage of attracting animals from places remote to the actual area investigated.

Extraction

There are no extraction methods devised that successfully extract all animals from a soil sample. The object, then, becomes to extract as many invertebrates of interest as possible, knowing the biases of the recovery method. Most investigators are familiar with the limitations imposed by different extraction techniques and new variations of established techniques are frequently reported (Barker et al. 1969, Bouche 1969, 1975, Huhta 1972, Beute 1974, Singh and Pillai 1976, Seastedt and Crossley 1978). The proliferation of these "new" techniques should be regarded not as being repetitive, but new advances, as investigators become more aware of the biases introduced by season, depth, soil type, flora, pH, etc. on both numbers of individuals and life stages extracted. This range of techniques provides an investigator with various options to evaluate prior to beginning a study. In many cases, a single technique or a combination of techniques may be chosen. Methods for extracting soil invertebrates are detailed by Phillipson (1971), Edwards and Lofty (1972) and discussed further by Petersen and Luxton (1982). Within the present chapter, we will not describe all the extraction methods and their recent adaptations, but instead note only advances and major changes.

Extraction methods for soil invertebrates are of two types: (1) Direct (Mechanical) - those by which the animals are physically separated from the soil or, (2) Indirect (Dynamic) - those by which the animal is driven from the soil, usually by a behavioral reaction.

Direct Extraction Procedures. These methods will extract active, inactive, dead and live animals and thus may lead to overestimates of population size if significant portions of dead individuals are collected. They do not depend upon the behavior of the animals.

Handsorting. This may appear as the simplest method of removing animals from the soil. However, the method is generally laborious, although it is widely used for earthworms (Satchell 1971), and rarely used for smaller invertebrates. Earthworms are separated from the surface soil layers by handsorting or by handsorting following sieving of mineral soil. The limitations include a poor recovery of larger deep burrowing species and of small specimens less than 0.2 g live mass. See Petersen and Luxton (1982) for a further discussion of limitations of this method for earthworms.

When the method has been used for smaller invertebrates, it has not always proven to be laborious. For example, Somme and Block (1982) examined collembola in Arctic soils and due to the lack of organic matter in this habitat, the organisms were removed using suction from a microaspirator and counted.

Flotation, centrifugation and sieving. These methods rely on differences in specific gravity between the soil invertebrates and a solution such as sugar, brine, magnesium sulfate or other solutions (Edwards and Fletcher 1971). The soil and solution are stirred, and as the animals are separated from the soil, they float to the surface and can be decanted onto a series of different size sieves. The use of centrifugation requires small soil samples. Both flotation and sieving methods have been used for earthworms (Satchell 1971). Modification of these methods for microarthropods include the Salt and Hollick specific gravity flotation method (Edwards and Fletcher 1971) and the grease film technique of Aucamp and Ryke (1964). In comparing several of these techniques with indirect methods for microarthropods, Edwards and Fletcher noted that flotation techniques collected both living and dead animals and were less efficient except for the extraction of collembola and larger arthropods, (Diptera, Thysanoptera), from sandy, fallow soils. Thus flotation, centrifugation and sieving are rarely used for microarthropods (Petersen and Luxton 1982).

It is interesting that both Edwards and Fletcher (1971) and Oostenbrink (1971) found centrifugation and flotation methods unsuitable for nematodes. Since 1971, however, there have been major advances in extraction methods for nematodes (Barker 1978). One of the most widely used methods is the semi-automatic elutriator (Byrd et al. 1976, Barker 1978) which processes four samples simultaneously. Although this technique generally has a poor extraction efficiency ranging from 20-80% depending on the size of the nematode, the popularity of this technique in the United States is a result of: 1) use of large soil samples, 500-1000 g being optimum, 2) the reduction of individual operator error by mechanical processing, thus allowing more accurate comparisons of abundance from different regions and soil types, 3) increased replications from the field due to rapidity of extraction (up to 250 samples can be processed in one day), and 4) the ability to also extract mycorrhizal spores and roots from the same soil samples. The disadvantages, namely cost, low efficiency and the extraction of both live and dead animals, appear to be minor when compared with the speed, ease of operation and increased replication. An advantage is the ability to obtain an estimate of efficiency for age structure by seeding different soil types with different size nematodes. This method is presently a major means of nematode extraction by many agricultural institutions in the USA, with some institutions having up to 3 semi-automatic elutriators. Mites, collembola and tardigrades have been observed in samples, but have never been quantified, which presents another challenge for future investigation.

<u>Indirect extraction procedures</u>. These techniques utilize the behavorial responses of the animal, mainly its activity, to remove it from soil. Eggs and non-motile stages are not extracted and different stages and species which vary in their activity, will not be extracted equally.

Wet extractors. The wet extractors include the Baerman funnel, used for nematodes, rotifers and tardigrades and its modifications for earthworms (Satchell 1971, Huhta and Koskennemi 1975) and for enchytraeids (O'Connor 1971). This method consists of a soil sample placed on a screen at the top of a water-filled funnel or bath and relies on the ability of the animal to move to the bottom of the funnel where it is collected. Petersen and Luxton (1982) noted that this method compared favorably with handsorting for earthworms. However, Sohlenius (1980) noted variations from 35-90% in extraction efficiencies. Variations in efficiency of the Baermann funnel have been detailed elsewhere by Freckman and Baldwin (1985) and include: 1) use of large soil samples (Sohlenius recommends 3 ml), 2) variability in nematode mobility which is related to species, 3) hatching of eggs or rehydration of anhydrobiotic forms giving a skewed distribution of juveniles, and 4) use of deionized water which decreases nematodes extracted.

Dry heat extractors. The simplest type of heat extractor is a Tullgren funnel, consisting of a light bulb hanging over soil contained in a wire mesh screen, suspended in a funnel. The animals fall from the soil to the bottom of the funnel. This extractor and the MacFadyen (1961) high-gradient extractor are the most frequently used for extraction of microarthropods. The MacFadyen high-gradient extractor maintains steep gradients of humidity and temperature in the sample and is a more sophisticated apparatus. It is generally accepted that temperature and humidity are the most important factors causing animals to emerge from soil or litter. Lasebikan (1974) discussed the reaction of different microarthropod species to these two environmental factors and found responses to temperature and/or humidity varied among species. For a description of the many modifications of these techniques and their limitations, in addition to comparisons of extraction efficiencies, see Edwards and Fletcher (1971) and Petersen and Luxton (1982). Although Edwards and Fletcher recommended the high gradient extractor as the more efficient technique for most soils and microarthropods, Petersen and Luxton, in their analysis of data from IBP investigators, noted little difference between the high-gradient extractor and the Tullgren funnel. The important consideration is knowing limits of the recovery method. Tamura (1976) noted that microarthropod adults are more efficiently extracted in a high-gradient soil arthropod extractor and thus, fewer individuals might be extracted from an actively reproducing population than from a population composed mainly of adults. Using this information, Lussenhop (1981) in a study of detrital processing in prairie soil, measured percent immatures to determine which of his treatments caused more rapid population growth. He noted that extraction might affect the interpretation of results if the procedure selected for certain life stages.

Attractants-repellents. The principle employed with this technique is to repel or attract the animal from soil or litter. The substances used are numerous and include chemicals, electrical stimulation and heat. Generally, these methods are not used for nematodes, enchytraeids, or microarthropods. Their major success has been with earthworms or larger invertebrates. The technique most used for earthworms is the formalin technique (Satchell 1971), or a combination of formalin and handsorting (Bouche 1975). Springett (1981) compared five extraction methods for lumbricid populations that occur mainly in the top 10 cm depth. Of the five methods, heat extraction in water, application of formalin in the field, handsorting in the lab, handsorting in the field, and placing soil cores into 0.2% formalin solutions in buckets in the field for 1 hour, the latter was most efficient, both in recovery and time expenditure. The new method extracted up to 90% of the earthworms compared to handsorting methods.

TECHNIQUES FOR MEASURING DECOMPOSITION AND NUTRIENT RELEASE

The decomposition of soil organic material is a combination of chemical and physical processes. It involves the degradation of complex molecules to simpler ones. During decomposition, nutrients are released from bound forms and rendered available to the plant/soil system. This liberation of nutrients occurs, in part, as a consequence of the biochemical breakdown of organic compounds, but also in part through physical distribution and dissolution. Nutrient release is therefore a function of organic decomposition but is not synonymous with it. A distinction is therefore made between these two processes in the following discussion.

Decomposition

Indirect methods. In order to achieve the molecular transformations required for chemical degradation, decomposer organisms, principally bacteria, fungi and actinomycetes, must expend energy. This, except in the cases of anaerobes and chemotrophs, requires the consumption of oxygen and the liberation of carbon dioxide. Rates of oxygen uptake and carbon dioxide evolution have been roughly equated with rates of decomposition of organic material and can often be employed as useful approximations. The measurement of carbon dioxide evolution is technically easier than the measurement of oxygen consumption, and is widely used today (Witkamp 1966, MacFadyen 1970, Anderson 1973, Minderman and Vulto 1973a,b). However, the evolution of CO_2 is complicated by solubility in solutions (Schlesinger 1977). These measurements are frequently assessments of total community respiration and, as such, do not distinguish the individual contributions made by microbes, fauna and plant roots (Swift et al. 1979). As a more discriminating alternative, but with attendant technical problems, respirometers designed to measure oxygen uptake from individual isolates can be used. Gilson and Warburg respirometers are considered suitable for measuring microbial activity in plant litter (Swift et al. 1979), while Cartesian divers (Linderstrom-Lang 1943) have been used to measure oxygen uptake by microarthropods and enchytraeids (Berthet 1963, O'Connor 1963, Healey 1967).

The oxidation of organic compounds is brought about by enzymatic activity, notably on the part of dehydrogenases (Swift et al. 1979). Techniques are available to measure dehydrogenase activity in soil (Howard 1972), but the difficulties involved in interpreting the results obtained are outlined by Swift et al. (1979).

These techniques are essentially indirect methods of assessment of decomposition in that they measure the activity of decomposer organisms as a function of gas exchange or enzymatic activity. The use of models (see Chapter 10) is presently being used as an indirect aid for assessing decomposition processes. Direct methods measure the rate at which the substrate is removed during the decomposition process and may be affected by physical transfer of substrate which may overestimate decomposition.

Direct methods. These may be classified as mechanical, visual chromatographic or gravimetric.

An example of the mechanical methods is the use of cotton strips, of known tensile strength, which are placed in the soil profile at appropriate depths for predetermined periods of time. After a particular time lapse, the strips are removed and their tensile strength recorded on a tensiometer (Latter and Howson 1977). The difference between original and final tensions, suitably calibrated, can provide an estimate of cellulose decomposition.

The decomposition of cellulose can also be estimated visually by the use of cellophane inserts into the soil profile at various levels (Harding 1967). These inserts can be removed from time to time and their decomposition measured as a function of surface area reduction. This approach has also been used to measure the decomposition of leaf disc (Heath et al. 1964); estimates can be made visually (by holding the discs up to the light, or comparing traces on graph paper) or photometrically – in the latter case up to 100 discs can be measured simultaneously.

Chromatography can be used to monitor microbial activity in the soil, since it permits the separation, identification and quantification of the volatile end products representative of such activity. Of course, many of the products are not volatile and they cannot be used. Gas chromatography techniques are appropriate for this purpose although their application to field situations is limited (Swift et al. 1979). To pursue this line of inquiry further takes us into the realm of nutrient release as a measure of decomposition rate and, as already noted, estimates of these two parameters are not easily reconcilable.

Gravimetric methods essentially involve recording the loss in weight of known quantities of litter after being exposed, for predetermined periods of time, to natural or artificial soil conditions. This is by far the most common method used by soil ecologists to assess rates of organic decomposition in soil systems; its advantages and disadvantages will be discussed in more detail later (see also Chapter 3).

Nutrient release

Nutrient release rates are related to rates of organic decomposition but they are not simple linear functions of the latter. Certainly, the biochemical decomposition of organic material releases nitrogen, phosphorus and potassium which would otherwise remain immobilized and unavailable in complex organic molecules. But the physical process of leaching soluble nutrients achieved principally through the agency of percolating soil solutions and fluctuating ground water levels, is also important. Techniques for determining rates of organic decomposition are, therefore, not appropriate for assessing rates of nutrient release. Crossley and Witkamp (1964) drew attention to this when they monitored the release of radioactive cesium (an analogue of potassium) from Cs-labeled leaf litter. They found rates of removal of this element from the litter were much greater than would be predicted from overall decomposition rates of the organic material. Labeling plant litter with radioisotopes is a convenient way of studying elemental loss by leaching and decomposition; the advantages and limitations of this technique have been reviewed by Swift et al. (1979).

Many of the techniques outlined above are not particularly new approaches to the study of soil biology – they have been available for fifteen or twenty years. But over this period of time, their limitations have become apparent, and modifications and refinements have been made to correct for these limitations. This period has also seen the development and elaboration of the 'microcosm' technique for the study of the processes associated with decomposition and nutrient release. Basically, there are two kinds of microcosm, field and laboratory (see also Chapter 8).

<u>Field microcosms</u>. Litter bags. Crossley and Witkamp (1964) confined the leaf litter they were monitoring in nylon mesh bags. In this way, `parcels' of organic material could be placed in the field and retrieved after varying periods of time (Fig. 1). The loss in mass of this organic

FIG. 1. Litter bags of the fiber-mesh type deployed on the surface in the Chihuahuan desert.

material over a specified time interval can then be used as an index of decomposition. The technique was developed by Bocock and Gilbert (1957) and is capable of various modifications. By varying the mesh size of the nylon bags, for example, certain size categories of the soil fauna can be excluded. This variant can be used to make separate assessments of the role played by microflora, microfauna and macrofauna in the decomposition process. Alternatively, the confined leaf litter can be `tagged' with radioisotopes, as in the Crossley and Witkamp study, and isotope loss monitored by scintillation techniques.

Litter bags deployed in the field provide a quasi-natural environment for the organic material confined within them. They represent a system that has a measure of isolation from its surrounding; a miniature world of its own. In completely natural conditions, the confined leaf litter would be fragmented and relocated deeper in the soil profile. But the litter-bag microcosm does not allow for this relocation of the products of organic decomposition. Furthermore, the litter bag creates its own environment - particularly with regard to moisture - which may encourage the growth of fungi. Another disadvantage of this technique is that it imposes a two-dimensional structure on the decomposing system, whereas under natural conditions this system operates in three dimensions. Leaf litter is not always flattened when it decomposes - it may be curled or folded. Such distortion adds a third dimension to the living space for decomposer organisms.

Clearly, these limitations of the litter-bag technique must be appreciated in analyzing and comparing the results obtained (Seastedt and Crossley 1980, St. John 1980). In truth we can echo the words of Anderson et al. (1979): "All microcosms, by design, place some limitation on the system by controlling the size or physical and chemical environment".

Synthetic 'logs' and litter boxes. Despite these reservations, the litter bag technique is a very useful tool for the study of the decomposition process and it is widely used by soil ecologists today. But it is by no means the only type of field microcosm that has been developed in recent years. For example, Fager (1968) used synthetic logs, oak boxes packed with sawdust, as field microcosms to study faunal succession during the course of wood decomposition. The nutrient quality of the sawdust filler was varied to evaluate the effect of substrate chemistry on decomposition.

A variant of this technique, which is appropriate for studying organic horizons, has been developed by Stevenson and Dindal (1981, 1982a,b) - the litter box. These boxes constructed from metal hardware cloth, can be of variable size - the authors used relatively small boxes (10 cm x 10 cm x 2 cm). Although these were used to investigate the community structure of soil fauna, they could equally well be used in decomposition studies. Synthetic logs and litter boxes remove one of the constraints imposed on the litter bag technique, namely that they provide a three-dimensional arena, rather than a two-dimensional one, in which decomposition processes can operate. The same effect can be produced by the use of screen cylinders.

Screen cylinders. This technique has been used by Whitford and his coworkers to investigate litter decomposition in the Chihuahuan desert (where it gives more repeatable results than fiber-mesh litter bags) (Santos et al. 1981). The screen cylinder microcosm (Fig. 2) has the advantage over the litter bag in that the decomposing leaf material is in direct contact with the soil. Known masses of leaf litter are placed on the mineral soil surface and a screen cylinder used to confine the sample and prevent it from being displaced or dispersed by wind, water or the activities of herbivorous animals. The cylinder allows free access of air and the percolation of soil solution such that the litter can mix with the mineral soil directly beneath it. Clearly, this method is only suitable for soils where the surface organic material is sharply demarcated from the mineral soil, so the former can be removed if necessary and replaced with the experimental samples. Many agricultural soils would qualify in this regard. Moreover, this technique, like that which uses litter bags, can be manipulated in various ways to address a variety of questions. Rates of decomposition of different litter types under the same temperature and moisture regime can be compared. Comparisons can also be made of the rate of decomposition of surface litter versus completely buried

FIG. 2. The screen cyclinder field microcosm under creosote bush (*Larrea tridentata*) in the Chihuahuan desert.

litter of the same type. Litter treated with biotic inhibitors, and enclosed in screen cylinders, can be used to assess the influence of selected groups of soil biota on decomposition as discussed below. This technique has also been used successfully to study the effects of simulated rainfall on the activities of soil microarthropods and nematodes (Whitford *et al*. 1981). There are, indeed, many extensions of this technique - it is, perhaps more flexible than the litter bag method in view of the greater reliance that can be placed on the results obtained.

Resin bags. One recent variant of the field microcosm which has been used successfully in organic soils to estimate available nutrients, particularly N and P, is the resin bag (Sibbesen 1977). This operates on the principle that resin confined in nylon bags, deployed in the field, provides a medium for ion exchange with the surrounding soil solution. Smith (1979) used a strongly basic, anion exchange resin to extract available phosphorus from sub-Antarctic soils. Elution with 0.4 N HCl separated the phosphorus from the resin and this solution was subsequently analyzed for phosphorus by a modification of the chlorophosphomolybdate method (Page 1982). Availability of nitrogen in forest soils was estimated through the use of a mixed (cation + anion) resin, the cation exchange capacity of which was loaded with mercuric chloride to inhibit microbial activity (Binkley 1982). An autoanalyzer was then employed to determine ammonium- and nitrate-nitrogen.

Laboratory microscosms

Field microcosms are subject to a number of environmental variables that are not always easily controlled. Monitoring rates of carbon dioxide evolution or oxygen consumption that can provide indices of overall rates of decomposition on the one hand, and animal participation in this decomposition on the other, are beset with problems when measured under natural conditions, or in field microcosm situations. Decomposition in most ecosystems is a function of environmental temperature and moisture content and it may not be possible to obtain continuous records of these variables in the soil throughout the year. The use of laboratory microcosms in ecological research has been discussed by Geisy (1980), Coleman et al. (1983, 1984) and Coleman (see Chapter 8).

Two examples of the method. Many environmental variables can be controlled and monitored by the use of microcosms developed in the laboratory. One such microcosm, used to study the interactions between bacteriophagous nematodes and bacteria, consisted of sterilized soil placed in flasks to which are added various nutrients and biotic amendments (Anderson et al. 1979). Nematodes are inoculated into this system from stock cultures.

A rather more sophisticated microcosm which can be used both for measurements of respiration and nutrient loss during decomposition has been devised by Anderson and Ineson (1982) (Fig. 3). Continuous recordings of carbon dioxide evolution can be made by the use of a conductivity cell incorporated in the lid (Fig. 3). Alternatively, a second type of lid, containing a gas absorbent, may be fitted (Fig. 3). Nutrient loss is monitored in leachate that drains

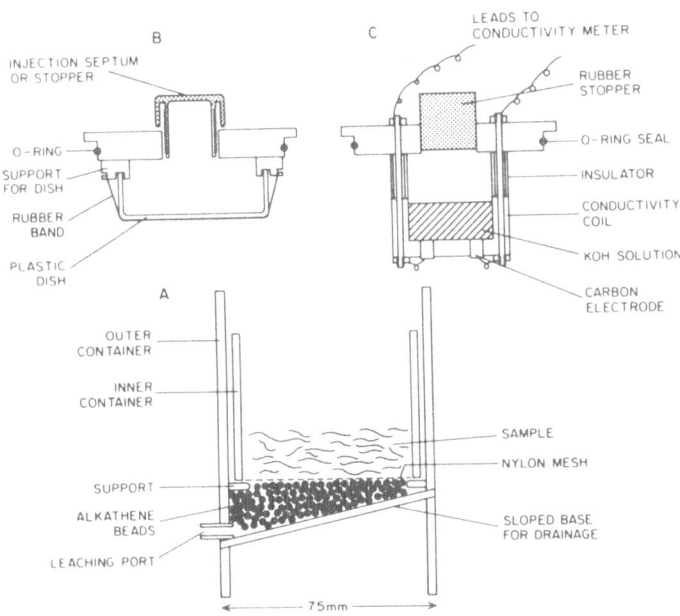

FIG. 3. A laboratory microcosm developed by Anderson and Ineson (1982). (a) General view. (b) A lid designed for continuous recordings of carbon dioxide evolution. (c) An alternative lide for absorbing emitted gases.

through the sample and is drawn off through a leaching port. As Anderson et al. (1979) point out, these microcosms place constraints on the complexity of the system, thus providing "an important tool in the interpretation and understanding of both organism associations and the effects of physical or chemical perturbations on individual species and systems". The microcosm system of Anderson and Ineson (1982) can be perturbed by the introduction (via the leaching port) of glucose treatments for determinations of microbial biomass, pesticides or other biological inhibitors. Using this technique, Ineson et al. (1982) were able to demonstrate that grazing of fungi by Collembola caused significant increases in the leaching of ammonium, nitrate and calcium, but not of potassium and sodium, from fragmented oak litter.

Biotic inhibitors

Much can be learned about decomposition and nutrient fluxes by experimental manipulation. Selective biotic inhibition, for example, can remove one particular trophic level in an animal/microbial or an animal/animal food chain, and the effect of this removal on decomposition rates of organic substrates can be measured.

Insecticides. This approach has been used by Santos et al. (1981) in studies on the bacteria/nematode/mite food chain associated with the decomposition of creosote bush litter (Larrea tridentata) in a Chihuahuan desert site (Fig. 4). Here, tydeid mites prey on bacteriophagous cephalobid nematodes that, in turn, feed on bacteria. Selective elimination of tydeid mites, through the experimental application of the insecticide chlordane, results in an up-surge of nematode numbers, a reduction in bacterial activity and, as a consequence, a retardation in the decomposition rate of Larrea litter. These investigations were conducted using field microcosms of the litter bag type.

Naphthalene. Seastedt and Crossley (1983), on the other hand, preferred to use naphthalene as a biotic inhibitor in experiments designed to investigate the role of soil microarthropods in nutrient retention of forest floor litter. This inhibitor was used in preference to an insecticide since it is an insoluble volatile. Naphthalene has been widely used in biotic inhibition studies since the time of Kurcheva (1960) and is still proving to be a valuable research tool (Best et al. 1978, Seastedt and Crossley, 1980).

Perturbation experiments

Chemical perturbations. The disruption of an ecological system by the application of an insecticide, bactericide, fungicide or some other biotic inhibitor is an example of a chemically-induced perturbation. Effectively, this disruption, if it is carried out properly, renders part of

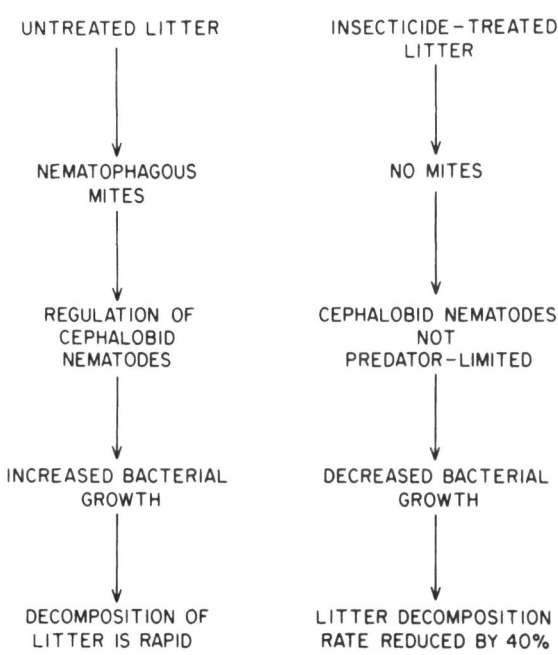

CREOSOTE-BUSH MICROCOSMS

UNTREATED LITTER	INSECTICIDE-TREATED LITTER
↓	↓
NEMATOPHAGOUS MITES	NO MITES
↓	↓
REGULATION OF CEPHALOBID NEMATODES	CEPHALOBID NEMATODES NOT PREDATOR-LIMITED
↓	↓
INCREASED BACTERIAL GROWTH	DECREASED BACTERIAL GROWTH
↓	↓
DECOMPOSITION OF LITTER IS RAPID	LITTER DECOMPOSITION RATE REDUCED BY 40%

FIG. 4. The results of biotic inhibition treatments on a bacteria/nematode/mite food chain in soil under creosote bush (Larrea tridentata) in the Chihuahuan desert.

the system sterile and therefore non-functional. In this category of perturbations must also be included irradiation techniques that sterilize the litter and soil so completely that decomposition processes can be followed from the initial stages. Irradiation at high levels does change the chemical structure of organic material (Coleman and MacFadyen 1966). Such techniques are not featured prominently in the literature but deserve consideration.

Physio-chemical perturbations. Disruptions under field conditions that are physical, rather than purely chemical, cannot be readily simulated in laboratory microcosms. Nevertheless, such physical disruptions can provide insights into the biological characteristics of the decomposition process and the particular pathways along which it is directed.

Lussenhop (1981) has examined this approach by reporting the effects of burning and raking on prairie soil fungi, while raking regimes promote decomposition pathways through bacteria/microarthropod food chains.

This type of approach requires more extensive field plots for experimental purposes since burning and raking effects cannot be studied on a small scale. The same is true of field experiments that are designed to ameliorate existing soil conditions, or to stimulate the growth of microbial and soil animal populations. It is not the purpose of this section to describe in detail the results of the various kinds of treatments that fall within this category, but rather to identify the techniques themselves.

One such technique that has been used for many years, and is still in vogue, is the adjustment of soil pH in the direction of neutrality by the addition of lime. This technique has been used recently in acid coniferous soils in Finland by Huhta et al. (1983), where it was compared with the efficacy of peat ash as an additive, for stimulating biological activity.

Experimental field plots have also been used to gauge the effects of waste disposal, municipal waste water, sewage,

and such agricultural by-products as cattle and pig slurry
and solids, on natural soil ecosystems. These wastes must
be deposited somewhere, and they can be placed in
terrestrial soil systems provided the latter can degrade
them and prevent toxic constituents from accumulating. Such
depositions represent perturbations to the natural
situation, altering the organic and elemental status of
soils. These perturbations are frequently
involuntarily-induced in the sense that they are the
products of economic necessity or convenience. However,
they can be exploited for scientific purposes provided
attention is paid to experimental design (see the collected
works edited by Hartenstein 1978, Huhta et al. 1979). Curry,
(Chapter 9) also discusses the possibilities of using
man-induced management practices as experimental
manipulations for studying ecosystem processes.

In a rather different vein, Brown and Harrison (1983)
used different mixtures of litter types to encourage the
growth and activity of earthworms in a spruce (Picea) forest
floor, with a view to increasing mineralization of N and P.
Similarly, amendments of nitrogen, in the form of urea, have
been used to stimulate microbial activity in black spruce
humus in Quebec (Behan et al. 1978).

Perturbation experiments, such as the ones described in
this section, provide useful insights into the factors that
influence the rate and type of decomposition of organic
material. Their applicability to agroecosystems where
emphasis is frequently placed on maximizing productivity in
the short term is obvious. However, they can also be
applied to natural ecosystems where the decomposition
process is more complicated.

SUMMARY

There are, at present, many adequate techniques for
quantifying the soil biota. The future for the development
of new techniques is promising. Direct observations of root
growth-fungal hyphae and other soil animals using fiber

optics in a mini-rhizotron system (Rush et al. 1984) will advance the study of ecological interactions in the soil. Computer imagery which could distinquish between morphological shapes could facilitate direct counts of fungal hyphae, bacteria or nematodes. The combination of direct and indirect methods, with the expanding use of new technology such as electron probe analysis or enzyme localization, will enhance the interpretation and integration of data from field studies. Field experimental manipulations in which small or large scale ecosystem responses are measured, should be designed to also include testing of the soil faunal-microfloral responses. Knowledge of biotic participation in controlling the critical ecosystem processes needs to be synthesized with overall models of decomposition. The combination of the new and various methods of studying soil biota will then focus on new avenues of research and thus our understanding of soil biology.

REFERENCES

1. Abrahamsen, G. 1972. Ecological study of enchytraeidae (Oligochaeta) in Norwegian coniferous forest soils. Pedobio. 12:26-82.
2. Anderson, J. M. 1973. Carbon dioxide evolution from two temperate, deciduous woodland soils. J. Appl. Ecology 10:361-378.
3. _____, and I. N. Healey. 1970. Improvements in the gelatin-embedding technique for woodland soil and litter samples. Pedobio. 10:108-120.
4. _____, and P. Ineson. 1982. A soil microcosm system and its application to measurements of respiration and nutrient leaching. Soil Biol. Biochem. 14:415-416.
5. Anderson, J. P. E., and K. H. Domsch. 1978. Mineralization of bacteria and fungi in chloroform-fumigated soils. Soil Biol. Biochem. 10:207-213.

6. Anderson, R. V., D. C. Coleman, C. V. Cole, E. T. Elliott, and J. F. McClellan. 1979. The use of soil microcosms in evaluating bacteriophagic nematode responses to other organisms and effects on nutrient cyling. Int. J. Environ. Stud. 13:175-182.

7. Andrassy, I. 1956. Die Rauminhalts and gewichtsbestimmung der Fadenwurmer (Nematoden). Acta. Zool. Acad. Sci. Hung. 2:1-15.

8. _____. 1962. The problem of number and size of sampling unit in quantitative studies of soil nematodes. Pages 65-67 in P. W. Murphy, editor. Progress in soil zoology. Butterworth's, London, England.

9. Anscombe, F. J. 1950. Soil sampling for potato root eelworm cysts. Ann. Appl. Biol. 37:286-295.

10. Aucamp, J. L., and P. A. J. Ryke. 1964. A preliminary report on a grease film extraction method for soil microarthropods. Pedobio. 4:77-79.

11. Axelsson, B., D. Gardetors, U. Lohm, and O. Tenow. 1971. Reliability of estimating standing crop of earthworms by handsorting. Pedobio. 11:338-340.

12. Ayoub, S. M. 1980. Plant nematology. An agriculture training aid. Department of Food and Agriculture, Sacramento, California, USA.

13. Barker, K. R. (Chairman). 1978. Determining nematode population responses to control agents. Pages 114-125 in E. I. Zehr, editor. Methods for evaluating plant fungicides, nematicides and bactericides. Am. Phytopathol. Soc., St. Paul, Minnesota, USA.

14. _____, and C. L. Campbell. 1981. Sampling nematode populations. Pages 451-474 in B. M. Zuckerman, and R. A. Rhode, editors. Plant parasitic nematodes. Vol. III. John Wiley and Sons, New York, New York, USA.

15. _____, C. J. Nusbaum, and L. A. Nelson. 1969. Seasonal population dynamics of selected plant-parasitic nematodes as measured by three extraction procedures. J. Nematol. 1:232-239.

16. _____, C. C. Carter, and J. N. Sasser, editors. 1985. An advanced treatise on Meloidogyne: Volume II, Methodology. University Graphics, North Carolina State University, Raleigh, North Carolina, USA (in press).

17. Baskerville, G. L., and P. Emin. 1969. Rapid estimation of heat accumulation from maximum and minimum temperatures. Ecology 50:514-517.

18. Behan, V. M., S. B. Hill, and D. K. E. McE Kevan. 1978. Effects of nitrogen fertilizers, as urea, on Acarina and other arthropods in Quebec black spruce humus. Pedobio. 18:249-263.

19. Berthet, P. 1963. Mesure de la consommation d' oxygene des Oribatides (Acariens) de la litiere des forets. Pages 18-31 in J. Doeksen, and J. Van der Drift, editors. Soil organisms. North Holland Publishing Company, Amsterdam, Holland.

432

20. _____. 1971. Mites. Pages 186-208 in J. Phillipson, editor. Methods of study in quantitative soil ecology: population, production and energy flow. IBP Handbook No. 18. Blackwell Scientific Publications, Oxford, England.

21. Best, G. R., J. V. Nabholtz, J. Ojasti, and D. A. Crossley. 1978. Response of microarthropod populations to napthalene in three contrasting habitats. Pedobio. 18:189-201.

22. Beute, M. K. 1974. A quantitative technique for the extraction of soil-inhabiting mites (Acarina) and spring tails (Collembola) associated with pod rot of peanut. Phytopathol. 64:571-572.

23. Binkley, D. 1982. Case studies of red alder and sitka alder in Douglas Fir plantations: nitrogen fixation and ecosystem production. Thesis. Oregon State University, Corvallis, Oregon, USA.

24. Bocock, K. L., and O. J. W. Gilbert. 1957. The disappearance of leaf litter under different woodland conditions. Plant Soil 9:179-185.

25. Bohlool, B. B., and E. L. Schmidt. 1968. Nonspecific staining: its control in immunofluorescence examination of soil. Science 162:1012-1014.

26. Boreham, P. F. L. 1979. Recent developments in serological methods for predator-prey studies. Misc. Publ. Entomol. Soc. Amer. 11:17-23.

27. Bouche, M. B. 1969. Comparison critique des methodes d'evaluation des populations de Lumbricides. Pedobio. 9:26-34.

28. _____. 1975. Fonction des lombriciens. III. Premieres estimations quantitative des stations francaise du P.B.I. coll. biologie du sol Montpellier 27. Mai-2. Juine 1973. Rev. Ecol. Biol. Sol. 12:25-44.

29. _____, and R. H. Gardner. 1983. Fonctions des lombricien (earthworm functions). VIII. Population estimation techniques. Rev. Ecol. Biol. Sol. 21:37-63.

30. Brown, A. H. F., and A. F. Harrison. 1983. The effect of tree mixtures on the earthworm populations and N and P status of spruce stands. Pages 101-107 in Ph. Lebrun, H. M. Andre, A. De Medts, C. Gregoire-Wibo, and G. Wauthy, editors. New trends in soil biology. Dieu-Brichart, Louvain-la-Neuve, Belgium.

31. Burges, A., and F. Raw, editors. 1967. Soil biology. Academic Press, London, England.

32. Byrd, D. W., Jr., K. R. Barker, H. Ferris, C. J. Nusbaum, W. E. Griffin, R. H. Small, and C. A. Stone. 1976. Two semi-automatic elutriators for extracting nematodes and certain fungi from soil. J. Nematol. 8:206-212.

33. Chao, Y. S. and A. A. Holland. 1970. Direct and indirect fluorescent antibody staining of Ophiobolus gramin Sacc. in culture and in the rhizosphere of cereal plants. Antonie van Leeuwenhoek 36:549-554.

34. Clarholm, M. 1981. Protozoan grazing of bacteria in soil - impact and importance. Microb. Ecol. 1:343-350.

35. _____, and T. Rosswall. 1981. Biomass and turnover of bacteria in a forest soil and tundra pest. Soil Biol. Biochem. 12:49-51.

36. Coleman, D. C., and A. MacFadyen. 1966. The recolonization of gamma-irridiated soil by small arthropods. Oikos 17:62-70.

37. _____, C. P. P. Reid, and C. V. Cole. 1983. Biological strategies of nutrient cycling in soil systems. Pages 1-55 in A. MacFadyen and E. D. Ford, editors. Advances in Ecological Research, Vol. 13. Academic Press, New York, New York USA.

38. _____, Ingham, R. E., J. F. McClellan, and J. A. Trofymow. 1984. Soil nutrient transformations in the rhizosphere via animal-microbial interactions. Pages 35-58 in J. M. Anderson, A. D. M. Rayner and D. W. H. Walton, editors. Invertebrate-microbial interactions. Cambridge University Press, New York, New York, USA.

39. Cromack, K. Jr., P. Sollins, W. C. Graustein, K. Speidel, A. W. Todd, G. Spycher, C. Y. Li, and R. L. Todd. 1979. Calcium oxalate accumulation and soil weathering in mats of the ectomycorrhizal fungus Hysterangium crassum (Tul. and Tul.) Fischer. Soil. Biol. Biochem. 11:463-468.

40. Crossley, D. A., and M. Witkamp. 1964. Some methods for assessing the activity of soil animals in the breakdown of leaves. C.R.I.er Congress Internat. d'Acarologie 137-146.

41. Cutler, D. W. 1920. A method for estimating the number of active protozoa in the soil. J. Agric. Sci. 10:135-143.

42. Edwards, C. A., and K. Fletcher. 1971. A comparison of extraction methods for terrestrial arthropods. Pages 150-185 in J. Phillipson, editor. Methods of study in quantitative soil ecology: population, production and energy flow. Blackwell Scientific Publications, Oxford, England.

43. _____, and J. R. Lofty. 1972. Biology of earthworms. Chapman and Hall, Ltd., London, England.

44. Elliott, J. M. 1971. Some methods for the statistical analysis of benthic invertebrates. Freshwater Biological Assoc. Scientific Publication No. 15, Ambleside, England.

45. Englemann, M. D. 1961. The role of soil arthropods in the energetics of an old field community. Ecol. Monog. 31:221-238.

434

46. Fager, E. W. 1968. The community of invertebrates in decaying oak wood. J. Anim. Ecol._37:121-142.

47. Fairbanks, B. C., L. E. Woods, R. J. Bryant, E. T. Elliot, C. V. Cole, and D. C. Coleman. 1984. Limitations of ATP estimates of microbial biomass. Soil Biol. Biochem._16:549-558.

48. Fichter, B. L., and W. P. Stephen. 1981. Time related decay in prey antigens ingested by the predator Podisus maculiventris (Hemiptera, Pentatomidae) as detected by ELISA. Oecologia 51:404-407.

49. Fogel, R., and G. Hunt. 1979. Fungal and arboreal biomass in a western Oregon Douglas-fir ecosystem: Distribution patterns and turnover. Can. J. For. Res._ 9:245-256.

50. Foster, R. C. 1981. Polysaccharides in soil fabrics. Science 214:665-667.

51. Frankland, J. C., A. D. Bailey, T. R. G. Gray, and A. A. Holland. 1981. Development of an immunological technique for estimating mycelial biomass of Mycena galopus in leaf litter. Soil Biol. Biochem._13:87-92.

52. Freckman, D. W. 1982a. Parameters of the nematode contribution to ecosystems. Pages 81-97 in D. W. Freckman, editor. Nematodes in soil ecosystems. Univ. Texas Press, Austin, Texas, USA.

53. _____, editor. 1982b. Nematodes in soil ecosystems. Univ. Texas Press, Austin, Texas, USA.

54. _____, and J. G. Baldwin. 1985. Soil nematoda in D. L. Dindal, editor. Soil biology. Wiley-Interscience. New York, New York, USA (in press).

55. _____, S. D. Van Gundy, and S. F. MacLean, Jr. 1977. Nematode community structure in Alaskan Soils. Nematol._9:268.

56. Geisy, J. P., editor. 1980. Microcosms in ecological research: selected papers from a symposium held in Augustus, Georgia. U.S. Dept. Energy, Washington, D.C.

56. Goodell, P. B. 1982. Soil sampling and processing for detection and quantification of nematode populations. Pages 179-198 in D. W. Freckman, editor. Nematodes in soil ecosystems. Univ. Texas Press, Austin, Texas, USA.

57. _____, and H. Ferris. 1980. Plant parasitic nematode distribution in an alfalfa field. J. Nematol._ 12:136-141.

58. Graustein, W. C., K. Cromack, Jr.,and P. Sollins. 1978. Calcium oxalate: occurrence of soils and effect on nutrient and geochemical cycles. Science 198:1252-1254.

59. Gray, T. R. G. 1967. Stereoscan electron microscopy of soil micro-organisms. Science 155:1668-1670.

60. Harding, D. J. L. 1967. Faunal participation in the breakdown of cellophane inserts in the forest floor. Pages 10-20 in O. Graff, and J. E. Satchell, editors. Progress in soil biology. Vieweg and Sohn, Braunschweig, Germany.

61. Harris, W. F, D. Santantonio, and D. McGinty. 1980. Dynamic belowground ecosystems. Pages 119-129 in R. H. Waring, editor. Forests: fresh perspectives from ecological analysis. Proc. 40th Biol. Colloq., Oregon State University Press, Corvallis, Oregon USA.

62. Hartenstein, R., editor. 1978. Utilization of soil organisms in sludge management. Soil microcommunity conference proc., Syracuse, New York, USA.

63. Healey, I. N. 1967. The population metabolism of Onychiurus procampatus Gisin (Collembola). Pages 127-137 in O. Graff, and J. E. Satchell, editors. Progress in soil biology. Vieweg and Sohn, Braunschweig, Germany.

64. _____. 1971. Apterygotes, Pauropods and Symphylans. Pages 209-232 in J. Phillipson, editor. Methods of study in quantitative soil ecology. IBP Handbook No. 18. Blackwell Scientific Publications, Oxford, England.

65. Heath, G. W., C. A. Edwards, and M. K. Arnold. 1964. Some methods for assessing the activity of soil animals in the breakdown of leaves. Pedobio. 4:80-87.

66. Howard, P. J. A. 1972. Problems in the estimation of biological activity in soil. Oikos 23:235-240.

67. Huhta, V. 1972. Efficiency of different dry funnel techniques in extracting arthropoda from raw humus forest soil. Ann. Zool. Fenn. 9:42-48.

68. _____, and A. Koskennemi. 1975. Numbers, biomass and community respiration of soil invertebrates in spruce forests in two latitudes in Finland. Ann. Zool. Fenn. 12:164-182.

69. _____, E. Ikonen, and P. Vilkamaa. 1979. Succession of invertebrate populations in artificial soil made of sewage sludge and crushed bark. Ann. Zool. Fenn. 16:223-270.

70. _____, A. Koskenniemi, R. Segersven, and P. Vilkamaa. 1983. Role of pH in the effect of fertilization on nematoda, oligochaeta and microarthropods. Pages 61-73 in Ph. Lebrun, H. M. Andre, A. de Medts, C. Gregoire-Wibo, and G. Wauthy, editors. New trends in soil biology. Dieu-Brichart, Louvain-la-Neuve, Belgium.

71. Hurlbert, S. H. 1984. Pseudoreplication and the design of ecological field experiments. Ecol. Monog. 54:187-211.

436

72. Ineson, P., M. A. Leonard, and J. M. Anderson. 1982. Effect of collembolan grazing upon nitrogen and cation leaching from decomposing leaf litter. Soil Biol. Biochem. 14:601-605.

73. Ingham, E. R., and D. A. Klein. 1982. Relationship between fluorescein diacetate-stained hyphae and oxygen utilization, glucose utilization, and biomass of submerged fungal batch cultures. Appl. Env. Microb. 44:363-370.

74. _____. 1984a. Soil fungi: measurement of hyphal length. Soil Biol. Biochem. 16:279-280.

75. _____. 1984b. Soil fungi: relationships between hyphal activity and staining with fluorescein diacetate. Soil Biol. Biochem. 16:273-278.

76. Jenkinson, D. S., and D. S. Poulson. 1976. The effects of biocidal treatments on metabolism in soil - V. A method for measuring soil biomass. Soil Biol. Biochem. 8:209-213.

77. Jones, P. C. T., and J. E. Mollison. 1948. A technique for the quantitative estimation of soil microorganisms. J. Gen. Microbiol. 2:54-69.

78. Kubiena, W. 1938. Micropedology. Iowa State University Press, Ames, Iowa USA. 79. Kurcheva, G. F. 1960. Role of invertebrates in the decomposition of oak litter. Soviet Soil Sci. (AIBS translation of Pochvovedeniye). 4:360-365.

80. Lasebikan, B. A. 1974. Preliminary communication on microarthropods from a tropical rain forest in Nigeria. Pedobio. 14:402-411.

81. Latter, P. M., and G. Howson. 1977. The use of cotton strips to indicate cellulose decomposition in the field. Pedobio. 17:147-155.

82. Lebrun, Ph. 1971. Ecologie et biocenotique de quelques Peuplements de 'arthropodes edaphiques. Inst. Royal Sci., Naturelles de Belgique. Memo. 165:1-203.

83. Lindemann, R. L. 1942. The trophic-dynamic aspect of ecology. Ecology 23:399-418.

84. Linderstrom-Lang, K. 1943. On the theory of the cartesian diver respirometer. C. R. Trans. Carlsberg, Ser. Chim. 24:333-398.

85. Lousier, J. D., and D. Parkinson. 1984. Annual population dynamics and production ecology of Testacea (Protozoa, Rhizopoda) in an aspen woodland soil. Soil Biol. Biochem. 16:103-114.

86. Lundkvist, H. 1978. A technique for determining individual fresh weights of live small animals with special reference to Enchytraeidae. Oecologia 35:365-367.

87. _____, 1982. Population dynamics of Cognettia sphagnetorum (Enchytraeidae) in a Scots pine forest soil in Central Sweden. Pedobio. 23:21-41.

88. Lussenhop, J. 1981. Microbial and microarthropod detrital processing in a prairie soil. Ecology 62:964-972.

89. Luxton, M. 1975. Studies on the Oribatid mites of a Danish Beechwood soil. II. Biomass calorimetry, and respirometry. Pedobio. 15:161-200.

90. MacFadyen, A. 1952. The small arthropods of a Molina fen at Cothill. J. Anim. Ecol. 21:87-117.

91. _____. 1961. Improved funnel-type extractors for soil arthropods. J. Anim. Ecol. 30:171-184.

92. _____. 1962. Soil arthropod sampling. Pages 1-34 in J. B. Cragg, editor. Adv. Ecol. Res., Academic Press, London, England.

93. _____. 1963. Animal Ecology - Aims and Methods. Sir Isaac Pitman & Sons Ltd., London, England.

94. _____. 1970. Simple methods of measuring and maintaining the proportion of carbon dioxide in air, for use in ecological studies of soil respiration. Soil Biol. Biochem. 2:9-18.

95. MacLean, S. F., Jr., G. K. Douce, E. A. Morgan, and M. A. Skeel. 1977. Community organization in the soil invertebrates of Alaska arctic tundra. Pages 90-101 in U. Lohm and T. Persson, editors. Soil organisms as components of ecosystems. Proceedings of the VI International colloquium on soil zoology. Ecol. Bull. (Stockholm) 25:90-101.

96. Merny, G., and J. DeJardin. 1970. Les nematodes phytoparasites des rizieres inondee de Cote d'Ivoire. II. Essai d' estimation de l'importance des populations. Cah. ORSTOM, Ser. Biol. 11:45-67.

97. Minagawa, N. 1979. Efficiences of two methods for extracting nematodes from soil. Appl. Entomol. Zool. 4:469-477.

98. Minderman, G., and J. C. Vulto. 1973a. Comparison of techniques for the measurement of carbon dioxide evolution from soil. Pedobio. 13:73-80.

99. _____. 1973b. Carbon dioxide production by tree roots and microbes. Pedobio. 13:337-343.

100. Mitchell, M. J. 1979. Energetics of Oribatid mites (Acari: Cryptostigmata) in an aspen woodland soil. Pedobio. 19:89-98.

101. Mitchell, M., and D. Parkinson. 1976. Fungal feeding of oribatid mites (Acari: Cryptostigmata) in an aspen woodland soil. Ecology 57:302-312.

102. Murphy, P. W., editor. 1962. Progress in soil zoology. Butterworths, London, England.

103. O'Connor, F. B. 1957. An ecological study of the enchytraeid worm population of a coniferous forest soil. Oikos 8:271-281.

438

104. _____. 1962. The extraction of the Enchytraeidae from soil. Pages 279-285 _in_ P. W. Murphy, editor. Progress in soil zoology. Butterworths, London, England.

105. _____. 1963. Oxygen consumption and population metabolism of some populations of Enchytraeidae from North Wales. Pages 32-48 _in_ J. Doeksen, and J. Van der Drift, editors. Soil organisms. North Holland Publishing Company, Amsterdam, Holland.

106. _____. 1971. The enchytraeids. Pages 83-106 _in_ J. Phillipson, editor. Methods of study in quantitative soil ecology: population, production and energy flow. IBP Handbook No. 18. Blackwell Scientific Publications, Oxford, England.

107. Odum, E. P. 1971. Fundamentals of ecology. W. B. Saunders Co., New York, New York, USA.

108. Olson, F. W. 1950. Quantitative estimates of filamentous algae. Trans. Amer. Micros. Soc. 69:272-279.

109. Oostenbrink, M. 1971. Comparison of techniques for population estimation of soil and plant nematodes. Pages 72-82 _in_ J. Phillipson, editor. Methods of study in quantitative soil ecology: population, production and energy flow. IBP Handbook No. 18. Blackwell Scientific Publications, Oxford, England.

110. Page, A. L., R. H. Miller, and D. R. Keeney. 1982. Methods of soil analysis. Part II: Chemical and microbiological properties. American Society of Agronomy, Madison, Wisconsin, USA.

111. Pande, Y. D., and P. Berthet. 1973. Comparison of the Tullgren funnel and soil section methods for surveying oribatid populations. Oikos 24:273-277.

112. Paris, O. H., and F. A. Piteka. 1962. Population characteristics of the terrestrial isopod _Armadillidium vulgare_ in a California grassland. Ecology 43:229-248.

113. Parker, L. W., D. W. Freckman, Y. Steinberger, L. Driggers, and W. G. Whitford. 1984. Effects of simulated rainfall and litter quantities on desert soil biota: soil respiration, microflora and protozoa. Pedobiol. 27:185-195.

114. Parkinson, D., T. R. G. Gray, J. Holding, and H. M. Nagel-de-Boois. 1971. Heterotrophic microflora. Pages 34-50 _in_ J. Phillipson, editor. Methods of study in quantitative soil ecology: population, production and energy flow. IBP Handbook No. 18. Blackwell Scientific Publications, Oxford, England.

115. Paul, E. A., and R. P. Voroney. 1984. Field interpretation of microbial biomass activity measurements. Pages 509-514 _in_ M. J. Klug and C. A. Reddy, editors. Current perspectives in microbial ecology. Amer. Soc. Microbiol., Washington, D.C., USA.

116. Peterson, H. 1978. Some properties of two high gradient extractors for soil microarthropods and an attempt to evaluate their extraction efficiency. Nat. Jutl. 20:95-121.

117. _____, and M. Luxton. 1982. A comparative analysis of soil fauna populations and their role in decomposition processes. Oikos 39:287-376.

118. Petrusewicz, K., and A. MacFadyen. 1970. Productivity of terrestrial animals, principles and methods. IBP Handbook No. 13. F. A. Davis Company, Philadelphia, Pennsylvania, USA.

119. Phillipson, J. 1971. Methods of study in quantitative soil ecology: population, production and energy flow. IBP Handbook No. 18. Blackwell Scientific Publications, Oxford, England.

120. Piearce, T. G., and B. Piearce. 1979. Responses of Lumbricidae to saline inundation. J. Appl. Ecol. 16:461-473.

121. Proctor, J. R., and C. F. Marks. 1975. The determination of normalizing transformations for nematode count data from soil samples and of efficient sampling schemes. J. Nematol. 20:395-406.

122. Richards, B. N. 1974. Introduction to the soil ecosystem. Longman Company, New York, New York, USA.

123. Robinson, A. F. 1984. Comparison of five methods for measuring nematode volume. J. Nematol. 16:343-347.

124. Rush, C. M., D. R. Upchurch, and T. J. Gerik. 1984. In situ observations of Phymatotrichum omnivorum with a borescope mini-rhizotron system. Phytopathol. 74:104-105.

125. Santos, P. F., J. Phillips, and W. G. Whitford. 1981. The role of mites and nematodes in early stages of buried litter decomposition in a desert soil. Ecology 62:667-669.

126. Satchell, J. E. 1969. Methods of sampling earthworm populations. Pedobio 9:20-25.

127. _____. 1971. Measuring population and energy flow in earthworms. Pages 261-267 in J. Phillipson, editor. Methods of studying soil ecology. UNESCO, Paris, France.

128. _____. 1974. Litter-interface of animate/inanimate matter. Pages XIV-XL in C. H. Dickenson, and G. J. F. Pugh, editors. Biology of Plant Decomposition. Academic Press, New York, New York, USA.

129. Schlesinger, W. H. 1977. Carbon balance in terrestrial ecosystems. Ann. Rev. Ecol. Syst. 8:51-81.

130. Seastedt, T. R. 1984. The role of microarthropods in decomposition and mineralization processes. Ann. Rev. Entomol. 29:25-46.

440

131. _____, and D. A. Crossley, Jr. 1978. Further investigations of microarthropod populations using the Merchant-Crossley high gradient extractor. J. Georgia Ent. Soc. 13:344-388.

132. _____. 1980. Effects of microarthropods on the seasonal dynamics of nutrients in forest litter. Soil Biol. Biochem. 12:337-342.

133. _____. 1983. Nutrients in forest litter treated with napthalene and simulated throughfall: A field microcosm study. Soil Biol. Biochem. 15:159-165.

134. _____, A. Kothari, and D. A. Crossely, Jr. 1980. A simplified gelatin embedding technique for sectioning litter and soil samples. Pedobio. 29:55-59.

135. Sibbesen, E. 1977. A simple ion-exchange resin procedure for extracting plant-available elements from soil. Plant Soil 46:665-669.

136. Singh, J., and K. S. Pillai. 1976. The use of a flotation method in the collection of microarthropods from arable soil in India. Rev. Ecol. Biol. Sol. 13:321-335.

137. Smith, V. R. 1979. Evaluation of a resin-bag procedure for determining plant-available P in organic volcanic soils. Plant Soil 53:245-249.

138. Sohlenius, B. 1979. A carbon budget for nematodes, rotifers and tardigrades in a Swedish coniferous forest soil. Holarc. Ecol. 2:30-40.

139. _____. 1980. Abundance, biomass and contribution to energy flow by soil nematodes in terrestrial ecosystems. Oikos 34:186-194.

140. Somme, L., and W. Block. 1982. Cold hardiness of Collembola at Signy Island, Maritime Antarctic. Oikos 38:168-176.

141. Southwood, T. R. E. 1975. Ecological Methods. Butler and Tanner, Ltd., London, England.

142. Springett, J. A. 1981. A new method for extracting earthworms from soil cores with comparison of four commonly used methods for estimating earthworm populations. Pedobio. 21:217-222.

143. St. John, T. V. 1980. Influence of litter bags on growth of fungal structures. Oecologia 46:130-132.

144. Standen, V. 1973. The production and respiration of an enchytraeid population in blanket bog. J. Anim. Ecol. 42:219-245.

145. _____. 1982. Associations of Enchytraeidae in experimentally fertilized grasslands. J. Anim. Ecol. 51:501-523.

146. Stevenson, B. G., and D. L. Dindal. 1981. A litter box method for the study of litter arthropods. J. Georgia Ent. Soc. 16:151-156.

147. _____. 1982a. Effect of leaf shape on forest litter collembola: community organization and microhabitat selection of two species. J. Georgia Ent. Soc. 17:369-376.

148. _____. 1982b. Effect of leaf shape on forest spiders: community organization and microhabitat selection of immature Enoplognatha ovata (Clerck) (Theridiidae). J. Arach. 10:165-178.

149. Swift, M. J., O. W. Heal, and J. M. Anderson. 1979. Decomposition in terrestrial ecosystems. Blackwell Scientific Publications. Oxford, England.

150. Takeda, H. 1979. On the extraction process and efficiency of MacFadyen's high-gradient extractor. Pedobio. 19:106-112.

151. Tamura, H. 1976. Biases in extracting collembola through Tullgren funnels. Rev. Ecol. Biol. Sol 13:21-34.

152. Terhivuo, J. 1980. Relative efficiency of handsorting, formalin application and combination of both methods in extracting Lumbricidae from Finnish soils. Pedobio. 23:175-188.

153. Thomas, J. O. M. 1979. An energy budget for a woodland population of orbatid mites. Pedobio. 19:346-378.

154. Van der Drift, J. 1951. Analysis of the animal community of a beech forest floor. Tijdschr. Ent. 94:1-168.

155. Van Straalen, N. M., and P. C. Rijninks. 1982. The efficiency of Tullgren apparatus with respect to interpreting seasonal changes in age structure of soil arthropod populations. Pedobio. 24:197-209.

156. Viglierchio, D., and R. V. Schmitt. 1983. On the methodology of nematode extraction from field samples: comparison of methods for soil extraction. J. Nematol. 15:450-454.

157. Waksman, S. A. 1932. Principles of soil microbiology. Williams and Wilkins Publishing Company, Baltimore, Maryland, USA.

158. Wallwork, J. A. 1970. Ecology of soil animals. McGraw-Hill Publishers, London, England.

159. _____. 1976. The distribution and diversity of soil fauna. Academic Press, London, England.

160. Warwick, R. M., and R. Price. 1979. Ecological and metabolic studies on free living nematodes from an estuarine mud-flat. Estuar. Coast. Mar. Sci. 9:257-271.

161. Whitford, W. G., D. W. Freckman, N. Z. Elkins, L. W. Parker, R. Parmalee, J. Phillips, and S. Tucker. 1981. Diurnal migration and responses to simulated rainfall in desert soil microarthropods and nematodes. Soil Biol. Biochem. 13:417-425.

162. Willard, J. R. 1972. Soil invertebrates. I. Methods of sampling and extraction. Matador Project. Technical Report F. Saskatoon, Canada.

163. Winfield, A. L., and D. A. Cooke. 1974. The ecology of Trichodorus. Pages 309-340 in C. E. Taylor, F. Lamberti, and J. W. Seinhorst, editors. Nematode vectors of plant viruses. Plenum Press, New York, New York, USA.

164. Witkamp, M. 1966. Rates of carbon dioxide evolution from the forest floor. Ecology 47:492-494.

165. Yeates, G. W. 1979. Soil nematodes in terrestrial ecosystems. J. Nematol. 11:213-229.

166. _____. 1981. Nematode populations in relation to soil environmental factors: a review. Pedobio. 22:312-338.

167. Zicsi, A. 1958. Determination of number and size of sampling unit for estimating lumbricid populations of arable soils. Pages 68-71 in P. W. Murphy, editor. Progress in soil ecology. Butterworths Publishing Company, London, England.

168. Zlotin, R. I. and K. S. Khodashova. 1980. The role of animals in biological cycling of forest-steppe ecosystems. Dowden, Hutchinson and Ross, Stroudsburg, Pennsylvania, USA.

THE ROLE OF MODELING IN RESEARCH ON MICROFLORAL AND FAUNAL
INTERACTIONS IN NATURAL AND AGROECOSYSTEMS

H. W. HUNT AND   W. J. PARTON*

INTRODUCTION

A variety of mathematical tools can be applied to
constructing models in biology. The distinction between
correlative and explanatory models (Gold 1977) is more
important for our purposes than the particular mathematical
tool employed. Explanatory models are structured to be
analogous to the real system under study and thus to embody
hypotheses about how the real system operates. The structure
of correlative models--regression models, for example--is
not constrained by what is known about mechanisms operating
in the real system. Both correlative and explanatory models
are expected to yield realistic predictions. Although both
types of models are needed in basic research, we restrict
our attention largely to explanatory models because they are
especially useful as an aid to understanding.

In this chapter, we emphasize recent developments and
topics considered particularly important, rather than attempt
a comprehensive overview of ecosystem modeling, such as
Wiegert (1975) provides. There are good general reviews of
crop models (Loomis et al. 1979), forest succession models
(Shugart and West 1980), and soil nitrogen models (Frissel
and Van Veen 1982). Roels and Kossen (1978) outlined general
methods for modeling microbial growth and reviewed models
for growth in pure culture. DeWit and Van Keulen (1972)
presented basic equations for transport of heat, salts,
ions, and water in soils.

*Natural Resource Ecology Laboratory, Colorado State
University, Fort Collins, Colorado 80523.

The following section expounds our modeling philosophy and gives a brief overview of the variety of models useful in ecosystem research. Second, we review modeling approaches for important ecosystem processes. Third, several particular models are described in detail, with emphasis on their use in research. The final section presents conclusions on the role modeling has played in natural and agroecosystem research and gives recommendations for future research applications.

Benefits of modeling

Woodmansee (1978) discussed the benefits of modeling to a research program. There are a variety of direct and indirect benefits, the most important of which is a better understanding of how the system operates.

Indirect benefits include the following. Constructing quantitative models requires a disciplined view of the system, and it reduces addlepated thinking. Models provide a framework for relating various kinds of information from different sources, and can serve to check the compatibility and internal consistency of data sets. In multidisciplinary research, modeling facilitates communication among participants by providing definite hypothetical statements about the system under study and helps scientists integrate their individual contributions into the whole study. Models may exhibit unexpected or unrealistic behavior during their development. Attempting to account for such model behavior in terms of a model's structure often leads to insights about the relationships among state variables, the linkage of cause and effect, and the consequences of simplifying assumptions. New insights may or may not be directly transferrable from model to real system, but they often constitute testable hypotheses. Altered perceptions of the relative importance of various components of the system can lead to changes in experimental design and protocol; thus, modeling can help direct a research program.

The direct benefits of modeling derive from models' formal role in the scientific method. We view models as complex hypotheses about the structure and operation of the system (Kowal 1971). Model structure and parameter values constitute the hypothesis, and model outputs are the predictions, or consequences, of the hypothesis. To test a model, its predictions are compared to data from the real world. When a model fails, some part of the complex hypothesis can be rejected, and the rejection of hypotheses is the main avenue for scientific advancement (Platt 1964). When a model succeeds, we have achieved to some degree the main objective of science--to find simple hypotheses accounting for the widest range of observations. To speed scientific progress, experiments should be designed to distinguish among competing hypotheses (Platt 1964).

Treating models as complex hypotheses helps to resolve the sometimes confusing issue of model validity. Hypotheses (models) may be rejected if they fail to mimic closely enough the real world phenomena they are intended to represent. Hypotheses (models) are never proven correct, but the greater the range of real world phenomena they mimic, the more confidence may be placed in them. The data sets used in formulating hypotheses (adjusting model structure and parameter values) cannot provide a test of the predictive power of the hypothesis (model), which requires an independent data set.

Some authors disagree with the above view of the formal role of simulation models in research (Romesburg 1981, Leigh 1968). There are technical problems in attempting to construct statistical tests of agreement between model output and real-world data, but model output can often be refuted confidently without formal tests. When a model fails, it may be difficult to decide which part of the complex hypothesis is at fault. This problem can be handled by systematically comparing alternative model structures (Hunt et al. 1984). Whether or not one treats simulation models as formal hypotheses, the indirect benefits discussed above are probably sufficient justification to include simulation modeling in a research program.

Given a model that adequately mimics the relevant aspects of behavior of a natural ecosystem or agroecosystem, that model can be used as a surrogate of the real system, and experiments performed on it. Thus, for example, a crop model can be used to evaluate the effects of varying the amounts and timing of a fertilizer application. Model results may be used to guide further research on fertilization, and they might be used to recommend fertilizer applications to farmers.

Levels of resolution and inclusiveness

The natural world has long been viewed as having an hierarchical structure (Bertalanffy 1968). The hierarchical levels of greatest interest in ecology are individual organisms, local species populations, guilds, communities, different communities within a landscape, and ecosystems. A model may be assigned a level of resolution according to the hierarchical level of the state variables and processes included. A wide range of levels of resolution may be appropriate, depending on the objectives of the modeling exercise and on the type of data available.

A generally adopted, but seldom discussed, approach to modeling is to attempt to account for system behavior at one hierarchical level by describing the elements of the next more detailed level. For example, crop models often attempt to account for the growth of a population of plants from information about processes such as photosynthesis, respiration, translocation, and nutrient uptake by individual plants (Loomis et al. 1979). The behavior of individual plants is often predicted from interactions among plant organs. Thus, it is assumed that population dynamics can be predicted from a knowledge of the physiology and behavior of individuals. In this paper, we illustrate a range of levels of resolution useful in studying natural and agroecosystems.

Most terrestrial ecosystem-level models deal with light, water, temperature, and biomass. Biomass commonly is expressed as dry mass, carbon, or chemical potential energy, all of which are closely related. Models of systems under significant nutrient limitations generally include N, P, or S. Fluxes of different elements may be handled more or less independently in different submodels (Innis 1978) or be integrated together (McGill et al. 1981).

There is wide variation in the degree of taxonomic diversity recognized in ecosystem models. For practical reasons, models of natural ecosystems either combine species into functional groups (e.g., warm-season grasses, chewing insects, microbivorous protozoa) or treat only small single-species patches, which presumably do not interact with adjacent patches. A major assumption of ecosystem-level models is that important aspects of system dynamics may be captured in models that either ignore taxa or aggregate taxa into functional groups. The relative success of some models that simplify biotic diversity indicates that such aggregation is legitimate. Nevertheless, the consequences of aggregating species is a subject of recent theoretical work (Gardner et al. 1982).

STRUCTURE AND PROCESSES

The next few sections survey the treatment of ecosystem structure and processes deemed particularly important for understanding microfloral and -faunal interactions: belowground primary production, decomposition, microflora, fauna, and N-cycle transformations. We have omitted from consideration factors controlling primary production, heat flow, water flow, and aboveground food webs.

The particular algebraic form of an equation for the rate of a process is seldom critical, because different mathematical forms can be made to approximate one another. It is far more important and difficult to correctly identify which variables affect a process and the general nature of interactions among controlling variables. Therefore, we have not attempted to compile a set of "off-the-shelf" equations

for processes but, instead, to identify the general features
necessary for a successful model.

Belowground primary production

Predicting aboveground production is relatively
straightforward (Hesketh and Jones 1980), and a great deal
is known about the kinds and amounts of plant material
coming into surface litter. In natural ecosystems, the
problems of estimating root biomass and of separating live
from dead roots have obstructed a comparable understanding
of the belowground system.

Singh et al. (1983) constructed "a structurally sound
but mechanistically simple simulation model of grassland
primary production . . ." and drew random samples from the
model output as if the model were the real system under
study. Various formulae for estimating root production were
applied to the model output and compared with the known
value from the model. The results indicated that estimates
of root production based on harvest techniques are woefully
inadequate and that production estimates based on root
harvest are probably overestimates. This conclusion applies
to the portion of root production related to release of
fragments detectable in soil cores and thus does not address
the question of release of exudates and sloughed cells. Singh
et al. (1983) further concluded that root production can be
estimated reliably only by tracer methods (Caldwell and Camp
1974), or perhaps by measuring separately the dynamics of
different functional categories or size classes of roots, as
done in forests (Harris et al. 1980). The model of Parton
et al. (1978) separated live roots into suberized, unsuberized,
and juvenile categories and simulated turnover in these
categories as a function of soil water and temperature
extremes. The model agreed with the qualitative observations
of Ares (1976) on the dynamics of the three root categories
in shortgrass prairie, but this study did not address the
release of exudates and sloughed cells.

Given our poor understanding of the partitioning of belowground production between large, slowly decomposing components detected in soil cores versus small particulates and dissolved components, it is not surprising that the controls on these production processes have not yet been accurately modeled. Gilmanov (1977) assumed that roots die fastest when conditions are worst for growth. McGill et al. (1981) reasoned that whole roots should be lost under stressful conditions of cold or drought, but that losses to herbivores would occur under more favorable conditions. They assumed that whole-root death occurred at a constant rate, while exudation and sloughing occurred fastest in warm, wet soil.

Scott et al. (1979) developed a simulation model for arthropod and nematode bioenergetics, which was used to predict rates of direct loss of roots to herbivores, given the level of each kind of herbivore. Total root loss, including loss of damaged plant tissue not directly consumed, was estimated to be about 100 g dry mass $\cdot$ m^{-2} $\cdot$ y^{-1}, which is almost 18% of annual belowground production [using the admittedly uncertain model value of production from Singh et al. (1983)]. Assuming that herbivores prefer smaller, metabolically more active roots, their impact might be considerable. Yet apparently no total-system models include belowground herbivores as state variables.

Mycorrhizal fungi are common in most natural ecosystems and may play an important role procurring P and perhaps other plant nutrients (Tinker 1982). Supplying organic C to the fungus may represent a significant cost to the plant, but mycorrhizae have not yet been included in ecosystem-level models.

Decomposition

We define decomposition as the breakdown of particulates into soluble material that can be taken up and used by microbes. Decomposition rate is regulated mainly by temperature, moisture, and the chemical properties of the substrate, although fauna and external supply of nutrients have an effect. The primary agents of decomposition--fungi, bacteria, and actinomycetes--appear to be inactive most of the time in some ecosystems but are so ubiquitous under natural conditions that decomposition rate might not be limited by their biomass (Hunt 1977).

Substrate quality. Although chemical reactions among inorganic substances in the soil solution may be an important part of ecosystem models (Dutt et al. 1972), organic matter is too heterogeneous to distinguish individual substances in a model. The most detailed approach attempted to date is to define classes of substances such as proteins, cellulose, etc. (Minderman 1968). Such an approach is difficult to apply, since the classes interact--for example, the effect of lignin on cellulose decomposition--and it is difficult to analyze changing chemical composition during decomposition. The least detailed approach is to treat substrates as homogeneous, which might be appropriate for models at a low level of resolution.

An empirical approach is to divide substrates into two or three arbitrary components with different decomposition rates (Hunt 1977). The initial division among components has been correlated with the original N and lignin content of the substrate (Melillo et al. 1982). Such an approach gives good predictions of the total amount of material remaining, but these models have little heuristic value, and the predicted dynamics of individual components usually cannot be verified. A similar but more elegant approach is to define, a priori, several components with different decomposition rates. For example, McGill et al. (1981) distinguished slowly decomposing structural materials, cell walls, from more labile metabolic materials, cytoplasm and nuclei. The initial division between structural and metabolic components was estimated from N content by assuming fixed C:N ratios for each component. This approach allows for different decomposition rates for plant, bacterial, and fungal matter and simplifies model structure by not requiring separate state variables for each kind of substrate.

McGill and Cole (1981) proposed an interesting classification of organic substances according to the covalent bonds present between C, N, and S (C-C, C-N, C-S, C-O-S, and C-O-P). The most important consequence of distinguishing bond classes is that the mechanisms for mineralizing C-bonded atoms (via intracellular enzymes) are different than for O-bonded atoms (via esterases, often extracellular). This leads to predictions about soil organic matter composition over a range of conditions (McGill and Cole 1981). Hunt et al. (1982) explored the consequences of chemical bond classes for ecosystem models, but the approach has not yet been applied.

Substrate heterogeneity presents a difficult modeling problem. There appear to be two viable approaches. One solution is empirically to define two or more pools with different turnover times. This is more appropriate in correlative models. The other solution is to define components based on their physiochemical properties and to determine how these relate to use by the soil biota. The latter solution is more appropriate for explanatory models.

Soil organic matter fractions. Hass et al. (1957) showed that long-term cultivation of soil can result in a decrease in soil organic matter, soil fertility, and crop yields. A variety of models has been used to represent the long-term change in soil organic N and C with cultivation. Jenny (1941) used the following equation to represent the decline of organic C and N in cultivated soils.

$$\frac{dX}{dt} = A - k \cdot X , \tag{1}$$

where A = the rate of annual addition of nitrogen or carbon, k = the decomposition constant (y^{-1}), and X = soil N or C. The major problem with this formulation is that $\underline{k}$ generally is assumed to be constant, and no allowance is made for changes in the composition of soil organic matter or effects of soil temperature and water.

Campbell et al. (1976) improved upon Eq. (1) by dividing soil organic matter into two different compartments: stable organic matter and relatively labile organic matter with turnover times (1/k) of 53 years and 1429 years, respectively. Paul and Van Veen (1978) made further improvements in the model by dividing the plant residue into "recalcitrant" and decomposable fractions and introducing the concept of physically protected and nonphysically protected soil organic matter. Critical assumptions in the latter model are that physically protected organic matter has a much lower decomposition rate than nonphysically protected organic matter and that the mechanical disruption of cultivation decreases physical protection. The model predicts that organic C decreases rapidly (by 26% of total C) during the first 20 years of cultivation.

Jenkinson and Rayner (1977) developed a model in which organic matter was separated into five compartments: decomposable plant material (0.24-y turnover time), resistant plant matter (3.33-y), soil biomass (2.44-y), physically stabilized organic matter (72-y), and chemically stabilized organic matter (2857-y). Decomposition of any of the five compartments results in the loss of CO_2 (76%) and the formation of soil biomass (8%), physically protected organic matter (12%), and chemically stabilized organic matter (4%).

The mechanisms of soil organic matter formation and breakdown are not well understood, as discussed in Chapter 4; and this is reflected in the somewhat empirical nature of soil organic matter models. Various fractions such as clay protected and physically protected (Van Veen and Paul 1981); active, slow, and passive (Parton et al. 1983b); humads and resistant soil organic matter (McGill et al. 1981); nonbiomass active, stabilized, and old organic matter (Juma 1981) have been defined, but in effect the different approaches merely divide organic matter into two or three pools with different turnover times. The actual mechanisms proposed in the various models are not well established and are the object of current research.

Effects of temperature and moisture on decomposition.
Many studies have shown that soil temperature, soil water,
and substrate quality are the primary factors controlling
decomposition in the soil. Decomposition rates are very
sensitive to temperature and increase rapidly with increasing
temperature over a certain range. The response to temperature
is typically expressed as the Q_{10} (rate at T°C divided by
rate at T - 10°C). Q_{10} values generally are assumed to be
around 2.0 but tend to be higher below 10°C and lower above
30°C (Giese, 1973). The effect of temperature on
decomposition rate can be represented by the Arrhenius
equation over the middle range of temperatures but not at
extreme temperatures (Swift et al., 1979). Malhi and McGill
(1982) showed that temperature optima for nitrification vary
appreciably with climate, suggesting adaptation of nitrifiers
to local conditions. Such local adaptation is probably
widespread among microbes, so that models will require
different temperature curves at different sites.

Decomposition rates are sensitive to changes in soil
moisture, and the response to soil moisture depends on
temperature. In general, decomposition is maximized with
soil water tension between -0.03 and -0.1 MPa and decreases
with lower and higher tension. Data from Wildung et al.
(1975) suggest that variation of soil water has little effect
at low soil temperatures (<10°C) but a large effect for
temperatures greater than 15°C.

Models representing the effect of soil temperature and
moisture on decomposition range from simple linear regression
models (Wildung et al. 1975) to mechanistic models having
different functions for the effect of temperature and moisture
on fungi and bacteria (McGill et al., 1981) (Fig. 1). In
most of these models the combined effect of moisture and
temperature on decomposer activity is represented as the
product of the moisture and temperature terms times the basic
decomposition rate, which is a function of substrate quality.
An alternative approach, Liebig's law of the minimum, assumes
that decomposer activity is a function of either soil

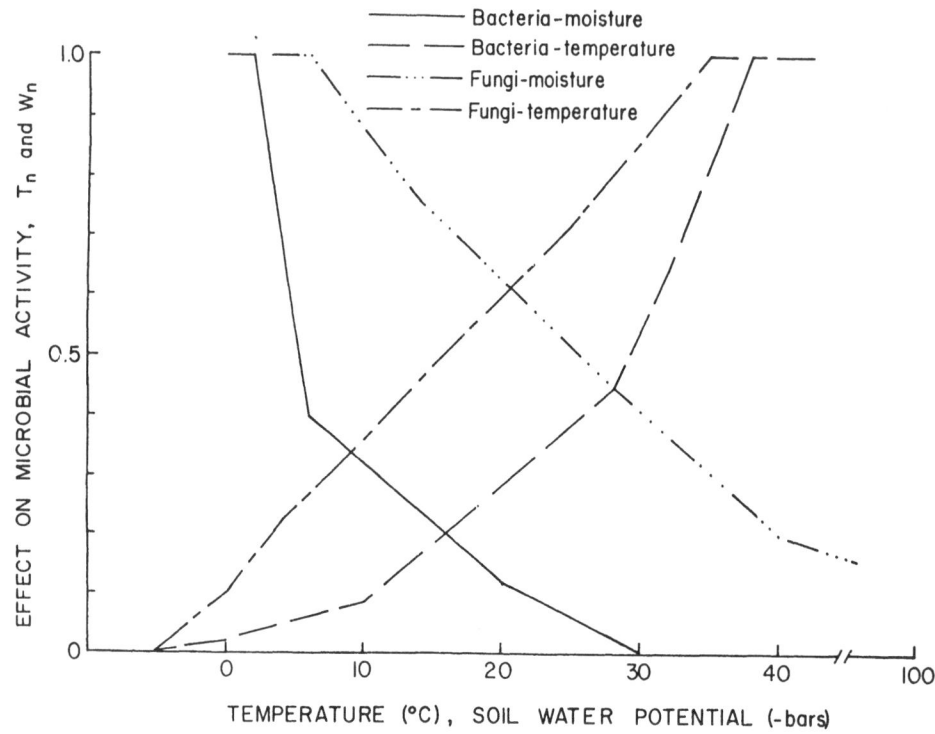

FIGURE 1. The effects of temperature and moisture on microbial activity (from McGill et al. 1981).

temperature or moisture effects times the basic decay rate (Parton et al. 1978). One of the simplest models is a linear regression developed by Wildung et al. (1975), in which soil respiration (CO_2 evolution) from a grassland was poorly correlated to moisture or temperature alone; however, the water-temperature interaction term (product of temperature and moisture) accounted for 70% of the variance. Heal and French (1974) developed a nonlinear regression for decomposer activity in tundra, using the following equation:

$$y = 11.62 + 0.0147 \cdot T \cdot W^2 + 0.000152 \cdot T \cdot W^3 , \qquad (2)$$

where y = percent mass loss (day^{-1}), T = temperature sum (of

°day > 0°), and W = soil moisture (% of dry mass). Nyhan (1976), Hunt (1972, 1977), and Bunnell et al. (1977) developed similar models in which decomposition is a function of the product of nonlinear moisture and temperature terms. Although the equations used by the various authors are quite different, the shapes of curves for temperature and moisture effects are similar. The models developed by Hunt (1972, 1977) and Nyhan (1976) are somewhat more general, since they use soil water tension instead of gravimetric water.

Microflora

Representing biomass. Some ecosystem models do not distinguish microbes from their substrates. This approach may be defended on the bases that microbial biomass is difficult to measure and that much of the data collected on substrate dynamics actually include microbes along with substrate. Also, useful generalizations about the C:N at which immobilization ceases and mineralization begins usually apply to the admixture of substrates and microbes. Paul and van Veen (1978) pointed out that apparent decomposition rates based on ^{14}C remaining in soil are underestimates because the remaining ^{14}C includes both microbes and substrate. Distinguishing microbes and substrates in a model facilitates the use of information on microbial physiology and microbial responses to environmental factors. It also makes possible an explicit treatment of the effect of microbivores. Microbes contain a significant amount of N and P potentially available for plant growth, and it probably is important to model microbial biomass in N- or P-limited systems.

Fungi generally are considered relatively more active than bacteria at cooler temperatures and at decreased moisture levels and more efficient in the decomposition of low-quality cellulosic materials. McGill et al. (1981) incorporated these assumptions into a shortgrass prairie model in order to determine the consequences of differences between the two groups. The model predicted that fungi are predominant, which agrees with our best information (Clark and Paul

1970). However, the McGill et al. model does not distinguish free-living from mycorrhizal fungi. It is not known what fraction of fungal biomass is mycorrhizal or whether mycorrhizal fungi have significant saprophytic ability. Thus, we are faced with the situation that the C source of fungi, the most abundant microbial group in most soils, might be live plants, or it might be litter. Clearly, this makes it difficult to model fungi.

There is no doubt that element ratios of microbes in general are variable; but because of the difficulty estimating biomass, the range of variation for soil microbes is poorly known. Some ecosystem models assume fixed element ratios, and others allow ratios to vary within bounds. A model adjusted to operate with variable element ratios may behave very differently when the range of variation is restricted (McGill et al 1981); but, if model parameters affecting other processes are adjusted along with those controlling element ratios, the effect of restricting variation may be much less important (Hunt et al. 1984). Thus, it is still not clear whether it is critical in ecosystem models to realistically represent variation in microbial element ratios.

The release of organic waste products by bacteria in liquid culture is well established and has been simulated (Hunt et al. 1977). Much less is known about release of organic matter by microbes in soil or those growing on particulate substrates. The rate of release of exoenzymes and their residence times in soils are not known quantitatively. Neither is it known at what rate fungi leave behind empty hyphae. Such processes might be critical to understanding turnover of soil organic matter, but there are no published ecosystem models including them.

Models for microbial growth under defined laboratory conditions (Roels and Kossen 1978) are more detailed than ecosystem models, and they have been shown to have some predictive power. However, there is insufficient information

on microbial biomass or processes to place confidence in an
ecosystem model's predictions about microbial biomass in soil.

   Respiration. Respiration rates of microbes, plants,
and animals commonly are related to two general kinds of
processes: growth and maintenance. The rate of
growth-associated respiration is affected by temperature,
moisture, the supply of substrate, and any other factor
influencing growth rate. Growth respiration can be modeled
by assuming that, for every x units of substrate C consumed,
a maximum $Y_m \cdot x$ is incorporated into new biomass, and the
rest is oxidized to CO_2 to provide energy for anabolism;
thus, Y_m is the maximum possible yield. Yield is less than
maximum because of the costs of homeostasis. Maintenance
respiration of active microbes generally is predicted as a
function only of temperature, but adverse conditions may
lead to physiological changes allowing survival in a
quiescent state. Resting stages have a respiration rate
lower than the maintenance rate of active microbes (Hunt
1977). The supply of substrate in natural systems may be
insufficient to maintain microbes at the level of activity
they are capable of in the laboratory. Thus, microbes in
grasslands must be inactive most of the time (Clark and Paul
1970). One approach to modeling respiration and activity is
to create separate state variables for active and resting
stages (Hunt 1977), which leads to consideration of the
factors controlling the transition between active and
inactive states. Another approach is to combine active and
resting stages into a single state variable but to use a
maintenance respiration rate lower than that observed in
laboratory cultures of active organisms (McGill et al.
1981), which is necessary to prevent excessive catabolism.
This approach finds support in field estimates of
maintenance rates, which are usually low, about 0.002 g
CO_2-C $\cdot$ (g microbial C $\cdot$ h)$^{-1}$. Most estimates of high
maintenance rates (> 0.02) are based on studies in liquid
culture, but Hunt et al. (1984) present evidence that
bacteria cultured in soil have high maintenance rates during

short periods of active growth. If high maintenance rates
are the rule for active soil microbes, there is strong
justification for distinguishing active and inactive microbes
in an ecosystem model.

Death. Death is among the least understood of microbial
processes in soil, and it is understandable that it is not
handled well in models. Most models assume that freezing
and drying kill microbes, and there is considerable
supporting literature. However, there are few good
quantitative data on death of mixed populations in the field
because of the difficulties of measuring biomass and
distinguishing live from dead biomass. McGill et al. (1981)
assumed no death from drying and some death from freezing
but that most death results from predation and other biotic
interactions during periods favorable for growth. Hunt
(1977) assumed that freezing and drying kill a high fraction
of active microbes, but most death in the model actually
resulted from starvation of active microbes failing to make
the transition to an inactive state.

Given the diversity of microbes in the field, it is
likely that different taxa vary considerably in their
susceptibility to freezing, drying, predation, and
starvation and that there are important interactions among
effects.

Uptake. In models that omit a soluble organic pool the
processes of uptake and decomposition are not distinguished.
For substances in solution, the nearly universal procedure
is to use the Michaelis-Menten expression for uptake as a
function of solution concentration:

$$U = \frac{V \cdot S \cdot B}{K + S} \, , \tag{3}$$

where U is rate of uptake per unit volume, S is the solution
concentration of substrate, B is microbial biomass per unit
volume, V is the maximum rate of uptake per unit biomass,
and K is the concentration of substrate allowing uptake at

one-half the maximum rate. This expression originally was derived for enzyme kinetics but can be extended to carrier transport kinetics (Giese 1973). Its use for uptake has been criticized (Bazin et al. 1976) because its derivation entails the steady state assumption, but the equation should still be appropriate in the likely situation when steady state between carrier and substrate is established quickly compared with the rate of change of the levels of carrier and substrate (Parker 1975). However, alternative expressions may fit uptake data better than Eq. (3) (Powell 1967).

For uptake from soil solution, several problems arise with the use of Eq. (3). First, adsorption to the solid phase is poorly known for some important substances, such as ammonium and soluble organic matter, so that "S" may not be known. Second, it is not known to what degree surface phenomena make it unrealistic to represent uptake as occurring from bulk solution. Third, there is indirect evidence that uptake in soil may be limited by diffusion (Hunt et al. 1984), which is not accounted for by Eq. (3). Adequate understanding of the dynamics of uptake in soil may well require knowledge of the spatial distribution both of active microbes and of various kinds of substrates in order to specify the length of diffusion gradients. For our current level of understanding, simple empirical equations for uptake from solution seem most appropriate.

Assuming that microbes directly attached to or within particulates are the primary agent of their decomposition, it is inappropriate to express the availability of particulates in terms of concentration per volume or area of soil. The term S in the Michaelis-Menten expression [Eq. (3)] is for the concentration of soluble substrates, for which the quantity of substrate can change without altering the volume of the system. In contrast, the mass of particulates in an ecosystem is not a measure of substrate concentration, since mass and volume of particulates tend to be proportional. For example, the substrate concentration

for a bacterium decomposing a leaf depends on the physical
relationship between bacterium and leaf, not on the total
mass of leaves. McGill et al. (1981) developed the following
equation for rate of decomposition [U_ℓ (g C $\cdot$ m^{-2} $\cdot$ d^{-1})] of
plant litter:

$$U_\ell = \frac{B_\ell}{1 + K_1 \left(\dfrac{B_\ell}{S_\ell}\right)^{K_2}} \quad , \qquad (4)$$

where B_ℓ is microbial biomass (g C/m^2), S is litter mass (g
C/m^2), and K_1, K_2 are parameters. In Eq. (4), unlike the
Michaelis-Menten equation, doubling the extent of the
substrate-microbe system (doubling both B_ℓ and S_ℓ), has no
effect on the specific rates of decomposition (U_ℓ/S_ℓ) or
microbial growth (U_ℓ/B_ℓ).

The treatment of substrate heterogeneity in a model has
an obvious effect on the representation of uptake. Hunt
(1977) assumed that the use of each of two kinds of substrate
proceeded independently. However, in models for more than
one chemical element, uptake must be regulated to avoid
predicting unreasonable element ratios. With a limited
array of substrates, it is a fairly simple matter to adjust
the uptake rates from various pools in order to keep element
ratios within bounds (McGill et al. 1981). With a wide
array of substrates, selection among acceptable combinations
may be required. The problem of modeling diet selection
(Ellis et al. 1976) more commonly arises in studies of
multicellular organisms than with microbes. However, the
phenomenon of diauxie, whereby microbes exhaust a favorable
C source before starting to use another, is well-established
in solution culture (Pastan and Perlman 1970). A Pseudomonas
species in soil largely exhausted added glucose before
utilizing glycol and probably exhausted a pool of soluble
material released during sterilization before switching to
glucose (Hunt et al. 1984). Of course, bacteria also may

simultaneously take up C, N, and P sources, as indicated by their ability to achieve steady-state growth on simple defined media in continuous culture.

We know of no general theory to predict when use of substrates is simultaneous and when sequential. Perhaps the point is moot in natural systems with mixed flora, where one might expect different taxa to specialize on different substrates. Thus, in ecosystem models, use of all substrates probably should be simultaneous; but it is not a trivial problem to predict the relative rates, especially if the mix of organisms changes over time.

Fauna

Classical population growth models--Lotka-Volterra equations and their derivatives--have played an important role in basic ecological research on birds, mammals, and invertebrates. The classical models are important in theoretical ecology, but they are too simple to predict dynamics of most natural populations (Wiegert 1977).

Detrital food webs are omitted from most total-system models. There are three important reasons for lack of progress. First, diets of many animals are poorly known and change as a function of food abundance, food quality, and animal age. Second, it is very laborious to estimate the populations of many species at once (see Chapter 10), and there is seldom enough data to provide a strong test of model output. The third difficulty is that there are many possible direct and indirect effects of trophic interactions, and little is known about their relative importance. The most obvious direct effect is the removal of microbes from the population. This removal, by itself, will reduce the level of microbes, but the actual rate of element transfer through microbes conceivably could either be increased or decreased depending on whether or not the element is a factor limiting microbial growth. Wastes released by microbivores are available to microbes, and mineralization of a growth-limiting element theoretically may have an important effect on microbial growth rate (Johannes 1965, Hunt et al. 1977). Microbivores release inorganic nutrients available to plants, and primary production may be increased (Ingham et al. 1984), which should in turn affect microbes. Porter (1976) showed that aquatic algae may survive passage through the guts of filter feeders and actually pick up additional nutrients in transit. Perhaps there is a similar phenomenon in soil.

Another effect of microbivores is to speed the distribution of microbes to favorable sites. This mechanism has been proposed to account for faster bacterial growth observed in sterilized soil in the presence of bacteriophagic nematodes (Trofymow and Coleman 1982). Presumably the phenomenon is less important in the field where resting stages of microbes will be well distributed, but we have not seen any research on the question. Comminution of particulates, thereby increasing the surface area of substrates and easing enzymatic attack, has been proposed as an important effect of larger animals. Finally, it is conceivable that microbivores release growth-stimulating substances or somehow "select for faster growing" microbes (Barsdate et al. 1974).

Over the past decade, attention has shifted from macrofauna such as earthworms and insect larvae to micro- and mesofauna, especially protozoans, nematodes, and mites, as the organisms most likely to have appreciable effects on element cycling through their impacts on microbial populations. Research at Colorado State University (T. Trofymow, D. Coleman, E. Ingham, R. Ingham, P. Reid, C. Cambardella, R. Ames, R. Hays, W. Hunt, T. Elliott, J. Frey, C. Morley; 1982, unpublished) has led to Fig. 2 as a working hypothesis about trophic structure in shortgrass prairie. Field experiments with selective biocides (E. R. Ingham, personal communication) will test the proposed structure and whether it is legitimate to disregard the effects of soil insects and other groups left out of the model.

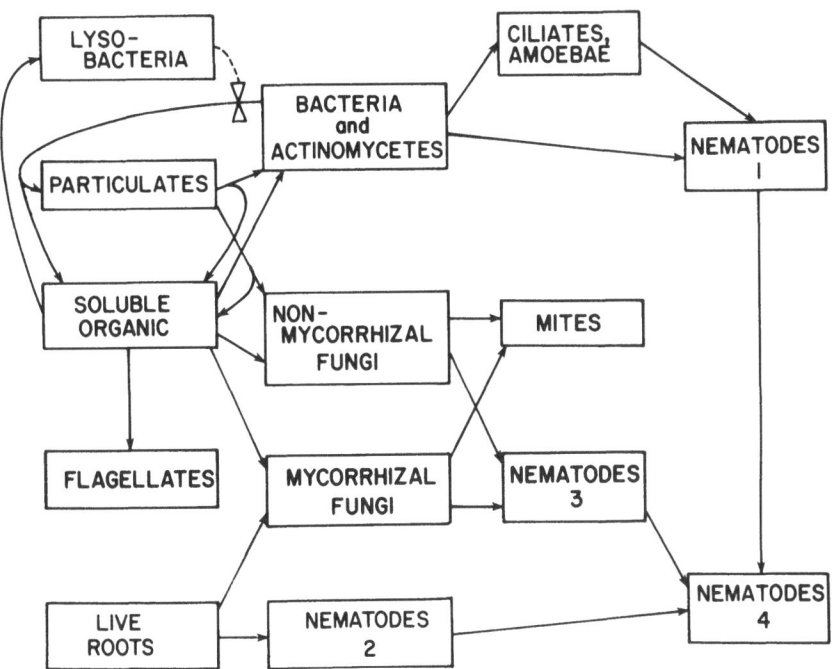

FIGURE 2. Provisional trophic structure for shortgrass
prairie detrital food web. Four groups of nematodes are
distinguished by their diets.

Theoretical work suggests the importance of correctly
representing the general features of the trophic web
(Fretwell 1977). Figure 3 shows a simple hypothetical food
chain. The model (Hunt, unpublished) may be run as a two or
three component chain by setting one or two state variables
to zero, and the behavior of the model is qualitatively
different for different chain lengths. For example, the
effect of enrichment (higher E) on the two-component system
is that bacteria increase, and nutrient level is unaffected.
In the three-component system, enrichment appears always to
increase nutrients and amoebae, with no effect on bacterial
biomass. In the complete four-component system, amoebae are
unaffected by enrichment, while the nutrient level may
actually decrease and bacteria and nematodes increase. The
stability properties of the three-component system have been
studied in detail. Bacteria and amoebae may become extinct

464

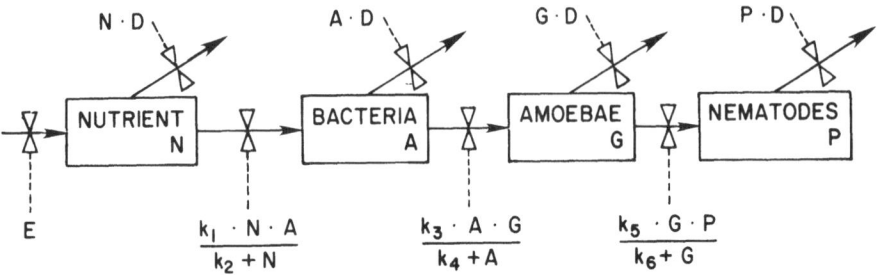

FIGURE 3. Hypothetical food chain. The levels of state variables are denoted N, A, G, and P. Math expressions for the transfers are given along the arrows. Parameters are k_i, i = 1, 2, ... 6. Each state variable is assumed to lose material from the system at specific rate D. The rate of input of nutrient into the system is E.

for certain combinations of parameter values and enrichment. In situations where both bacteria and amoebae persist, they may stay constant or exhibit damped or sustained oscillations. Allowing for the return of nutrients by amoebae does not substantially alter the effects of enrichment on stability of the three-component system. These results illustrate that even simple models may exhibit surprising and complicated behavior. Wiegert (1977) showed that adding a "space control"--negative feedback of consumer density on rate of consumption--to a model very similar to Fig. 3 allows even more complicated model behavior, including alternate stable steady states with all components persisting. Incorporating feeding thresholds, such as a minimum level of bacteria required for amoebae to feed (Hunt et al. 1984) tends to stabilize the model.

It seems obvious from modeling and other theoretical work that food web structure should have important system level consequences, and this conclusion appears to hold in aquatic ecosystems (Lynch 1979, Hurlbert et al. 1972). It has been argued, although not convincingly, that food chain length is an important variable in terrestrial ecosystems (Fretwell 1977), but it is not known whether decomposer food webs in temperate terrestrial ecosystems vary in overall structure. Apparently tillage has direct adverse effects on some fauna, which presumably could alter nutrient cycles.

Perhaps because of the potential complexity of their effects, microbivores have been included in few ecosystem models. Patten (1972) presented a 23-component detrital food web for a grassland model, but the structure of the model was not described in detail and no model output was shown. One of Patten's objectives was to show that we lack the information needed to model such a complicated system. Witkamp and Frank (1970) included millipedes in a model for ^{137}Cs transfers during leaf decomposition, but did not distinguish microbial biomass from substrate, so the effects of microbivory were not explicit. McBrayer et al. (1974) constructed a model of litter decomposition that distinguished fungi, fungivores, detritivores, and predators.

It appears to us that the main values of including the soil fauna in a system model are to investigate the consequences of various hypotheses about their roles and to discover new explanations for patterns already observed in the field. It would be difficult to argue that the fauna must be included in models intended for practical applications.

N-cycle transformations

Nitrification, denitrification, and NH_3 volatilization are important processes frequently included in nitrogen cycling models. See Chapter 6 for a general discussion of these processes. A variety of models has been used for nitrification. Van Veen and Frissel (1981) developed a fairly complex model treating both nitrification steps, using Monod kinetics and considering the effect of temperature and moisture. McGill et al. (1981) represented nitrification as

a one-step process which is a function of soil moisture,
temperature, and NH_4^+ concentrations. A double Michaelis-Menten
relationship was used to represent the effect of NH_4^+
concentration. Tanji et al. (1981) represented nitrification
as a first order process and did not consider the effect of
soil temperature and moisture. Recent evidence shows that
NH_4^+ oxidation results in the formation of N_2O. Ritchie and
Nicholas (1972) suggest that N_2O formation is primarily a
result of reduction of NO_2^- formed from NH_4^+. Mosier et al.
(1982) developed a model for predicting N_2O losses from the
soil as a function of NO_3^-, NH_4^+ and the soil water content.
The model has separate equations for predicting N_2O loss
resulting from nitrification and denitrification.

For denitrification, Bowman and Focht (1979) developed
a model which uses double Monod kinetics for the effects of
NO_3^- or NO_2^- concentrations, and the rate is affected by soluble
organic carbon. McGill et al. (1981) used a single
Michaelis-Menten expression for the effect of NO_3^- concentration,
and the rate was modified by soil moisture (related to O_2
levels) and soil temperature terms. Tanji et al. (1981)
developed a simpler model using first-order kinetics for the
effect of NO_3^- concentration and modifying the rate as a
function of the soil water content. Mosier et al. (1982)
also developed a model predicting N_2O loss from
denitrification as a function of NO_3^- concentration and soil
water content. With all of these models, the effect of O_2
on denitrification is represented in the soil moisture term.
The theory for modeling the diffusion of O_2 in soil is well
understood; however, it is difficult to define the physical
environment (i.e., geometry of soil aggregation and spatial
variability).

Significant losses of nitrogen from soil via NH_3
volatilization are primarily associated with addition of
large amounts of urea or NH_4^+ fertilizer. Volatilization of
NH_3 occurs mainly from surface soils and is controlled by
physicochemical processes. Parton et al. (1981) developed a
mostly mechanistic model for NH_3 volatilization. The model

represents the equilibrium between NH_4^+ in solution and adsorbed to soil colloids, chemical equilibrium between NH_4^+ and NH_3 in solution, and the equilibrium between NH_3 in solution and in the soil atmosphere. Ammonia in the soil atmosphere, calculated by solving two equilibrium equations, increases with increasing NH_4^+ solution concentration, pH, and soil temperature. Diffusion of NH_3 gas among six soil layers and from the top soil layer to the atmosphere decreases with increasing soil water content. Van Veen and Frissel (1981) developed a model with similar structure but with only one soil layer.

CASE STUDIES

The sections above covered special topics and reviewed a variety of modeling approaches that have been or could be taken. This and the next two sections take a different approach, describing particular models in detail. This will demonstrate the wide range of levels of resolution possible and illustrate the roles modeling can play in ecological research.

NAB model

In this section we give an historical account of the NAB model development to show how modeling can influence the design of experiments. One of the outcomes of the International Biological Program (IBP) was a greater appreciation of the importance of soil processes (Caldwell 1979), especially decomposition, mineralization, and microbial dynamics. Post-IBP projects were initiated at Colorado State University to carry out basic research on these processes under controlled laboratory conditions. Simulation models were constructed in order to better understand the data, to help direct the research, and to facilitate exchange of ideas among members of a multidisciplinary team.

Our first model, for bacterial growth in liquid culture (Hunt et al. 1977), was impractical for soil because it required too many assumptions about microbial physiology and too much data on chemical composition of biomass. Thus, we took a simpler approach in the NAB model (Cole et al. 1984) (Fig. 4), representing each species by a single state variable and including two kinds of C source--added glucose and the aggregate of native soluble organic matter and the large amount ($\sim$1500 µg C/g soil) of glycol from the propylene oxide sterilant used. Bacteria and amoebae were assumed to have refuge levels below which they were safe from predators (Elliott et al. 1980). Uptake rates from substrate pools were according to Michaelis-Menten expressions. A fraction of eaten bacteria and amoebae was digestible, the rest becoming particulates, which were assumed to decompose at a constant specific rate. Respiration was separated into growth and maintenance components.

The NAB experiment (Coleman et al. 1978a, 1978b; Anderson et al. 1978; Cole et al. 1978; Woods et al. 1982) included glucose amended and unamended treatments combined

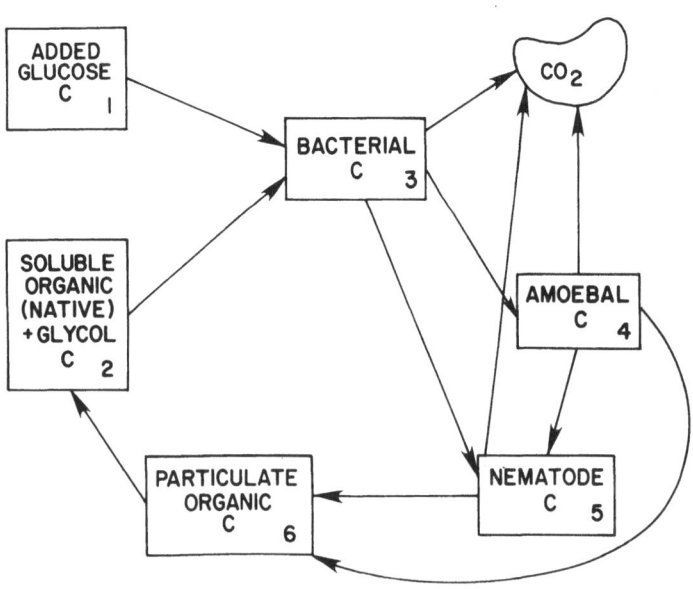

FIGURE 4. Carbon flows and compartments considered in NAB model (from Coleman et al. 1978b).

with four biotic treatments: bacteria alone, bacteria with
amoebae, bacteria with nematodes, and all three together.
The modeling strategy was to adjust bacterial growth
parameters using the data for bacteria alone and to keep the
same parameter values for growth in the presence of a
predator. We soon met an impasse. It was not possible to
adjust the model to fit the final (561-h) observed cumulative
CO_2 evolution by bacteria alone without grossly
underestimating CO_2 at 64 h. It was not obvious why the
model failed. Therefore, a second experiment was designed
to examine in greater detail bacterial growth and respiration
(BGR experiment, Elliott et al. 1983) during the first three
days of incubation, when growth was fastest and conditions
were rapidly changing. A variety of simple models (BGR
models, similar to NAB) were tested for their ability to
mimic the BGR data (Hunt et al. 1984), and it was concluded
that rapid growth in the first 20 to 30 h was almost solely
at the expense of a pool of nitrogenous organic material
released from biomass by sterilization. Subsequent growth
utilized added glucose, which was exhausted before glycol
was utilized. The only important difference between the NAB
and BGR models is that the latter use separate state variables
for glycol and for native soluble organic N, which allows
the BGR models to correctly predict early CO_2 evolution in
the NAB experiment (Fig. 5).

It may appear obvious that material released by
sterilization must be distinguished from glycol, if one
considers methods designed to estimate soil biomass by CO_2
evolution after fumigation (Jenkinson and Powlson 1976).
Nevertheless, we probably would never have progressed to our
present understanding of the use of various substrates in
soil if we had not built the NAB model, questioned why the
model failed to mimic the NAB CO_2 data, and carried out the
BGR experiment. Furthermore, a traditional statistical
analysis of the BGR data (Elliott et al. 1983) did not lead
to any conclusions about the importance of separating

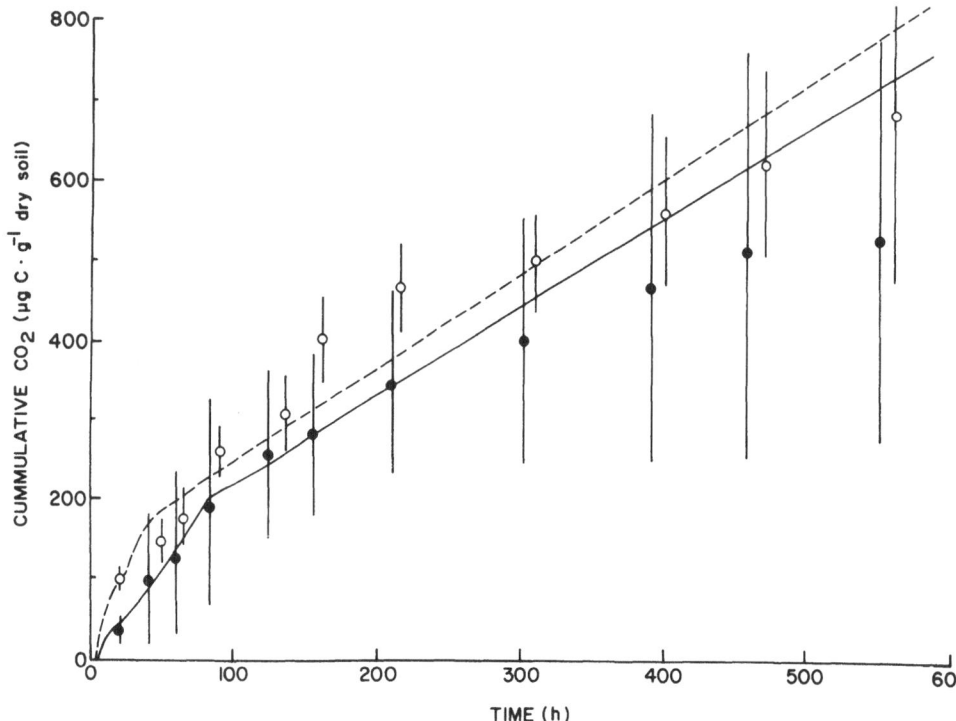

FIGURE 5.   Predicted (lines) and observed (mean ±95%
confidence interval) CO_2 evolution by bacteria inoculated
into sterilized soil (NAB data).   The O and dashed line
represent the C-amended treatment, while ● and solid line
are for unamended.   The model was identical to the BGR
fixed-ratio model of Hunt et al. (1984) except that the
yield (0.57) and maintenance (0.0031 h-¹) parameters were
lower, and glycol uptake proceeded at a maximal specific
rate of 0.045 h-¹ with half-saturation constant of 2 μg
glycol C/g soil.

substrate pools.   Thus, the BGR models aided substantially
in data interpretation.

The fit of the NAB model to final biomass data is
presented in Table 1.   The model is within 95% confidence
intervals for the data in 11 of 16 cases.   We consider this
an acceptable fit, given the difficulty of modeling biotic
interactions, but the NAB model still cannot be accepted as
an adequate theoretical statement about interactions among
substrates, bacteria, and predators.

Table 1. Comparison of measured and simulated values* of bacterial, amoebal and nematode biomass after 24-day incubation of soil (NAB experiment).

Treatment[+]		Biomass (µg C/g soil)		
		B	A	N
Glucose amended	B	1131 ±708 (847)		
	AB	570 ±509 (56)	223 (281)	
	NB	702 ±310 (610)		18 (24)
	NAB	283 ± 47 (72)	40 (56)	26 (30)
Unamended	B	669 ±439 (492)		
	AB	186 ±125 (55)	178 (156)	
	NB	281 ±256 (327)		12 (18)
	NAB	218 ±256 (73)	42 (55)	16 (18)

*Data given with 95% confidence intervals, which were available only for bacteria. Simulated value in parentheses.

[+]B = bacteria alone, AB = bacteria + amoebae, NB = bacteria + nematodes, NAB = all three together.

Our ultimate objective is to evaluate the variety of possible predator-prey interactions by determining which mechanisms are sufficient to account for our results. Modeling is an integral part of this research.

PHOENIX model

This model (McGill et al. 1981) is nearly the opposite of the NAB model along the spectrum of possible levels of complexity for ecological models. PHOENIX simulates C and N transfers in a natural sward of blue grama grass. The model was constructed by scientists sharing a general interest in modeling and N cycling but not working together on a common research project. There were no particular observations or ecosystem phenomena that the model was especially constructed to account for. Rather, the stated objectives for modeling were "to explore . . . the relationships between plant processes and microbial processes and their effect on plant production, microbial secondary production, and nitrogen cycling" (McGill et al. 1981).

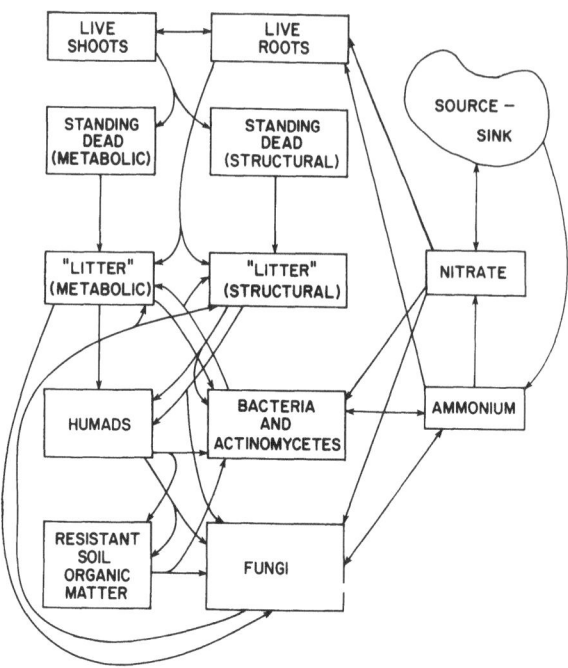

FIGURE 6.   PHOENIX N-flow diagram (McGill et al. 1981).

Figure 6 shows model structure for N. The C-flow model
has all the same state variables, except for inorganic N,
and C flows are similar to N flows. PHOENIX has a number of
distinctive features. Dying shoots, roots, and microbes are
divided between two substrate pools: the structural component,
consisting of cell walls, and the metabolic component,
including nucleus and vacuole, if present, and cytoplasm.
Ammonium, humads (see below), and the metabolic component
all are assumed to maintain steady-state distributions
between adsorbed and dissolved phases, predicted either from
a Langmuir or Freundlich isotherm. Uptake from the solution
phase is according to modified Michaelis-Menten equations:

$$U = \frac{V \cdot S \cdot B}{K + S} \cdot E_T \cdot E_W \quad , \tag{5}$$

where U is rate of uptake [μg substrate (g soil · day)$^{-1}$], S is solution concentration (μg substrate/ml), V is the maximal rate (day^{-1}), and K is the half-saturation constant (μg substrate/ml). B is biomass of either bacteria or fungi (μg C/g soil). Two nondimensional factors, E_T and E_W, each ranging from zero to one, reduce uptake according to temperature and soil water potential (Fig. 1). Uptake of NH_4^+ or NO_3^- is further modified by microbial demand for N, reflected in microbial C:N ratio. Uptake of inorganic N proceeds at the maximal rate for C:N ratios above an upper threshold, at zero for ratios below a lower threshold, and at intermediate rates for intermediate ratios. This mechanism ties immobilization directly to microbial biomass and indirectly to the relative supplies of organic C and N.

The decomposition rate of the structural component is predicted according to Eq. 4, further modified by temperature and water factors, as in Eq. (5), and by the effect of microbial C:N ratio. Thus, decomposition of N-poor structural material may be speeded by suppling either inorganic or soluble organic N. Fungi and bacteria differ in C:N ratios and in their responses to temperature, water, substrate quality, and substrate quantity.

The term humad, a contraction of the words humic and adsorbed, "designates materials . . . believed to be of a diverse, heterogeneous chemical nature which are stabilized by humification and adsorption" (McGill et al. 1981). In the model, humads are formed by the adsorption of dissolved metabolic material onto preexisting humads, an entirely abiotic process, and by the release of a small fraction of decomposing structural material representing phenolics. Microbes decompose humads, but a residual portion is retained as organic matter with a long turnover time.

Nitrifer populations are modeled using previously published equations developed from laboratory experiments; the effect of a hypothetical nitrification inhibitor released by roots is also included. PHOENIX also simulates microbial

death, microbial respiration, denitrification, leaching, and
atmospheric N inputs. Plant growth is treated in less detail
than are decomposition, N immobilization, and mineralization.
Processes modeled include photosynthesis; uptake of inorganic
N; translocation of C and N between roots and shoots; shoot
death from aging, freezing, and dry soil; transfer of
standing dead shoots to litter; root exudation and sloughing;
and root death.

Judging from our experience with other comparable
ecosystem-level models; and considering that relatively little
effort was put into iteratively adjusting model structure
and parameters, we feel that PHOENIX generated surprisingly
realistic output. Shoot production, CO_2 evolution from soil,
shoot:root ratio, inorganic N, and microbial turnover all
were close to independently derived estimates. Perhaps more
impressive was the model's ability to predict qualitatively
the response of the grassland to fertilization, soil
fumigation, organic residue amendment, and cultivation even
though the model response to these perturbations was never
explicitly considered during model development.

At the time PHOENIX was developed, there were no reliable
data for soil inorganic N levels at our research site.
Figure 7 compares recent information (Schimel 1982) to model
output. In the original PHOENIX model nitrification was
largely inhibited, and NO_3^- levels were grossly underestimated.
The predictions in Fig. 7 were generated by a model identical
to PHOENIX, except that (1) Parton's (1978) heat and water
flow models were used to allow dynamic interaction between
soil water and plant growth, (2) the maximal rate of net
photosynthesis was reduced from 0.4 to 0.2 day^{-1}, and (3)
changes were made to several parameters which affect
nitrification, but have little other effect on the model
output. The model slightly underestimated the general level
of inorganic N (Fig. 7), but correctly predicted higher levels
in spring than in fall. However, the model failed to predict
the high level observed in early September 1980 and the abrupt
increase during December 1980. The model also predicted that

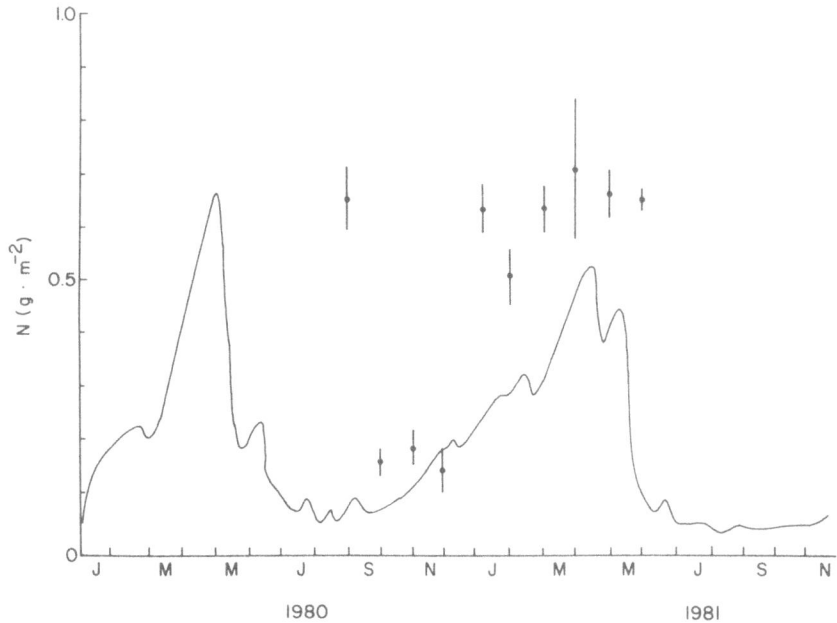

FIGURE 7. Predicted (line) and observed (points) inorganic N (NH_4^+ + NO_3^- + NO_2^-) in the upper 10 cm on an upland site at the Pawnee grassland. Data (Schimel 1982) are shown with 95% confidence intervals.

NH_4^+/NO_3^- ratios varied from 0.1 to 2.0, where the data showed much less variation (0.38 to 0.80). Furthermore, the model predicted seasonal trends in NH_4^+/NO_3^- not found in the data.

Thus, several versions of PHOENIX making different assumptions about nitrification have failed to predict observed NH_4^+/NO_3^- ratios. Of course, this ratio also will be affected by rates of ammonification and by the relative rates of uptake of NH_4^+ and NO_3^-. Thus, it will be no easy task to determine the reason for model failure.

The metabolic component of litter was included in PHOENIX to represent the heterogeneous nature of plant matter. The decision to include the metabolic component as a state variable led to its appearance in several key processes.

476

Dying microbes were divided between structural and metabolic.
A role of soluble organic matter in humification was
postulated. Fig. 8 shows accumulated C flows involving the
metabolic pool in a 1-year control run of the model. The
metabolic pool accounted for almost half of the C supply of
bacteria but was relatively less important as a C source for
fungi. The metabolic pool was more important as a source
of N (not shown) than C, at least for fungi. Surprisingly,
microbes supplied more C to the metabolic pool than did
plants. This is possible because microbes may grow with a
high yield, and organic C may be cycled several times before
being lost as CO_2. Although it constituted no more than
0.3% of total substrate, the metabolic pool was very dynamic,

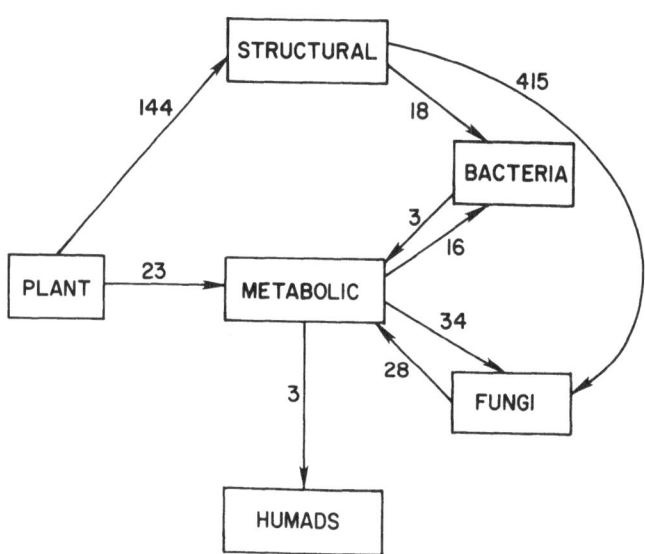

FIGURE 8. Transfers ($g\ C\ m^{-2}\ yr^{-1}$) involving metabolic pool
in PHOENIX (McGill et al. 1981). Not all transfers are
indicated for the other state variables.

ranging from about 0.2 to 1.7 μg C/g dry soil in the top 30 cm. It had an average turnover time of less than five days and probably much less when conditions were favorable for decomposition.

Thus a small pool of soluble organic material with a high turnover rate plays a key role in model dynamics. One may object that the model was bound to give these results, because the very structure of the model assumes the existence and importance of such a pool, and that the model is not telling us anything novel. However, the model was never adjusted to generate output corresponding to preconceptions on the relative contributions to the metabolic pool, its use by bacteria and fungi, or its level and dynamics in soil. We cannot accept model predictions as statements about how a metabolic pool in the real world behaves, but the modeling exercise makes it seem plausible that a small dynamic soluble organic matter pool is important.

Fungi account for 93% of C uptake (Fig. 8), which results from the model's assumption that fungi are more efficient than bacteria at breaking down the structural component, the most plentiful substrate, and fungal tolerance of low temperatures and dry soil. Bacteria assimilate the metabolic component faster per unit biomass than fungi, which should give bacteria an important competitive advantage under N limitation. Additional model runs are being made to investigate the importance of differences between bacteria and fungi.

The PHOENIX model has been part of several research efforts since 1976. The decomposition processes were part of a larger model used to evaluate strip-mine reclamation practices (Ellis and Parton 1977) and was used as an aid to research on the effects of atmospheric SO_2 on Montana grasslands (Coughenour et al. 1980), and agroecosystem research in Sweden (K. Paustian, personal communication). The plant model was revised to respond to belowground and aboveground herbivory (Bachelet et al. 1983), and the new model was used to compare the effects of herbivory by nematodes, grasshoppers, and cattle (Bachelet 1983). A

model for the detrital food web (Fig. 2) is being developed
for inclusion in PHOENIX, and the resulting model will be
used to explore the effects of fauna.

SOM model

Parton et al. (1983) developed a comprehensive model to
simulate the effect of different cultivation practices on
soil organic matter levels (SOM, N and C) and crop yields.
The model is designed to simulate changes over long periods
(100-1000 y). It has a fairly simple structure compared with
PHOENIX, which would be too expensive to run for such long
periods. The SOM model (Fig. 9) is based on a conceptual
model developed by Anderson (1979) in which dead plant and

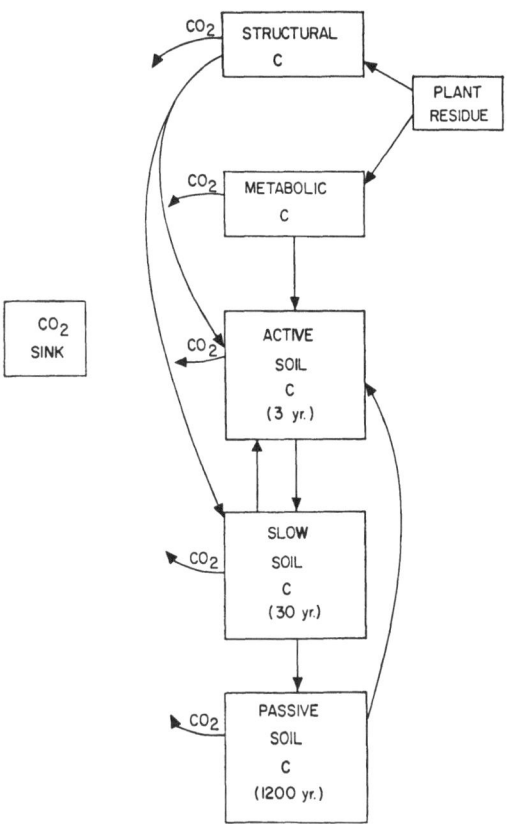

FIGURE 9. Carbon flow diagram for SOM model (from Parton et al.
1983a).

microbial residues decompose into relatively simple substrates
for humus formation. Polymerization of these compounds results
in high molecular weight molecules with many side chains.
These side chains participate in the formation of organomineral
aggregates because of chemical and physical bonding with clay
minerals and amorphous mineral colloids. This physically
stabilized soil organic matter decomposes at an expectedly slow
rate and yields lower-molecular-weight products, which are
generally more aromatic.

From this conceptual model, a simulation model has been
constructed containing three soil organic fractions: (1) an
active fraction of soil C and N consisting of live and dead
microbial structural and metabolic substances with a 3-year
turnover time, (2) a pool of C and N that is physically
protected and/or chemically resistant to decomposition,
with an intermediate turnover time (30 y), and (3) a fraction
that is chemically recalcitrant and that may also be
physically protected, with the longest turnover time (1200 y).
The model assures that the C:N ratio remains fixed at 80, 5,
8, 10, and 11, respectively, for the structural, metabolic,
active, slow, and passive soil organic fractions. As in
PHOENIX, plant residue was divided into structural and
metabolic components as a function of C:N ratio of the plant
residue.

Decomposition of each state variable was calculated
using the following equation:

$$\frac{dC_i}{dt} = K_i \cdot M \cdot T \cdot C_i , \qquad (6)$$

where C_i = the carbon in the state variable (i = 1, 2, 3, 4,
5, for structural and metabolic litter and active, slow, and
passive soil fractions); K_i = the decomposition rate parameter
(y^{-1}) for the ith state variable (K_i = 4, 8, 2.0, 0.20,
0.004); M = the effect of soil moisture on decomposition;
and T = the effect of soil temperature on decomposition.
The soil moisture term was a function of the ratio of annual

precipitation to the potential evapotranspiration rate, and the temperature term a function of average 10-cm soil temperature, calculated from aboveground plant biomass and the average growing-season air temperature. One of the major effects of cultivation is to modify soil moisture and temperature. The model assumed that 60% of decomposing structural material and soil organic fractions was lost as microbial respiration and that the remaining C, representing net microbial production (NMP), was transferred to the active and slow pools. Decomposition of metabolic material results in the loss of 80% of the C as microbial respiration and 20% to NMP. For structural residue, 90% of NMP was transferred to active soil organic matter, and 10% was transferred to slow soil organic matter. Metabolic C decomposes to active soil organic matter; active soil organic matter decomposes to slow organic matter; and slow organic matter decomposes to active (92%) and passive (8%) organic matter. Passive organic matter decomposes to active organic matter.

Since N is bonded mostly to C, N flows were assumed to be stochiometrically related to C flows (Fig. 10). The N-flow rate was equal to the product of the carbon-flow rate and the fixed N:C ratio of state variable receiving the C. The N attached to the carbon lost in microbial respiration (60 or 80% of the C) was assumed to be mineralized. Analysis of the flow diagram (Fig. 10) and the assumed C:N ratios for state variables shows that decomposition of metabolic residue and active, slow, and passive soil organic matter results in a net mineralization of N, while decomposition of structural residue results in immobilization of N. Net mineralization of N equals the difference between the mineralization and immobilization flows. Plant production was calculated as a function of the mineral-N pool.

The SOM model was developed from data in the United States and tested with data from a long-term (25 y) field experiment in Sweden. The five treatments modeled include a fallow treatment (no cropping), control (cropping with no addition of aboveground C or N), nitrogen treatment (cropping

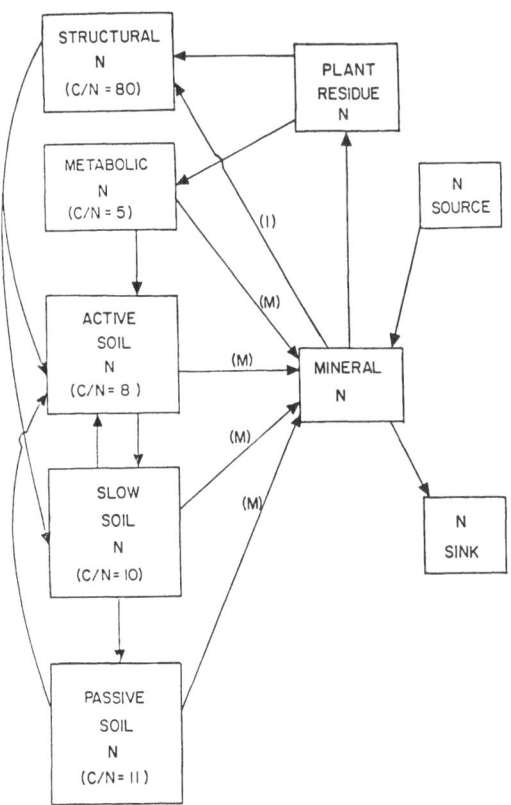

FIGURE 10. Nitrogen flow diagram for SOM model (from Parton et al. 1983a).

with N fertilization), straw treatment (cropping with incorporation of straw into the soil), and an N-plus-straw treatment (annual cropping with N fertilizer and straw additions). The four cropping treatments were planted annually, and 97% of the aboveground plant material was removed from the sites. One hundred eighty-three g C m^{-2} were added annually for the straw-incorporation treatment. C:N ratio of the straw was 60. The N fertilization treatments added 8 g N m^{-2} to the soil each year.

A comparison of the observed and simulated total soil organic matter (0-20 cm) for the different treatments (Fig. 11) shows that the straw-plus-nitrogen treatment has the highest soil organic matter level, followed by the straw, nitrogen, control, and fallow treatments. The results suggest

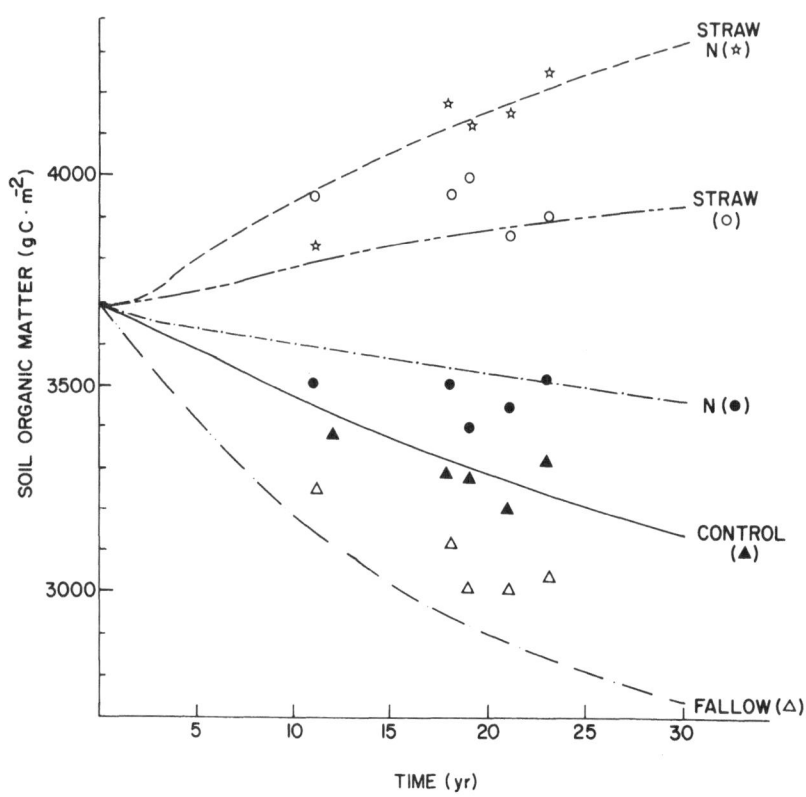

FIGURE 11. Comparison of observed and simulated total soil organic matter for the control, nitrogen, fallow, straw, and straw-plus-nitrogen treatments (from Parton et al. 1983a).

that the model correctly represented the major effects of the treatments on soil organic matter (SOM) and that the difference in SOM levels among treatments was primarily a result of differential C inputs and decomposition rates. The model also correctly simulated the effect of the different treatments on crop yields. The Swedish data were not used to develop the model; however, in preliminary runs the model overestimated the immobilization of N in the straw-addition treatments. The preliminary results plus recent data from Kassim et al. (1981) prompted us to increase the fraction of structural residue lost to CO_2 from 60% to 80%. This change

corrected the problem and shows how testing a model over a
variety of different systems and treatments leads to
improvements in the model and our understanding of the system.

We currently are developing a more mechanistic version
of the SOM model using monthly time steps. Changes are being
made to better represent the different effects of no-till,
minimal-till, and conventional tillage practices on soil
organic matter. Both versions of the model are being used
to help guide agroecosystem research at Colorado State
University and elsewhere. Many of our current experiments
were designed to answer questions raised as a result of
modeling.

CONCLUSIONS

The status of ecological modeling

The better understood a system, the more satisfactory
are simulation models of the system. Thus, models for
laboratory experiments often are more successful than those
for ecosystems. The most successful models of a field system
are probably those for the aboveground parts of fertilized
and irrigated crops. Nutrient and water limitations and the
presence of several plant species introduce complexities that
have not been completely solved for natural systems and have
not been simulated well.

The predictive power of ecological models appears to be
limited by an inadequate understanding of important processes
and perhaps by the failure of modelers to use all available
information. In addition to the roles of mycorrhizae and
microbivores, discussed in previous sections, the following
areas appear to us to offer the greatest opportunities for
improving ecosystem and agroecosystem models.

Soil properties. Dr. Francis E. Clark, at the close of
a nitrogen-cycling workshop in Fort Collins in April 1982,
asked how our simulation models represent the effects of
physical and chemical soil properties on microbial processes.
The answer, unfortunately, appears to be that soil properties
may determine the values needed for many parameters affecting

growth rate, survival, and soil organic matter transformations.
The need for numerous site-specific parameters in models
applied at more than one locale probably is partly a reflection
of the effects of soil properties. The amount and mineralogy
of clay affect the adsorption and availability of soluble
organic matter and may directly affect survival of bacteria,
apart from the profound effect on soil water relations
(Hattori and Hattori 1976). Texture, porosity, and aggregate
size may affect predator-prey relationships by restricting
predator movement (Elliott et al. 1980) and affect oxygen
diffusion and the development of anaerobic microsites.
Cation-exchange capacity and pH will affect the relative
availability of the wide array of organic and inorganic
nutrients in the soil solution. What is needed is a
mechanistic understanding of how soil properties affect
microbial processes, so that changes in soil physicochemical
properties can be automatically accounted for in a few key
parameters.

Spatial heterogeneity. Soil properties generally are
studied using bulk samples, for obvious practical reasons;
yet pH, water, O_2, and nutrient content might vary over a
distance of millimeters or centimeters, a scale large
compared with most soil organisms. In addition to differences
among soil horizons, heterogeneity would be associated with
soil aggregates, dead plant fragments, and rhizosphere zones,
especially in heavy textured soil, where roots may repeatedly
grow down the same channels. Thus, it might be misleading
to apply laboratory results using sieved and homogenized soil
to a field situation. Research on soil aggregate structure
(Tisdall and Oades 1982) and the differences between
rhizosphere and nonrhizosphere soil (Chapter 5, Smucker and
Safir) point to better appreciation of small scale spatial
heterogeneity.

At a larger spatial scale, different decomposer
communities may be associated with different plant species,
which would provide organic matter of different kinds and
amounts and at different seasons. The degree of patchiness

in plant distribution should determine the intensity of competition among plant species for light, water, and nutrients. Botkin et al. (1972) implicitly allowed for such spatial effects in a forest succession model by representing individual 10- × 10-m plots with competition only within plots. Nutrient-cycling models of desert ecosystems have distinguished shrub and nonshrub zones, and some crop models separate row and between-row areas (Morgan et al. 1983), but most nutrient-cycling models have assumed uniformly mixed plant communities.

Representing spatial heterogeneity increases computer run time, but the continued trend towards faster computers should allow inclusion of spatial effects in more ecological models. Until more data are collected on spatial heterogeneity, the role of modeling will be mainly to raise the possibility of significant effects and to influence the design of experiments.

Disease. Simulation models have long been important in epidemiology, but the effect of microbial diseases of plants, animals, and microbes in agroecosystems, and especially in natural ecosystems, has received scant attention from ecologists. Bacteriophage may be abundant in soil and potentially could affect bacterial numbers (Tan and Reanney 1976), but phage is seldom, if ever, considered in explaining microbial biomass dynamics. The die-off of a population of microbes, with subsequent regrowth on dead microbes, could lead to N mineralization under conditions otherwise favoring immobilization.

Levels of resolution

The models described above under Case Studies differ widely in scope. The BGR models predict growth of a single bacterial species under defined conditions in the laboratory. The NAB model adds the dynamics of amoebae and nematodes. NAB and BGR both are for time periods of a few days or weeks and do not include processes important in senescent cultures. PHOENIX emphasizes plant and microbial processes in native prairie, taking into account seasonal but not diurnal environmental changes. The model is comprehensive in that it includes most nutrient-cycling processes, but it treats only one plant speices and disregards all animals. PHOENIX includes soil organic matter pools with long turnover times and, in principle, could simulate changes occurring over hundreds of years, but such long runs would be prohibitively expensive. The SOM model treats some of the same processes as PHOENIX but in much less detail, so that changes occurring over hundreds of years can be simulated inexpensively. The version of SOM described above runs with a 1-year time step and makes no predictions about seasonal dynamics.

The NAB, PHOENIX, and SOM models have very different objectives and levels of resolution, and each has played a role in research on ecosystems and agroecosystems. However, the kinds of benefits derived are somewhat different for different levels of resolution. As discussed above under Benefits of Modeling, the more complicated the model, the more difficult it becomes to determine the reason for model failure. Thus, the direct benefits of modeling (hypothesis testing) are realized only with simple models. This suggests that one should construct simple models and add complexity only as needed to ensure that the model meets its objectives. On the other hand, indirect benefits (disciplined view of system, etc.) apply at all levels of resolution. Indeed, it may be that indirect benefits are maximized by considering as comprehensive a model as possible, so that researchers critically examine all aspects of the system under study. Models of the same system at different levels of resolution have different uses and complement each other, so that modeling probably should proceed simultaneously at several levels of resolution.

Applications of ecological models

Our experience is largely with models intended to aid basic research. The "application" of such models occurs mostly during their development, although a completed model may be subjected to various perturbations and parameter changes to help generate insights about the behavior of the ecosystem. These insights actually apply only to the model, and must be tested in the real ecosystem to establish their validity. In our opinion, the fact that a simulation model behaves in a certain way can never be taken as evidence that the ecosystem behaves similarly. Interpretation of model behavior sometimes is difficult and always requires an intimate knowledge of model structure. It may be necessary to make special runs to determine which aspects of model structure are responsible for particular aspects of model behavior.

It commonly happens that ecological models are tuned (adjusted by iterative changes in structure and parameter values) to fit one set of data closely but then fail under slightly different conditons, such as a different weather pattern. This phenomenon occurs because there is often more than one way to alter a model to make it fit a particular observed pattern, and many of the possibilities are unrealistic in that they do not strengthen the analogy between model and ecosystem. That is, models may be right for the wrong reasons, and this may not be discovered until the model is tested under new conditions.

It is appropriate that ecological models built to aid basic research often include a few poorly understood processes or unestablished mechanisms. The model provides a means to evaluate alternate hypothetical formulations for such mechanisms.

Research models can be used to investigate management problems, but it must be remembered that research models may have some theoretical aspects included for heuristic value and that they not uncommonly fail to predict system behavior adequately under new conditions. Results pertaining to management should be treated the same way as other results from research models--i.e., as predictions that need to be tested experimentally.

Models constructed solely for practical applications ultimately are judged by their success at predicting ecosystem dynamics, without regard for how well their structure parallels the real system. Thus, validation criteria should be different for management models and research models (Welch et al. 1981); the former emphasizes only output variables of direct economic importance. However, it may be argued that mechanistic models are intrinsically better at prediction than black-box models, especially for conditions beyond the data base used for model development. Consequently, some management models (Morgan et al. 1982) are as mechanistic as research models. However, the distinction between research models and management models should be preserved, since an apparently realistic structure is no guarantee that a model is predictive. Management models, no matter how plausible their structure, should be tested thoroughly across a wide range of conditions before being recommended to users (Welch et al. 1981).

Many modelers worry that users of either management or research models may uncritically accept inadequately tested models. The danger of misuse of ecological models will be minimized if modelers and users alike will cultivate the critical attitude that good scientists take towards theory in a science developing so rapidly as ecology.

REFERENCES

1. Anderson, D. W. 1979. Processes of humus formation and transformation in soils of the Canadian Great Plains. J. Soil. Sci. 30:77-84.
2. Anderson, R. V., E. T. Elliott, J. F. McClellan, D. C. Coleman, C. V. Cole and H. W. Hunt. 1978. Trophic interactions in soils as they affect energy and nutrient dynamics. III. Biotic interactions of bacteria, amoebae, and nematodes. Microb. Ecol. 4:361-371.
3. Ares, J. 1976. Dynamics of the root system of blue grama. J. Range Manage. 29:208-213.
4. Bachelet, D. 1983. Simulation of carbon and nitrogen distribution and utilization in blue grama to represent various kinds of herbivory and their effect on plant processes. Ph.D. dissertation. Colorado State University, Fort Collins.
5. Bachelet, D., H. W. Hunt, J. K. Detling, and D. W. Hilbert. 1983. A simulation model of blue grama biomass dynamics, with special attention to translocation mechanisms. Pages 457-466 in W. K. Lauenroth, G. V. Skogerboe, and M. Flug, editors. Analysis of ecological systems: State-of-the-art in ecological modelling. Elsevier Scientific Publ. Co., Amsterdam, The Netherlands.

6. Barsdate, R. J., R. T. Prentki, and T. Fenchel. 1974. Phosphorus cycle of model ecosystems: Significance for decomposer food chains and effect of bacterial grazers. Oikos 25:239-251.

7. Bazin, M. J., P. T. Saunders, and J. I. Prosser. 1976. Models of microbial interactions in the soil. C.R.C. Critical Reviews in Microbiology 4:463-498.

8. Bertalanffy, L. von. 1968. General System Theory. George Braziller, New York.

9. Botkin, D. B., J. F. Janak, and J. R. Wallis. 1972. Some ecological consequences of a computer model of forest growth. J. Ecol. 60:849-872.

10. Bowman, R. A., and D. D. Focht. 1974. The influence of glucose and nitrate concentrations upon denitrification rates in sandy soils. Soil Biol. Biochem. 6:297-301.

11. Bunnell, F. L., D. E. N. Tait, P. W. Flanagan, and K. Van Cleve. 1977. Microbial respiration and substrate weight loss. Soil Biol. Biochem. 9:33-40.

12. Caldwell, M. M. 1979. Root structure: The considerable cost of belowground function. Pages 408-432 in O. T. Solbrig, S. Jain, G. B. Johnson and P. H. Raven, editors. Topics in Plant Population Biology. Columbia Univ. Press, New York.

13. Caldwell, M. M., and L. B. Camp. 1974. Belowground productivity of two cool desert communities. Oecologia (Berl.) 17:123-130.

14. Campbell, C. A., E. A. Paul, and W. B. McGill. 1976. Effect of cultivation and cropping on the amounts and forms of soil N. Pages 7-101 in Proc. of Western Canada Nitrogen Symp., Alberta Soil Sci. Workshop, Calgary, Alberta.

15. Clark, F. E., and E. A. Paul. 1970. The microflora of grasslands. Adv. Agron. 22:375-435.

16. Coughenour, M. B., W. J. Parton, W. K. Lauenroth, J. L. Dodd, and R. G. Woodmansee. 1980. Simulation of a grassland sulfur-cycle. Ecol. Model. 9:179-213.

17. Cole, C. V., E. T. Elliott, H. W. Hunt, and D. C. Coleman. 1978. Trophic interactions in soils as the affect energy and nutrient dynamics. V. Phosphorus transformations. Microb. Ecol. 4:381-387.

18. Cole, C. V., H. W. Hunt, D. C. Coleman, R. V. Anderson, and L. E. Woods. 1984. Trophic interactions in soils as they affect energy and nutrient dynamics. VII. Simulating transfers of carbon, nitrogen, and phosphorus among bacteria, amoebae, and nematodes in soil microcosms. (In prep.).

19. Coleman, D. C., R. V. Anderson, C. V. Cole, E. T. Elliott, L. Woods, and M. K. Campion. 1978a. Trophic interactions in soils as they affect energy and nutrient dynamics. IV. Flows of metabolic and biomass carbon. Microb. Ecol. 4:373-380.

20. Coleman, D. C., C. V. Cole, H. W. Hunt, and D. A. Klein. 1978b. Trophic interactions in soils as they affect energy and nutrient dynamics. I. Introduction. Microb. Ecol. 4:345-349.

21. Dutt, G. R., M. J. Shaffer, and W. J. Moore. 1972. Computer simulation model of dynamic bio-physicochemical processes in soils. Tech. Bull. 196. Dept. Soils, Water and Engineer, Agric. Exp. Sta., Univ. Arizona, Tucson.

22. Elliott, E. T., R. V. Anderson, D. C. Coleman, and C. V. Cole. 1980. Habitable pore space and microbial trophic interactions. Oikos 35:327-335.

23. Elliott, E. T., C. V. Cole, B. C. Fairbanks, L. E. Woods, R. Bryant, and D. C. Coleman. 1983. Short-term bacterial growth, nutrient uptake, and ATP turnover in sterilized inoculated and C-amended soil: The influence of N availability. Soil Biol. and Biochem. 15:85-91.

24. Elliott, E. T., D. C. Coleman, R. V. Anderson, C. V. Cole, H. W. Hunt, L. E. Woods, W. D. Gould, and J. F. McClellan. 1980. Microbial trophic structure and habitable pore space in soil. Pages 1050-1070 in J. P. Giesy, Jr., editor. Microcosms in Ecological Research. Tech. Info. Center, U.S. Dept. Energy. CONF-781101.

25. Ellis, J. E., and W. J. Parton, editors. 1977. Impact of strip-mine reclamation practices: A simulation study. A report to the Western Energy and Land Use Team, Office of Biological Services, U.S. Fish and Wildlife Service.

26. Ellis, J. E., J. A. Wiens, C. F. Rodell, and J. E. Anway. 1976. A conceptual model of diet selection as an ecosystem process. J. Theoret. Biol. 60:93-108.

27. Fretwell, S. D. 1977. The regulation of plant communities by the food chains exploiting them. Perspect. Biol. Medicine 20:169-185.

28. Frissel, M. J., and J. A. Van Veen. 1982. A review of models for investigating the behaviour of nitrogen in soil. Phil. Trans. R. Soc. Lond. B 296:341-349.

29. Gardner, R. H., W. G. Cale, and R. V. O'Neill. 1982. Robust analysis of aggregation error. Ecology 63:1771-1779.

30. Giese, A. C. 1973. Cell Physiology. 4th ed. W. B. Saunders Co., Philadelphia.

31. Gilmanov, T. G. 1977. Plant submodel in the holistic model of a grassland ecosystem (with special attention to the belowground part). Ecol. Model. 3:149-163.

32. Gold, H. J. 1977. Mathematical modeling of biological systems--an introductory guidebook. John Wiley & Sons, New York. 357 pp.

33. Harris, W. F., D. Santantonio, and D. McGinty. 1980. The dynamic belowground ecosystem. Pages 119-129 in R. H. Waring, editor. Forests: Fresh Perspectives from Ecosystem Analysis. Oregon State Univ. Press, Corvallis.

34. Haas, H. J., C. E. Evans, and E. F. Miles. 1957. Nitrogen and carbon changes in Great Plains soils as influenced by cropping and soil treatments. U.S. Dep. Agric. Tech. Bull. 1164. 111 p.

35. Hattori, T., and R. Hattori. 1976. The physical environment in soil microbiology: An attempt to extend principles of microbiology to soil microorganisms. CRC Crit. Rev. Microbiol. 4:423-461.

36. Heal, O. W., and D. D. French. 1974. Decomposition of organic matter in tundra. Pages 279-310 in A. J. Holding, O. W. Heal, S. F. MacLean, Jr., P. W. Flanagan, editors. Soil organisms and decomposition in tundra. IBP Tundra Biome.

37. Hesketh, J. D., and J. W. Jones. 1980. Predicting photosynthesis for ecosystem models. Vol. I and II. CRC Press. Boca Raton, Florida.

38. Hurlbert, S. H., J. Zedler, and D. Fairbanks. 1972. Ecosystem alteration by mosquitofish (Gambusia affinis) predation. Science 175:639-641.

39. Hunt, H. W. 1972. Decomposer section. Pages 138-164 in J. C. Anway, editor. ELM: Version 1.0. US/IBP Grassland Biome Tech. Rep. No. 156, Fort Collins, Colo.

40. Hunt, H. W. 1977. A simulation model for decomposition in grasslands. Ecology 58:469-483.

41. Hunt, H. W., C. V. Cole, and E. T. Elliott. 1984. Models for growth of bacteria inoculated into sterilized soil. Soil Sci. (under revision).

42. Hunt, H. W., C. V. Cole, D. A. Klein, and D. C. Coleman. 1977. A simulation model for the effect of predation on bacteria in continuous culture. Microb. Ecol. 3:259-278.

43. Hunt, H. W., J. W. B. Stewart, and C. V. Cole. 1982. A conceptual model for the basis of interactions among carbon, nitrogen, sulfur, and phosphorus transformations. In B. Bolin, editor. SCOPE Report on Interactions of the Biogeochemical Cycles (in press).

44. Ingham, R. E., J. A. Trofymow, E. R. Ingham, and D. C. Coleman. 1984. Interactions of bacteria, fungi and their nematode grazers: Effects on nutrient cycling and plant growth. Ecol. Monogr. (submitted).

45. Innis, G. S., editor. 1978. Grassland Simulation Model. Springer-Verlag, New York.

46. Jenkinson, D. S., and D. S. Powlson. 1976. The effects of biocidal treatments on metabolism in soil. V. A method for measuring soil biomass. Soil Biol. Biochem. 8:209-213.

47. Jenkinson, D. S., and J. H. Rayner. 1977. The turnover of soil organic matter in some of the Rothamsted classical experiments. Soil Sci. 123:298-305.

48. Jenny, H. 1941. Factors of soil formation. McGraw-Hill, New York.

49. Johannes, R. E. 1965. Influence of marine protozoa on nutrient regeneration. Limnol. Oceanogr. 10:434-442.

50. Juma, N. G. 1981. Dynamics of soil and fertilizer nitrogen. Ph.D. Thesis, Univ. of Saskatchewan, Canada. 185 p.

51. Kassim, G., J. P. Martin, and K. Haider. 1981. Incorporation of a wide variety of organic substrate carbons into soil biomass as estimated by fumigation procedure. Soil Sci. Soc. Am. J. 45:1106-1112.

52. Kowal, N. E. 1971. A rationale for modeling dynamic ecological systems. Pages 123-194 in B. C. Patten, editor. Systems Analysis and Simulation in Ecology.

53. Leigh, E. G., Jr. 1968. Making ecology an applied science. Review of Ecology and Resource Management by K. E. F. Watt. Science 160:1326-1327.

54. Loomis, R. S., R. Rabbinge, and E. Ng. 1979. Explanatory models in crop physiology. Ann. Rev. Plant Physiol. 30:339-367.

55. Lynch, M. 1979. Predation, competition, and zooplankton community structure: An experimental study. Limnol. Oceanogr. 24:253-272.

56. Malhi, S. S., and W. B. McGill. 1982. Nitrification in three Alberta soils: Effect of temperature, moisture and substrate concentration. Soil Biol. Biochem. 14:393-399.

57. McBrayer, J. F., D. E. Reichle, and M. Witkamp. 1974. Energy flow and nutrient cycling in a cryptozoan food-web. Publ. No. 575, Environmental Sci. Division, Oak Ridge Natl. Lab. EDFB-IBP-73-8.

58. McGill, W. B., and C. V. Cole. 1981. Comparative aspects of cycling of organic C, N, S, and P through soil organic matter. Geoderma 26:267-286.

59. McGill, W. B., H. W. Hunt, R. G. Woodmansee, and J. O. Reuss. 1981. PHOENIX, a model of the dynamics of carbon and nitrogen in grassland soils. In F. E. Clark and T. Rosswall, editors. Terrestrial Nitrogen Cycles. Ecol. Bull. (Stockholm) 33:49-115.

60. Melillo, J. M., J. D. Aber, and J. F. Muratore. 1982. Nitrogen and lignin control of hardwood leaf litter decomposition dynamics. Ecology 63:621-626.

61. Minderman, G. 1968. Addition, decomposition and accumulation of organic matter in forests. J. Ecol. 56:355-362.

62. Morgan, J. A., D. N. Baker, W. J. Parton, C. V. Cole, W. O. Willis, D. E. Smika, A. Bauer, and A. L. Black. 1983. Simulation of climatic and management effects on wheat production. Pages 517-524 in W. K. Lauenroth, G. V. Skogerboe, and M. Flug, editors. Analysis of ecological systems. Proc. 3rd Int. Conf. on State-of-the-art in Ecol. Model.

63. Mosier, A. R., W. J. Parton, G. L. Hutchinson. 1982. Modelling nitrous oxide evolution from cropped and natural soils. Ecol. Bull. 35:(in press).

64. Nyhan, J. W. 1976. Influence of soil temperature and water tension on the decomposition rate of carbon-14 labeled herbage. Soil Sci. 121:288-291.

65. Parker, M. 1975. Similarities between the uptake of nutrients and the ingestion of prey. Verh. Internat. Verein. Limnol. 19:56-59.

66. Parton, W. J. 1978. Abiotic section of ELM. Pages 31-53 in G. S. Innis, editor. Grassland Simulation Model. Springer-Verlag, New York.

67. Parton, W. J., W. D. Gould, F. J. Adamson, S. Torbit, and R. G. Woodmansee. 1981. NH$_3$ volatilization model. Pages 233-244 in M. J. Frissel and J. A. Van Veen, editors. Simulation of nitrogen behavior of soil-plant systems. Pudoc, Wageningen.

68. Parton, W. J., J. S. Singh, and D. C. Coleman. 1978. A model of production and turnover of roots in shortgrass prairie. J. Appl. Ecol. 47:515-542.

69. Parton, W. J., J. Persson, and D. W. Anderson. 1983a. Simulation of organic matter changes in Swedish soils. Pages 511-516 in W. K. Lauenroth, G. V. Skogerboe, and M. Flug, editors. Analysis of ecological systems: State-of-the-art in ecological modelling. Elsevier Publ. Co. Amsterdam, The Netherlands.

70. Parton, W. J., D. W. Anderson, C. V. Cole, and J. W. B. Stewart. 1983b. Simulation of soil organic matter formation and mineralization in semiarid agroecosystems. In R. L. Todd, editor. (In press).

71. Pastan, I., and R. Perlman. 1970. Cyclic adenosine monophosphate in bacteria. Science 169:339-344.

72. Patten, B. C. 1972. A simulation of the shortgrass prairie ecosystem. Simulation 19:177-186.

73. Paul, E. A., and J. Van Veen. 1978. The use of tracers to determine the dynamic nature of organic matter. Trans. 11th Int. Congr. Soil Sci. 3:61-102.

74. Platt, J. R. 1964. Strong inference. Science 146:347-353.

75. Porter, K. G. 1976. Enhancement of algal growth and productivity by grazing zooplankton. Science 192:1332-1334.

76. Powell, E. O. 1967. The growth rate of microorganisms as a function of substrate concentration. Pages 34-55 in E. O. Powell, C. G. T. Evans, R. E. Strange and D. W. Tempest, editors. Microbial Physiology and Continuous Culture.

77. Ritchie, G. A. F., and D. J. D. Nicholas. 1972. Identification of the sources of nitrous oxide produced by oxidative and reductive processes in Nitrosomonas europae. Biochem. J. 126:1181-1191.

78. Roels, J. A., and N. W. F. Kossen. 1978. On the modelling of microbial metabolism. Progress in Industrial Microbiology 14:95-203.

79. Romesburg, H. C. 1981. Wildlife science: gaining reliable knowledge. J. Wildl. Manage. 45:293-313.

80. Schimel, D. S. 1982. Nutrient and organic matter dynamics in grasslands: Effects of fire and erosion. Ph.D. dissertation. Colorado State Univ., Ft. Collins.

81. Scott, J. R., N. R. French, and J. W. Leetham. 1979. Patterns of consumption in grasslands. Pages 89-105 in N. R. French, editor. Perspectives in grassland ecology. Springer-Verlag, New York.

82. Shugart, H. H., Jr., and D. C. West. 1980. Forest succession models. BioScience 30:308-313.

83. Singh, J. S., W. K. Lauenroth, H. W. Hunt, and D. M. Swift. 1983. Bias and random errors in estimators of net root production: A simulation approach. Ecology (under revision).

84. Swift, M. J., O. W. Heal, and J. M. Anderson. 1979. Decomposition in terrestrial ecosystems. Blackwell Sci. Publ., London.

85. Tan, J. S. H., and D. C. Reanney. 1976. Interactions between bacteriophages and bacteria in soil. Soil Biol. Biochem. 8:145-150.

494

86. Tanji, K. K., M. Mehran, and S. K. Gupta. 1981. Water and nitrogen fluxes in the root zone. Pages 51-66 in M. J. Frissel and J. A. Van Veen, editors. Simulation of nitrogen behaviour of soil-plant systems. Pudoc, Wageningen.

87. Tinker, P. B. 1982. Mycorrhizas: The present position, Trans. 12th Int. Congr. Soil Sci., Vol 5, New Delhi, pp 150-166.

88. Tisdall, J. M., and J. M. Oades. 1982. Organic matter and water-stable aggregates in soils. J. Soil Sci. 33:141-163.

89. Trofymow, J. A., and D. C. Coleman. 1982. The role of bacterivorous and fungivorous nematodes in cellulose and chitin decomposition in the context of a root/rhizosphere/soil conceptual model. Pages 117-138 in D. W. Freckman and J. A. Wallwork, editors. Nematodes in soil ecosystems. Univ. Texas Press, Austin.

90. Van Veen, J. A., and M. J. Frissel. 1981. Simulation model of the behaviour of N in soil. Pages 126-144 in M. J. Frissel and J. A. Van Veen, editors. Simulation of nitrogen behaviour of soil-plant systems. Pudoc, Wageningen.

91. Van Veen, J. A., and E. A. Paul. 1981. Organic carbon dynamics in grassland soils. 1. Background information and computer simulation. Can. J. Soil Sci. 61:185-201.

92. Welch, S. M., B. A. Croft, and M. F. Michels. 1981. Validation of pest management models. Environ. Entomol. 10:425-432.

93. Wiegert, R. G. 1975. Simulation models of ecosystems. Annu. Rev. Ecol. Syst. 6:311-338.

94. Wiegert, R. G. 1977. Population models: Experimental tools for analysis of ecosystems. In D. J. Horn, R. Mitchell, and G. R. Stans, editors. Proc. colloquium on analysis of ecosystems. Ohio State Univ. Press, Columbus.

95. Wildung, R. E., T. R. Garland, and R. L. Buschbom. 1975. The interdependent effects of soil temperature and water content on soil respiration rate and plant root decomposition in arid grassland soils. Soil Biol. Biochem. 7:373-378.

96. de Wit, C. T., and H. van Keulen. 1972. Simulation of transport processes in soils. Centre for Agricultural Publ. and Documentation, Wageningen, The Netherlands.

97. Witkamp, M., and M. L. Frank. 1970. Effects of temperature, rainfall, and fauna on transfer of ^{137}Cs, K, Mg and mass in consumer-decomposer microcosms. Ecology 51:465-474.

98. Woodmansee, R. G. 1978. Critique and analyses of the grassland ecosystem model ELM. Pages 257-281 in G. S. Innis, editor. Grassland Simulation Model. Springer-Verlag, New York.

99. Woods, L. E., C. V. Cole, E. T. Elliott, R. V. Anderson, and D. C. Coleman. 1982. Nitrogen transformations in soil as affected by bacterial-microfaunal interactions. Soil Biol. Biochem. 14:93-98.

INDEX